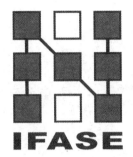

第六届结构工程新进展国际论坛文集

Proceedings of the 6th International Forum on Advances in Structural Engineering (2014)

结构抗震、减震新技术与设计方法

Seismic resistance of structures: new technologies and design methods

徐正安 任伟新 丁克伟 朱兆晴 王静峰 主编

Editors: XuZhengan, RenWeixin, DingKewei, ZhuZhaoqing, WangJingfeng

中国建筑工业出版社

China Architecture & Building Press

图书在版编目（CIP）数据

结构抗震、减震新技术与设计方法/徐正安等主编.
北京：中国建筑工业出版社，2014.9
（第六届结构工程新进展国际论坛文集）
ISBN 978-7-112-17142-2

Ⅰ.①结… Ⅱ.①徐… Ⅲ.①建筑结构-抗震结构-防震设计-文集 Ⅳ.①TU973

中国版本图书馆CIP数据核字（2014）第179957号

本书汇集11位国内外专家最新研究成果的专辑。为第六届结构工程新进展国际论坛特约报告文集。本届论坛的主题是"结构抗震、减震新技术与设计方法"。11位特邀报告主题涵盖了：大型复杂建筑结构、复杂多塔隔震结构设计、屈曲约束支撑减震技术、钢板剪力墙底层边界柱耐震设计、工程结构黏滞阻尼减振技术、城市高架桥抗震性能评估方法研究、约束钢管混凝土柱的低周疲劳性能、面向抗震设计的钢板件和钢截面分类、基于高性能计算的工程抗震与防灾、公路桥的地震分析、抗震工程在高压变电站中的应用。

责任编辑：赵梦梅　刘婷婷
责任设计：张　虹
责任校对：张　颖　党　蕾

第六届结构工程新进展国际论坛文集
结构抗震、减震新技术与设计方法
徐正安　任伟新　丁克伟　朱兆晴　王静峰　主编
Editors：XuZhengan，RenWeixin，DingKewei，ZhuZhaoqing，WangJingfeng
*
中国建筑工业出版社出版、发行（北京西郊百万庄）
各地新华书店、建筑书店经销
北京红光制版公司制版
北京盛通印刷股份有限公司
*
开本：787×1092毫米　1/16　印张：17¼　字数：420千字
2014年9月第一版　　2014年9月第一次印刷
定价：**56.00元**
ISBN 978-7-112-17142-2
　　（25926）

版权所有　翻印必究
如有印装质量问题，可寄本社退换
（邮政编码　100037）

前 言 Preface

改革开放三十年以来，尤其是近15年，中国正在持续地从事着世界最大规模的土木工程建设，而这一建设高潮还会伴随着我国各项事业的发展和新型城镇化的推进而持续一段时间。持续的、大规模的工程建设，为结构工程技术的发展、学科的建设、人才的培养、队伍的壮健，特别是创新能力的提高，提供了历史性的机遇。正是基于这种时代背景，在建设部的支持下，由中国建筑工业出版社、同济大学《建筑钢结构进展》编辑部、香港理工大学《结构工程进展》（Advances in Structural Engineering）编委会联合主办，安徽省建筑设计研究院有限责任公司、合肥工业大学、安徽建筑大学联合承办的第六届"结构工程新进展国际论坛（The 6nd International Forum on Advances in Structural Engineering)"在合肥举行。

本次论坛的主题是"结构抗震、减震新技术与设计方法"。近年来，大地震频发已引起人们的高度关注，目前，建筑工程的抗震、隔震和耗能减震控制技术已成为国际学术界和工程界关注的热点、研究的前沿。在本次论坛中我们荣幸地邀请了15位特邀报告人，他们的报告主题涵盖了近年来抗震、隔震、耗能减震的最新研究成果、设计方法、控制理论以及相应新型材料及构件的应用；阐述了在这些领域内的最新发展信息；同时也向与会者提供了一个与专家互动并获取宝贵经验的机会。

感谢特邀报告人，他们不仅在大会上做了精彩的主题报告，而且还奉献了精心准备的论文，使得本书顺利出版。

感谢论坛自由投稿作者以及参加本次论坛的所有代表，正是大家的积极参与配合，才使得本次论坛能够顺利进行。

感谢住建部执业资格注册中心、中国建筑工业出版社、同济大学《建筑钢结构进展》编辑部、香港理工大学《结构工程进展》编辑部对本次论坛的指导、支持和帮助。

感谢安徽省建筑设计研究院有限责任公司、合肥工业大学、安徽建筑大学对本次论坛成功主办的努力和付出。

目 录 Contents

SEISMIC ASSESSMENT OF HIGHWAY BRIDGES SUBJECTED TO
  COMBINED HORIZONTAL AND VERTICAL MOTIONS/
    Shan Li and S. K. Kunnath …………………………………………………… 1
大型复杂建筑结构创新与实践/傅学怡 ……………………………………………… 21
复杂多塔隔震结构设计/郁银泉 邓 烜 曾德民 高晓明 肖 明 …………… 31
屈曲约束支撑减震技术的研究与应用/李国强 胡大柱 郭小康 孙飞飞 …… 50
钢板剪力墙底层边界柱耐震设计、分析与试验/蔡克铨 李弘祺 李昭贤 苏 磊 …… 83
工程结构黏滞阻尼减振技术的研究与应用/李爱群 陈 鑫 张志强 黄 镇 …… 109
城市高架桥抗震性能评估方法研究/任伟新 陈 亮 王佐才 ………………… 135
Low Cycle Fatigue Behavior of Confined Concrete Filled Tubular Columns/
    Y. Xiao and P. Qin ……………………………………………………… 160
面向抗震设计的钢板件和钢截面分类/童根树 付 波 ……………………… 169
基于高性能计算的工程抗震与防灾：从单体到城市/
    陆新征 卢 啸 许 镇 熊 琛 韩 博 叶列平 ……………………… 196
PERFORMANCE-BASED EARTHQUAKE ENGINEERING APPLIED
  TO HIGH-VOLTAGE SUBSTATIONS USING REAL-TIME
  HYBRID SIMULATION AND PEER METHODOLOGY/
    Khalid M. Mosalam, Selim Günay and Qiang Xie ………………… 244
第六届论坛特邀报告论文作者简介 ………………………………………………… 268

SEISMIC ASSESSMENT OF HIGHWAY BRIDGES SUBJECTED TO COMBINED HORIZONTAL AND VERTICAL MOTIONS

Shan Li [1] and S. K. Kunnath [1,2]

[1] Civil and Environmental Engineering, University of California, Davis, CA 95616, USA;
[2] College of Engineering, Hunan University, Changsha 410082, China
Email: skkunnath@ucdavis.edu

ABSTRACT: A detailed evaluation of two typical bridge configurations is carried out to assess the suitability of linear response spectrum analysis to combined horizontal and vertical response spectra for predicting moment demands in the bridge girder. The process consists of first selecting the corresponding Caltrans ARS curve based on the earthquake magnitude and peak rock acceleration at the bridge site. This spectrum is applied in the longitudinal direction of the bridge. The vertical spectrum is built up using the procedures described in the report by Kunnath et al. (2008). The moment demands of the bridge girder are then computed from elastic response spectrum analysis considering sufficient number of modes using the longitudinal and vertical spectra. These results are compared to demands estimated with time history simulations using nonlinear models of the bridge columns. It is concluded that response history analysis using the horizontal and vertical design spectra is a valid preliminary approach to estimate the effects of vertical ground motions on ordinary highway bridges. The estimates from response spectrum analysis are typically more conservative for multi-span bridges than for 2-span overcrossings.

KEYWORDS: Nonlinear time history; response spectrum analysis; vertical ground motions

INTRODUCTION

In a recent study by Kunnath et al. (2007) investigating the effect of vertical ground motions on the seismic response of ordinary highway bridges, strong vertical accelerations have been found to have significant effects on (i) the axial force demand in columns; (ii) moment demands at the face of the bent cap, and (iii) moment demands at the middle of the span. The last issue was identified as the primary issue to be considered for the analysis and design of typical short-span column supported bridge configurations. This is because, in the absence of vertical effects, the design of the mid-span section is governed by positive moments whereas strong vertical motions can cause significant negative moments in the mid-span that can yield the top reinforcement.

The effects of vertical ground motions on structural response have also been investigated by many researchers in the past. Saadeghvaziri and Foutch (1991) conducted one of the early studies on the effect of vertical ground motions and found that the energy-dissipating capacity of bridge columns was reduced and the section shear capacity was influenced by the variation of axial forces due to vertical excitations. Broekhuizen (1996) and Yu et al. (1997) investigated the response of several overpasses on the SR 14/15 interchange after the 1994 Northridge earthquake. The study by Broekhuizen indicated that the vertical accelerations could significantly increase tensile stresses in the deck while the latter study found increases of about 20% in axial force demand and only a marginal change in the longitudinal moment when vertical motions were considered in the evaluation. The evaluation of 60 prestressed box-girder bridges by Gloyd (1997) indicated that the dynamic response from vertical acceleration was larger than dead load effects. Papazoglou and Elnashai (1996) reported analytical and field evidence of the damaging effect of vertical ground motions on both building and highway bridge structures. They state that strong vertical motions induced significant fluctuations in axial forces in vertical elements leading to a reduction of the column shear capacity. In certain cases, compression failure of columns was also reported to be likely. Numerical simulations were carried out to confirm these observations. Later Elnashai and Papazoglou (1997) and Collier and Elnashai (2001) worked on simplified procedures to combine vertical and horizontal ground motions. Both papers focus on near-fault ground motions that have been recorded within 15 km of the causative fault since these ground motions were observed to possess significant vertical components. Moreover, it was suggested to limit the damping ratio of elements susceptible to vertical effects to 2% because vertical ground motions are associated with higher frequency oscillations. Secondly, there are limited hysteretic energy dissipation mechanisms for vertical inelastic response than in the case of transverse response. Button et al. (2002) examined several parameters including ground motion and structural system characteristics. However, most of their studies were limited to linear response spectrum and linear dynamic analyses. Finally, Veletzos et. al. (2006) carried out a combined experimental-analytical investigation on the seismic response of precast segmental bridge superstructures. Among other issues, they also examined the effects of vertical ground motions. Their numerical analyses indicated that the prestressing tendons above the piers of one of the bridge structures yielded under positive bending. The median positive bending rotations were found to increase by as much as 400% due to vertical ground motions. Lee (2012) carried out an experimental and analytical investigation of reinforced concrete columns subjected to horizontal and vertical ground motions. In the experiments conducted on the UC-Berkeley shaking table, the shear behavior of two 1/4- geometrical scale specimens was examined under combined vertical and horizontal components. The experimental results confirmed that vertical accelerations can induce tensile strains which result in shear strength degradation of RC bridge columns.

For ordinary standard bridges constructed on sites where the peak rock acceleration is expected to be more than 0.6g, SDC-2006 (Caltrans 2006) requires consideration of vertical effects but does not require analysis of the structure under combined horizontal and vertical components of the ground motion. Instead, it stipulates the check of the nominal capacity of the structure designed considering horizontal effects only under an equivalent vertical load with a magnitude of 25% of the dead load (DL) of the structure applied separately in the upward and downward directions to account for vertical effect.

This study is a direct extension of the research completed by Kunnath et al. (2008) which concluded that if the expected peak rock acceleration is less than 0.4g, the vertical components of ground motions may be ignored and the design can proceed in accordance with existing SDC (Caltrans 2006) guidelines. If the expected peak rock acceleration is higher than 0.4g, Kunnath et al (2008) recommend a 3D linear response spectrum analysis as an initial first step to estimating seismic demands, particularly the mid-span moments, from vertical ground motions.

In the present study, a detailed evaluation of two typical bridge configurations is carried out to assess the suitability of linear response spectrum analysis to combined horizontal and vertical response spectra for predicting moment demands in the girder. The process consists of first selecting the corresponding Caltrans ARS curve based on the earthquake magnitude and peak rock acceleration at the bridge site. This spectrum is applied in the longitudinal direction of the bridge (longitudinal spectrum). The vertical spectrum is built up using the procedures described in the report by Kunnath et. al. (2008). The moment demands of the bridge girder are then computed from elastic response spectrum analysis considering sufficient number of modes using the longitudinal and vertical spectra. These results are compared to demands estimated with time history simulations using nonlinear models of the bridge columns.

MODELING OF BRIDGES

Two types of bridges will be considered in the study: single bent, two span overpasses and multi-span single frame bridges. A segment of an existing bridge in California, the Camino DelNorte Bridge, is selected as the prototype of an overpass bridge, whereas the Amador Creek Bridge is selected as the prototype multi-span column supported. Several configurations of each bridge are generated from the base configuration of each system without violating the specifications in SDC-2006 (Caltrans 2006) on allowed dimensional and balanced stiffness requirements to cover a practical range of fundamental periods

Modeling of a typical overcrossing

With the objective of selecting a typical ordinary standard bridge that was representative of a reinforced concrete over-crossing designed according to post-Northridge Caltrans specification, we selected a portion of the widening project of Camino Del Norte Bridge. The selected system is a single bent reinforced concrete bridge with two spans of 31.0 and 30.5 meters in length. The single bent is composed of two octagonal columns with spiral reinforcement. Line model is used in the numerical simulations for two-column bent over-crossings. The nonlinear simulations will be performed using the open-source software, OpenSEES (2009).

The typical two-column overcrossing and section detailing are shown in Figure 1. The column height and superstructure spans are identified in Figure 2. The column diameter is 1.68m. The column is reinforced with 25 #36mm bars and transverse reinforcement consisted of #25mm bars at a spacing of 0.1 m. Material properties are based on 27.6 MPa compressive strength with an ultimate strain of 0.006 concrete (unconfined) and 413.69 MPa steel for both longitudinal and spiral reinforcement. The confined concrete is modeled using Mander's model. A bilinear model with a post-yield stiffness of 1.33% of the initial stiffness was used to model the reinforcing steel. Each pier is modeled using a force-based nonlinear beam column element with fixed hinge lengths at element ends. The inelastic behavior of the hinge region is simulated with a discretized fiber section model. Using a fiber section is of critical importance in the present evaluation since it enables the consideration of the variations in the column moment capacity due to changes in the axial force in the columns as a result of vertical ground motions. The simulation model used in the nonlinear time history analyses is displayed in Figure 2.

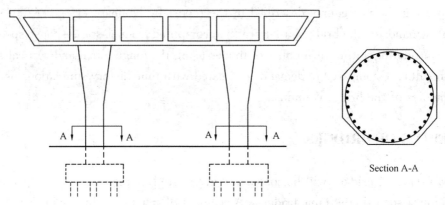

Figure 1 Configuration of two-column bent

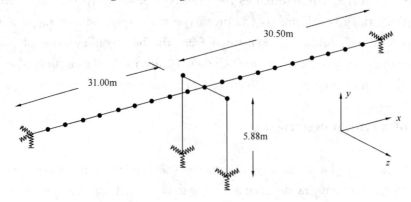

Figure 2 Simulation model of two-column bent

Following the SDC-2006 (Caltrans 2006) guideline which uses a capacity design approach to limit the inelastic behavior to the column element, the superstructure is designed and modeled to remain elastic under seismic motion. Based on this assumption, elastic elements are uses to model bridge girders. The elastic properties that used for the girder are summarized in Table 1.

The end conditions both at the abutments and at the bottom of the columns are modeled using spring elements to simulate the flexibility of the soil-pile-foundation system. SDC-2006 (Caltrans 2006) provides guidelines to determine these spring constants. The abutment stiffness in the longitudinal direction was computed as $K = K_i \omega h/5.5$, where K_i is the initial stiffness of the abutment and is taken to be equal to 0.69 kN/m per unit length of the abutment, ω and h are the width and height of the diaphragm abutment. For the abutment stiffness in the transverse direction and foundation stiffness in both translational directions, an empirical value, based on recommendations in the Caltrans guidelines, equal to 4.55 kN/m per pile is used.

Elastic properties of the box girder of Camino Del Norte Bridge Table 1

Parameter	Value
Area, A	7.32 m²
Moment of Inertia, I_x	3.51 m⁴
Moment of Inertia, I_y	141.92 m⁴
Torsional constant, J	1.50 m⁴

Based on preliminary dynamic analysis, the fundamental periods of the base configuration of Camino Del Norte Bridge were determined to be 0.55, 0.32 and 0.19 seconds in the longitudinal, transverse and vertical directions, respectively. In order to cover a wide range of fundamental periods, especially in the vertical direction, the base bridge configuration was modified to develop the additional bridge configurations. Care is taken not to violate the limits imposed by SDC-2006 (Caltrans 2006) on the geometry and dimensional restrictions for Ordinary Standard Bridges. Accordingly, only the span lengths of the bridge are modified from the original values, which in turn alter the mass of the bridge, and correspondingly the fundamental period of the bridge in all three directions.

The selected Camino Del Norte Bridge has been modeled in OpenSEES and SAP2000. In OpenSEES, material properties and modeling procedure are described as above, while in SAP2000 a line model with linear material properties is used. Similar period and model properties are found in both models. The span lengths and bridge periods in longitudinal, transverse, and vertical directions of final configurations are summarized in Table 2.

Properties and periods of highway overcrossings considered in the study Table 2

Configuration	Original	Config1	Config2	Config3	Config4	Config5
Left span (m)	31.0	20.9	41.0	46.0	51.0	36.0

续表

Configuration	Original	Config1	Config2	Config3	Config4	Config5
Right span (m)	30.5	20.5	40.5	45.5	50.5	35.5
T_L (s)	0.55	0.46	0.64	0.68	0.75	0.59
T_T (s)	0.32	0.27	0.38	0.39	0.45	0.34
T_V (s)	0.19	0.13	0.30	0.36	0.43	0.24

The moment curvature relationship of bridge column for each configuration is shown in Figure 3. The loading case is gravity only. The differences among each curves origins from the change of weight of superstructure. Configuration 4 and configuration 1 are the cases that have longest and shortest span length, respectively. Due to different gravity load that carried by bridge column, column section of configuration 4 shows highest moment capacity compared to other configurations at same curvature, while configuration 1 gives the lowest moment capacity.

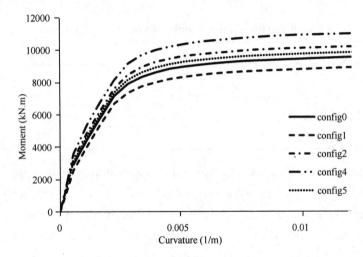

Figure 3 Moment curvature relationships of the column sections of the Camino Del Norte Bridge under gravity loads

Modeling of multi-span single frame bridges

For the case of a prototype bridge that represents a typical multi-span bridge, the Amador Creek Bridge which is a multi-frame pre-stressed concrete bridge built by Caltrans according to post-Northridge design practice was selected. It is a three-bent, four span bridge with a total length of 208.8 m. The elevation view and column details are shown in Figure 4. The elastic properties of the superstructure are presented in Table 3. The spring properties used to model footings have been summarized in Table 4, and the structural dynamic properties are summarized in Table 5. The moment curvature relationship of bridge column for each configuration is shown in Figure 5.

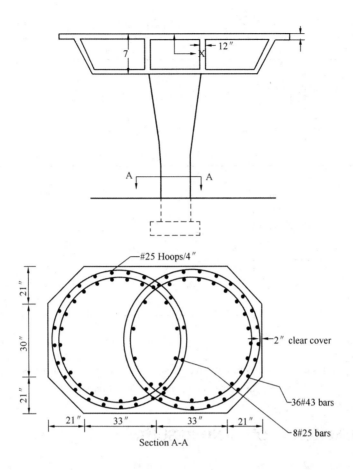

Figure 4 Elevation view and column details of Amador Creek Bridge

Elastic properties of the Amador Creek Bridge superstructure Table 3

Parameter	Value
Area, A	6.73 m²
I_x	4.56 m⁴
I_y	73.70 m⁴
J	30.52 m⁴

Elastic properties of springs to model footings Table 4

Spring Direction	Value
Translation, x	5.18×10^6 kN/m
Translation, y	6.01×10^6 kN/m
Translation, z	4.99×10^6 kN/m
Rotation, x	1.05×10^8 kN·m/rad
Rotation, y	1.16×10^8 kN·m/rad
Rotation, z	5.30×10^7 kN·m/rad

Dynamic properties of selected multi-span bridges   Table 5

Configuration	Config1	Config2	Config3
Side span (m)	40.5	34.0	25.0
Middle span (m)	54.0	45.0	35.0
T_L (s)	2.50	2.29	1.98
T_T (s)	2.33	2.12	1.77
T_V (s)	0.52	0.37	0.23

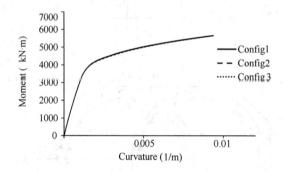

Figure 5 Moment-curvature relationships of column sections
of the Amador Creek Bridge under gravity loads

NUMERICAL SIMULATIONS AND RESULTS

In order to investigate the effectiveness of simplified elastic response spectrum analysis (RSA) to estimate moment demands in the girder for the typical ordinary bridges, the response parameters of interest computed by RSA will be compared with results from nonlinear time history analysis. The moment demands at the mid-span and interior supports are identified as the main response parameters of interest. For the response spectrum analysis, the following three cases will be considered: (1) the use of the ARS design spectrum and the developed vertical spectrum by Kunnath et al. (2008); (2) Intensity scaled (to match the spectral acceleration of the ARS spectra at the fundamental period) response spectrum of each selected ground motion used in the nonlinear time-history (NTH) simulations, record by record, both horizontal and vertical components; (3) Mean horizontal and vertical spectrum of the response spectra of the NTH ground motion set. In addition to comparing moment demands in the bridge girder, the probability density distribution of the girder moment demands estimated by both methods will also be compared.

Ground motions used in simulation study

A site in San Francisco was selected as the location for all the bridge configurations investigated in the study. The Caltrans ARS spectrum for the site was developed to be applied in the longitudinal direction of the bridge. Caltrans ARS Online (version 2.1.05) (2013) is the web-based tool that calculates both deterministic and probabilistic acceleration response spectra for any location in California based on criteria provided in Appendix B of Caltrans Seismic Design Criteria (2006). In the report by Kunnath et al. (2008), a procedure was introduced to construct a vertical response spectrum based on site characteristics and horizontal spectrum. The V/H ratio (ratio of spectral accelerations at each period) is calculated using Equation 1. The coefficients used for this equation are summarized in Kunnath et al. (2008) and listed in Table 6.

$$\ln(V/H) = b_1 + b_2 \times (M-6) + b_3 \times \ln(D+5) + b_4 + b_5 \\ \times (\ln(PGA_{rock} + 0.05)) \times \ln(vs_{30}) + \sigma \quad (1)$$

In the above expression, PGA_{rock} is taken as the horizontal peak ground spectral acceleration in g, M is the maximum moment magnitude of the site, and D is the closest rupture distance to the rupture plane in km. In this study, the standard deviation for the selected ground motions is assigned as $\sigma = 0.5$.

List of parameters for V/H model　　　　　　　　　　Table 6

Period (sec)	b1	b2	b3	b4	b5
0.01	0.644	−0.039	−0.15	−0.073	0.0135
0.02	0.534	−0.058	−0.118	−0.046	0.024
0.03	1.185	−0.079	−0.118	−0.115	0.003
0.05	2.135	−0.076	−0.168	−0.241	0.0219
0.1	1.89	0.03	−0.111	−0.245	0.023
0.15	1.63	0.064	−0.191	−0.248	−0.003
0.2	0.488	0.048	−0.144	−0.11	−0.002
0.3	−1.03	0.051	−0.083	0.059	−0.012
0.4	−1.536	0.041	−0.068	0.107	−0.022
0.5	−2.264	0.033	−0.006	0.191	−0.02
0.75	−3	0.05	−0.015	0.287	−0.041
1	−2.83	0.053	−0.068	0.292	−0.042
1.5	−3.29	0.094	−0.116	0.398	−0.047
2	−3.39	0.103	−0.113	0.434	−0.04
3	−2.86	0.217	−0.092	0.338	−0.038

The site characteristics are as follows: shear wave velocity $V_{S30} = 500$ m/s and nearest rupture distance = 14.32 km. Using the data in the web-based tool Caltrans ARS Online (version 2.1.05) (2013), the horizontal spectrum of site is generated and is shown in Figure 6.

Based on the specifications set forth in the Caltrans Seismic Design Criteria (2006), the Design Spectrum (DS) is defined as the greater of:

- Probabilistic spectrum based on a 5% in 50 years probability of exceedance (or 975-year return period);
- Deterministic spectrum based on the largest median response resulting from the maximum rupture (corresponding to M_{max}) of any fault in the vicinity of the bridge site;
- Statewide minimum spectrum defined as the median spectrum generated by a magnitude 6.5 earthquake on a strike-slip fault located 12 kilometers from the bridge site.

Thus, design spectrum is chosen as the probabilistic spectrum based on the USGS spectrum that corresponds to anexceedance probability of 5% in 50 years, with near fault factor applied. For the design spectrum used in this paper, the maximum magnitude (M_{max}) is 7.9 and peak rock acceleration in vertical direction is 0.58g which is larger than 0.4g. Hence a full 3D elastic bridge model is required for the corresponding response spectra analysis, based on SDC regulations (2006).

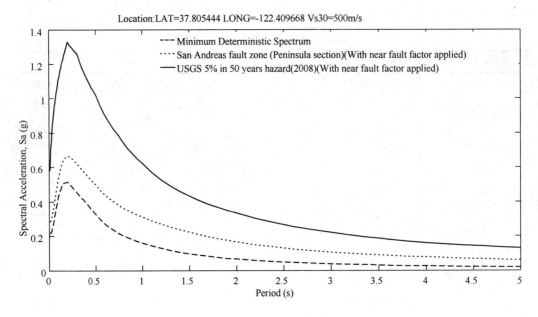

Figure 6 Site-specific spectra for selected bridge location

Twenty ground motions (see Table 7) with peak ground acceleration (PGA) of one or both horizontal components larger than 0.25g and relatively high vertical-to-horizontal PGA ratios are selected from PEER NGA database (2011). The ground motions are scaled to match the spectrum acceleration value on the design spectrum (the Caltrans ARS spectrum, see Figure 6) at the fundamental longitudinal period of bridge models, for each of the configurations considered in the study. For consistency, the same scale factor is applied to both horizontal and vertical directions of a ground acceleration record. The mean horizontal spectra of the twenty scaled ground motions for configuration 0 of both the model of Camino Del Norte Bridge and Amador Creek Bridge are displayed in Figure 7 and

8 respectively. The "target" spectrum – the Caltrans ARS design spectrum and the developed vertical spectrum are also plotted.

Selected ground motions for the study Table 7

No.	EQ name	Station	V/H ratio
1	Gazli, USSR	Karakyr	1.76
2	Imperial Valley-06	El Centro Array #5	1.03
3	Coalinga-05	Oil City	0.66
4	Nahanni, Canada	Site 1	1.90
5	N. Palm Springs	North Palm Springs	0.63
6	N. Palm Springs	Whitewater Trout Farm	0.77
7	Baja California	Cerro Prieto	0.42
8	LomaPrieta	Capitola	1.02
9	LomaPrieta	LGPC	0.92
10	Cape Mendocino	Cape Mendocino	0.50
11	Landers	Lucerne	1.04
12	Northridge-01	Jensen Filter Plant	0.81
13	Northridge-01	Jensen Filter Plant Generator	0.81
14	Northridge-01	Pacoima Dam (upper left)	0.78
15	Northridge-01	Rinaldi Receiving Sta	1.01
16	Northridge-01	Sylmar - ConverterSta	0.65
17	Chi-Chi, Taiwan	TCU071	0.69
18	Northridge-06	Rinaldi Receiving Sta	0.92
19	Chi-Chi, Taiwan-06	TCU079	0.75
20	Chi-Chi, Taiwan-06	TCU080	0.89

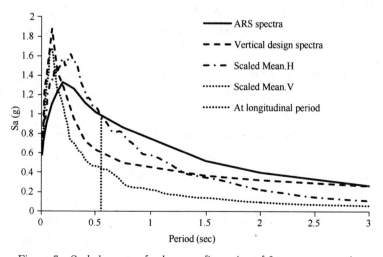

Figure 7 Scaled spectra for base configuration of 2-span overcrossings

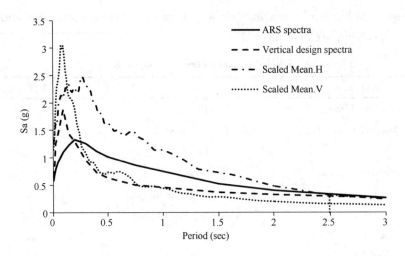

Figure 8　Scaled spectra for base configuration of multi-span bridge

Results of simulations for two-span overcrossing systems

The nonlinear time history (NTH) analysis of six configurations listed in Table 2 subjected to the 20 selected ground motions (Table 7) is accomplished using OpenSEES. For comparison, the response spectrum analysis (RSA) is performed using SAP2000 (Computers and Structures, 2012). In the results presented in this section, the girder positive moment indicates tension on the bottom face. In general, in the absence of vertical motions, the peak mid-span moments are positive moments and the peak support (girder-column region) are negative moments. However, as demonstrated in the previous research by Kunnath et al. (2008), one of the significant effects of strong vertical motions is to induce large negative moments in the mid-span which may result in yielding of the top reinforcement. The ability of RSA to capture the amplification of the mid-span moments is examined in this section. In order to facilitate better interpretation of results, the maximum values of girder moments are normalized by the dead load moment and expressed as moment ratios.

In Figure 9, the moment ratios at different critical sections are plotted as a function of the vertical periods of each configuration. The moment demands computed by RSA and NTH are compared in these plots. In Figure 9a and Figure 9b, the maximum and minimum moment demands at the middle of the left span of the over-crossing are compared. The mean of the NTH demands for vertical periods corresponding to 0.13 sec and 0.19 sec are slightly higher than the predictions using RSA. At longer periods, the mean computed demands for both methods are practically similar.

Similar plots for the right mid-span moment are displayed in Figure 9c and Figure 9d. Similar trends are evident in this case too-with mean NTH demands being slightly higher than those predicted using RSA. These differences are not significant considering the fact that the girder is modeled as an elastic element. The objective of the study is to verify if the predicted demands using RSA is sufficient to provide engineers with information on whether yielding is likely in the girder. It is clear from the results that RSA can provide this preliminary information, following which more detailed nonlinear simulations can be conducted.

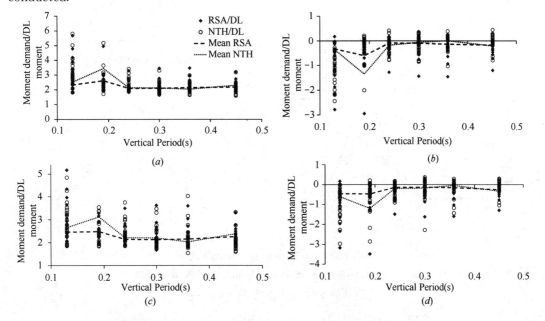

Figure 9 Variation of maximum and minimum moment demands using NTH and RSA as a function of vertical period for scaled 20 ground motions:
(a) Maximum moment demands at left mid-span; (b) Minimum moment demands at left mid-span;
(c) Maximum moment demands at right mid-span; (d) Minimum moment demands at right mid-span

A general observation for all cases is that as the vertical periods get longer (or span length becomes larger), the peak demands decrease and the differences between RSA and NTH are negligible. As the vertical period increases, the system moves away from the predominant period of the vertical motions which are typically in the range of 0.1 – 0.25 seconds.

In Figure 10 to Figure 12, the probability density distributions (PDF) of moment ratios at critical sections have been generated. This provides a measure of the dispersion in the demands so as to gain additional confidence that the demands computed from RSA are adequate. Since all configurations in which the vertical period was higher than 0.2 seconds produced similar mean demands in both RSA and NTH, only Configurations 0 and 1 that provided relatively higher demands have been selected for generating the probability distributions. Based on observations from results in Figure 9 to Figure 12, we can conclude that a linear Response Spectrum Analysis using Ground Motion spectra or the mean spectrum of a set of ground motions is adequate to get an estimate of girder demands due to vertical effects.

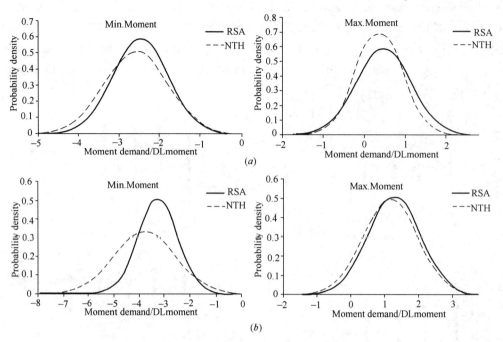

Figure 10 Probability density distribution of moment demands in
bent cap of 2-span overcrossing [Camino Del Norte bridge]
(a) Configuration 0 ; (b) Configuration 1

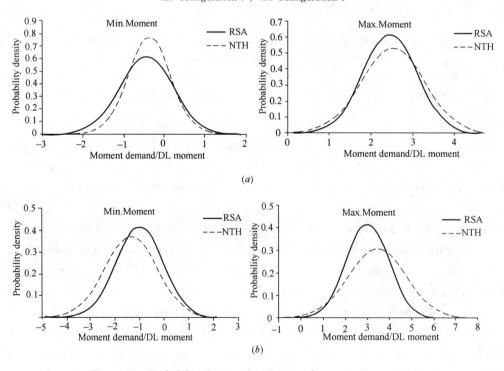

Figure 11 Probability density distribution of moment demands in
left mid-span of 2-span overcrossing [Camino Del Norte bridge]
(a) Configuration 0; (b) Configuration 1

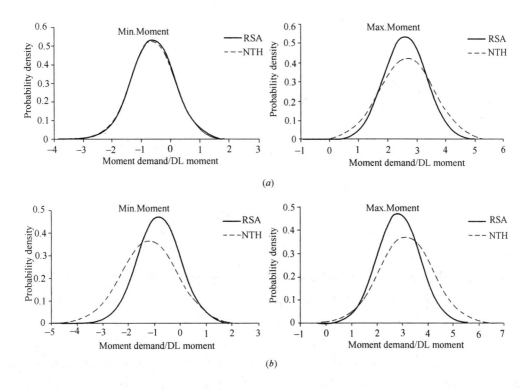

Figure 12 Probability density distribution of moment demands in right mid-span
of 2-span overcrossing [Camino Del Norte bridge]
(a) Configuration 0; (b) Configuration 1

Results of simulations for multi-span systems

For the case of multi-span bridges, the peak demands at various critical sections (mid-spans) are plotted in Figure 13 as a function of the vertical period of the bridge. All demand values are normalized by the dead load moment values at the particular section so as to provide a sense of the magnitude of the demands due to vertical effects.

Figure 13 (a) and (b) show the peak maximum and minimum demands at the mid-span of the exterior span. The estimate of RSA, on average, are much higher than the predictions by NTH. The model with medium vertical period has minimum demand ratio. The RSA estimates are significantly higher than NTH estimates. This suggests that the peak demands occur during the elastic phase of response of the columns.

The peak demands are compared for the interior mid-span in Figure 14. Estimates from nonlinear time-history analysis indicate a nearly constant demand (on average) for the three configurations. The RSA results show a trend with increasing demands as the vertical period increases. In fact, the mean RSA estimates are slightly lower than NTH for a vertical period of 0.23 secs, almost the same at the intermediate period and higher than NTH for a vertical period of 0.55 seconds. This observed trend is reversed for the mini-

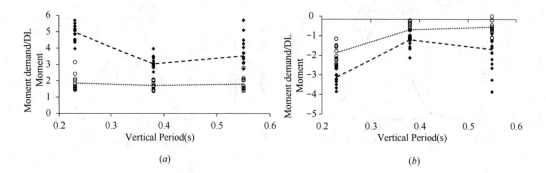

Figure 13 Comparison of moment demands between RSA and NTH at critical sections for all configurations as a function of vertical periods:
(a) Exterior mid-span (maximum); (b) Exterior mid-span (minimum)

mum moment demands (the critical demand parameter that induces tension in the top surface of the girder) with RSA estimates being higher than NTH at a vertical period of 0.23 seconds and lower than NTH at a vertical period of 0.55 seconds.

However, the variation in the minimum demands should be viewed with care. In most cases, the minimum values are still positive indicating that the vertical ground motions did not cause a reversal of dead load effects from positive to negative moments. Therefore, the only values of concern are the interior mid-span moments where negative demands are estimated in some cases. In all of the cases where the minimum moments become negative, the estimates from RSA are higher than NTH. Hence the use of RSA for evaluating the effects of vertical ground motion is adequate.

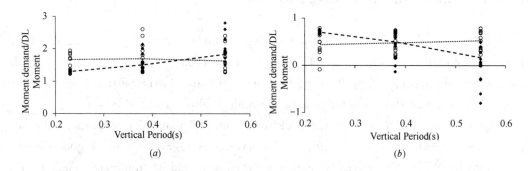

Figure 14 Comparison of moment demands between RSA and NTH at critical sections for all configurations as a function of vertical periods:
(a) Interior mid-span (maximum); (b) Interior mid-span (minimum)

The modeling of the multi-span bridges was accomplished by the use of gap springs at the abutments. A compression-only spring with a gap of 0.1 meter is provided in the longitudinal direction corresponding to realistic dimensions of the gap between the bearing pad and abutment. Once the superstructure of the bridge moves in the longitudinal direction due to the imposed ground motion, the gap is closed and significant stiffness is provided in the longitudinal direction to restrain further movement. This has an effect of decreasing the stiffness of the bridge in the direction of motion. To evaluate the effect of gap closing, an earthquake record was selected to generate deck displacements that exceeded the gap. The analysis was then repeated by scaling down the magnitude of the peak acceleration to prevent the gap from closing. The time history response of the column shear from these two simulations is shown in Figure 15.

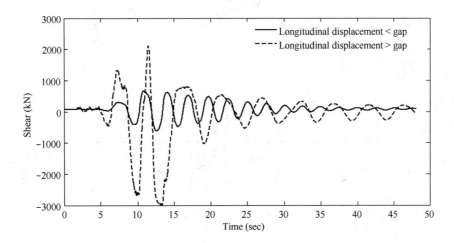

Figure 15   Shear force time history of column under two scenarios: with and without closure of gap between girder and abutment

The response from the lower intensity ground motion is seen to have a frequency that is equal to the longitudinal period of the bridge (the response was essentially elastic). However, when the ground acceleration is large enough to cause longitudinal displacements that exceed the available gap, pounding of the deck with the abutment occurs. The present analysis does not have the ability to exactly simulate pounding effects, however, the large compression stiffness provided upon contact of the deck with the abutment approximately accounts for the change in system period due to contact. The large shear demands imposed on the column due to girder impact with the abutment is evident in the response. The overall lengthening of the longitudinal period of the bridge due to nonlinearity is also observed. It must be emphasized that the response only approximately captures the effect of gap closing. Shear yielding of the column was not explicitly modeled in the analysis because none of the columns reached their yield capacity in shear.

Figure 16 shows the peak moment demands along the span of the girder. Figures include moment demands at the supports and at mid-spans of the girder. The values are normalized with respect to the moments resulting from dead load only. It is observed that the estimates from RSA consistently predict higher demands than NTH. This indicates that a simple elastic response spectrum analysis (RSA) provides conservative demands estimates of girder moment demands.

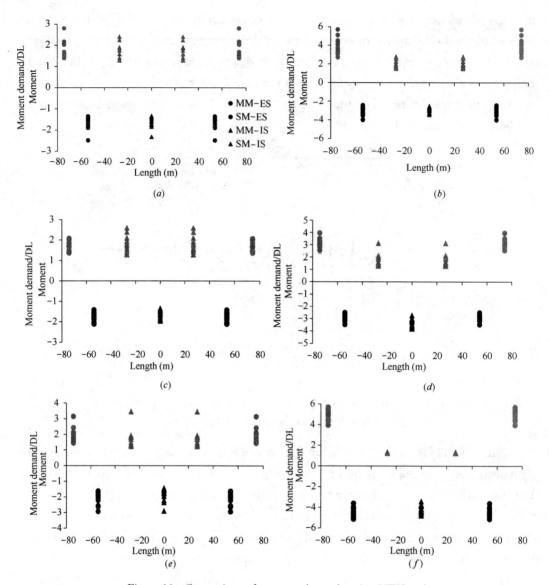

Figure 16 Comparison of moment demands using NTH and RSA along the deck of Amador Creek Bridge:
(a) Configuration 1 - NTH; (b) Configuration 1 - RSA; (c) Configuration 2 - NTH;
(d) Configuration 2 - RSA; (e) Configuration 3 - NTH; (f) Configuration 3 - RSA

CONCLUSIONS

The issue of the significance of vertical components of ground motions on structural response has continued to be a matter of debate since it is difficult to establish direct evidence of damage from vertical motions. Past research has clearly identified several potential issues that deserve attention - the moment demands in the girder that may result in inelastic behavior (this is an important consideration since highway bridges in the US are designed to keep the girder elastic and restrict inelastic action in the columns). The main objective of this study was to assess if an elastic response spectrum analysis (RSA) is sufficient to estimate critical demands in the girder. Such an analysis would only be a necessary first step to caution engineers that vertical effects should be considered and additional detailed nonlinear simulations would be necessary to determine the seriousness of the problem.

A comprehensive series of simulations was carried out on a wide range of over-crossings and multi-span bridge configurations to evaluate the effectiveness of estimating girder moment demands by response spectrum analysis (RSA) and comparing the estimated demands with results from nonlinear time history analysis. Results of these analyses reveal that response history analysis using the horizontal and vertical design spectra is a valid preliminary approach to estimate the effects of vertical ground motions on ordinary highway bridges. The estimates from RSA are typically more conservative for multi-span bridges than for 2-span overcrossings.

ACKNOWLEDGMENTS

The authors gratefully acknowledge the financial support provided for this study by the California Department of Transportation (Caltrans) under Contract No. 59A0688.

REFERENCES

Abrahamson, N. A. and Silva, W. J. (1997), "Empirical Response Spectral Attenuation Relations for Shallow Crustal Earthquakes."*Seism. Soc. Am.*, 68(1), 94-127.

ACI-318 (2008), Building Code Requirements for Structural Concrete and Commentary, ACI 318-08, American Concrete Institute, Farmington Hills, MI.

Broekhuizen, D. S. (1996). "Effects of vertical acceleration on prestressed concrete bridges."MS thesis, Univ. of Texas at Austin, Texas.

Button, M. R., Cronin, C. J. and Mayes, R. L. (2002). "Effect of Vertical Motions on Seismic Response of Bridges." ASCE *Journal of Structural Engineering*, Vol. 128 (12), 1551-1564.

California Department of Transportation (Caltrans) (2006), "Seismic design criteria", SDC-2006, Sacramento, CA

Caltrans ARS Online, http://dap3.dot.ca.gov/shake_stable/v2/index.php.

Collier C. J. andElnashai A. S. (2001), "A procedure for combining vertical and horizontal seismic action effects." *Journal of Earthquake Engineering*, 5(4), 521-539.

Elnashai, A. S. and Papazoglou, A. J. [1997] "Procedure and Spectra for Analysis of RC Structures Subjected to Strong Vertical Earthquake Loads," *Journal of Earthquake Engineering* 1, 121-155.

Gloyd, S. (1997). "Design of ordinary bridges for vertical seismic acceleration." Proc., FHWA/NCEER Workshop on the National Representation of Seismic Ground Motion for New and Existing Highway Facilities, Tech. Rep. No. NCEER-97-0010, National Center for Earthquake Engineering Research, State Univ. of New York at Buffalo, NY, pp. 277 – 290.

Kunnath, S. K., Erduran, E., Chai, Y. H. and Yashinsky, M. (2007). "Near-Fault Vertical Ground

Motions on Seismic Response of Highway Overcrossings." ASCE *Journal of Bridge Engineering*, Vol. 13, No. 3, 282-290.

Kunnath, S. K., Abrahamson, N., Chai, Y. H., Erduran, E. and Yilmaz, Z. (2008), "Development of Guidelines for Incorporation of Vertical Ground Motion Effects in Seismic Design of Highway Bridges", Technical Report, California Department of Transportation, Sacramento, CA.

Lee, H. (2011), "Experimental and Analytical Investigation of Reinforced Concrete Columns Subjected to Horizontal and Vertical Ground Motions", Ph. D. Dissertation, University of California, Berkeley.

OpenSees (2009), Open System for Earthquake Engineering Simulation, http://opensees.berkeley.edu/

Papazoglou A. J. and Elnashai A. S. (1996). "Analytical and field evidence of the damaging effect of vertical earthquake ground motion", *Earthquake Engineering and Structural Dynamics*, 25, 1109-1137

PEER-NGA Database (2011), http://peer.berkeley.edu/nga/

Saadeghvaziri, M. A., and Foutch, D. A. (1991). "Dynamic behavior of R/C highway bridges under the combined effect of vertical and horizontal earthquake motions." *Earthquake Engineering and Structural Dynamics*, 20, 535 - 549.

Veletzos M. J, Restrepo, J. I, Seible F. (2006). "Seismic response of precast segmental bridge superstructures" Report submitted to Caltrans - SSRP-06/18, University of California, San Diego

Yu, C-P., Broekhuizen, D. S., Roesset, J. M., Breen, J. E., and Kreger, M. E. (1997). "Effect of vertical ground motion on bridge deck response." Proc., Workshop on Earthquake Engineering Frontiers in Transportation Facilities, Tech. Rep. No. NCEER-97-0005, National Center for Earthquake Engineering Research, State Univ. of New York at Buffalo, NY, pp. 249 - 263.

大型复杂建筑结构创新与实践 *

傅学怡

(悉地国际（深圳）设计顾问有限公司，广东 深圳 518023)

摘 要：针对近几年所主持设计的 6 项国内外重大建筑工程，简要阐述各项目的结构设计过程及技术难点、结构体系与方案、分析计算方法、关键技术等，着重介绍其中采用的新结构体系、新结构设计方法，新分析计算技术等。

关键词：大型复杂建筑结构；创新与实践

中图分类号：TU375

INNOVATION AND PRACTICE OF LARGE-SCALE AND COMPLICATED BUILDING STRUCTURES

X. Y. Fu

(CCDI，Shenzhen，518023 China)

Abstract：This paper is based on the topic of "structure design", by combining the major constructions in domestic and overseas which mainly designed by the author. The details of the technological difficulty, structural system and scheme, methods of analysis and calculation among the above constructions are systematically introduced. The new type of structural systems, design methods and analysis techniques are specially emphasized in this paper.

Key words：large-scale and complicated building structure；innovation and practice

引言

笔者长期从事高层、大跨空间建筑结构的设计与研究[1]。并有幸与国内外同行专家经常广泛切磋探讨有关高层、大跨建筑结构的一些关键技术问题。实践使我们一致深感，在科学技术尤其是计算机科学技术高度发展的今天，在各种计算软件已广泛应用于高层、大跨空间建筑结构设计计算的今天，在新颖复杂高层、大跨建筑不断涌现的今天，如何从概念和整体上把握住高层、大跨空间建筑结构的基本力学特性、设计方法，如何去理解并掌

---

* 基金项目：国家自然科学基金重大研究计划（90715012）。

作者：傅学怡（1945—），男，研究员，主要从事复杂大跨、超高建筑结构设计研究，E-mail：fu _ xue _ yi@aliyun. com

握新型建筑结构设计技术，进而达到使高层、大跨建筑结构设计得更加适用、经济安全，乃是当前高层、大跨建筑结构设计迫在眉睫的重大课题。

以下针对近几年所主持设计的 6 项国内外重大建筑工程，简要阐述各项目的结构设计过程及技术难点、结构体系与方案、分析计算方法、关键技术等，着重介绍其中采用的新结构体系、新结构设计方法，新分析计算技术等。

1. 卡达尔多哈塔[2,3]

卡塔尔多哈塔位于卡塔尔首都多哈，南邻多哈湾，地下 4 层，地面以上 44 层钢筋混凝土主体结构，顶盖为一直径 36m 钢结构穹顶，上设 27m 高桅杆，总高 231m，总建筑面积约 10 万 m^2，由世界著名法国建筑师 John Novel 创意设计，整体建筑形态简洁淳朴富有阿拉伯民族特色。2008 年底项目结构主体施工封顶，2010 年 3 月整个项目竣工。

该建筑采用了世界首创钢筋混凝土交叉柱外网筒结构，项目的结构标书设计由法国工程师完成。受总承包方中国建筑工程总公司委托，中建国际设计顾问有限公司与深圳大学土木工程学院联合团队于 2005 年 12 月开始介入该项目，通过对该项目标书结构设计文件的详细分析和研究，发现项目结构标书设计不仅存在着设计不经济的情况而且存在着交叉柱节点承载力不足、混凝土斜柱变形差异将引起主楼倾斜、影响电梯运行等重大安全隐患。项目团队对该结构设计的理念、混凝土徐变、施工模拟、部分预应力、连续倒塌、交叉柱节点、温度收缩效应及屈曲稳定等多方面开展研究分析，提出了多项重大结构调整和优化建议如下：

（1）进行结构自重下完整的全过程施工模拟，逐层找平，逐层找正。计算表明，总重力载荷标准值作用下环梁最大拉力可由原标书设计未做施工模拟的 7500KN 降到 3698KN。

（2）将原标书设计楼盖与外网筒脱开改为恢复楼盖与网筒连接，提高结构整体性，楼盖参与环梁共同工作，环梁最大拉力进一步减小 20% 左右。

（3）将原标书设计环梁超预应力设计（无拉应力）改为环梁设计采用部分预应力适度强化的理念，允许环梁出现适量拉力，允许环梁在正常工作状态出现裂缝，控制裂缝宽度 <0.1mm，即 BS8110 CLASS3 的标准，从而使环梁预应力筋数量大幅度减小 90% 左右，预应力筋可只布置在环梁截面内，悬臂采用梁板结构，板厚调整为 130mm，悬臂结构自重大为减小，仅此一项，全楼结构总重减小可达 70MN，有利于减小南侧斜柱的轴力及其与北侧斜柱轴力的差异。

（4）将北侧下部 1~28 层柱实心圆截面改为空心圆截面。在保留建筑师同层斜柱截面相同和清水混凝土的设计前提下，平衡南、北两侧斜柱重力载荷作用下的压应力水平的差异；在满足承载力要求前提下，调整斜柱配筋率，从 4.5% 下调至 2.77%，与此同时，北侧 28 层以上斜柱配筋率适当调低至 2%。徐变效应计算分析表明，20 年后结构顶点的水平位移可由 282mm 降至 145mm；南北侧斜柱顶点竖向位移差可由 114mm 降至 64mm，巧妙地解决了结构南倾的重大隐患。

调整设计不仅消除了结构安全隐患，还取得了很好的经济效益，大大方便了施工，获得了业主、原标书设计法国工程师、审查单位独立复核第三方中国建筑科学研究院的认可。

创新点主要有以下 7 项：
- ◆ 实现世界首创现浇混凝土交叉柱外网筒结构；
- ◆ 引入混凝土徐变收缩模式计算分析重力荷载下结构长期变形，揭示原标书设计将导致新"比萨斜塔"出现；
- ◆ 创新提出交叉柱节点核心区钢板凳加强；
- ◆ 创新提出考虑结构自重、温度、混凝土收缩徐变、预应力、后浇带全面施工模拟技术；
- ◆ 创新提出并采用了整体结构线性屈曲稳定分析确定受压构件计算长度的方法；
- ◆ 创新提出预应力张拉施工模拟的设计方法；
- ◆ 创新提出和采用了保留施工模拟重力荷载下结构实际工作状态，引入"倒塌荷载"的结构抗连续倒塌设计方法。

该项目获 2009 年第六届全国优秀建筑结构设计一等奖、获 2010 年国际混凝土协会 FIB 特别贡献大奖、2011 年华夏建设科学技术二等奖，2012 年国际高层建筑都市委员会杰出大奖。

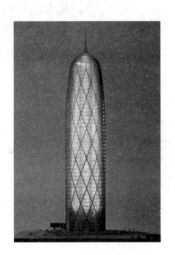

图 1 多哈塔项目效果图

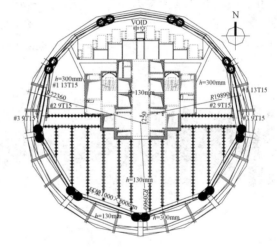

图 2 多哈塔标准层结构布置图

2. 深圳大梅沙万科中心[4]

大梅沙·万科中心项目位于深圳盐田区大梅沙度假区。与美国 STEVEN HOLL 建筑事务所合作设计。总用地面积 61730 m²，总建筑面积 137116 m²。自建筑落成以来获得了广泛的关注和好评。

建筑设计将多个功能体以水平几何形态连接在一起，并将整个建筑抬起，将基地最大程度地还原给自然并大大提升了建筑内部的景观。

"斜拉桥上盖房"，结构设计以独特的结构形式诠释了建筑理念。落地筒体、实腹厚墙、柱支承离地 10~15m 的上部 4~5 层结构，中间跨度 50~60m，端部悬臂 15~20m，底部形成了连续的大空间。采用世界首创落地钢筋混凝土筒体、墙—斜拉索—首层钢结构楼盖—上部钢筋混凝土框架混合结构,上部结构重力荷载由预应力拉索、首层钢结构楼盖和上部各层混凝土结构楼盖整体协同工作传递到竖向落地构件——筒体、墙、柱;水平荷载通过各层楼屋盖传递到落地筒体。索通过铸钢节点与落地竖向构件连接,索的连接过渡区埋入型钢,前期承受索张拉应力,后期参与工作。

该体系特点如下:(1)结构悬挑达 15~20m,中部跨越达 50~60m;(2)结构自重大,索力大,索径粗,所用 D7×499 索为国内最大直径索;(3)拉索张拉后上部结构逐层施工,索力增长大。

本工程整体结构平面狭长且多支,为加强各筒体之间的协同工作能力,保证斜索拉力的有效传递,在结构底层和顶层楼层平面内加设水平交叉斜撑,加强楼屋盖面内刚度和承载力,同时兼作承重梁。中间楼层楼盖采用主次梁结构,以利于减轻结构自重,减轻索的负荷。

在该项目中提出并实现斜拉索跨越支承多层建筑创新结构。采取自配重自平衡设计理念,实现大直径索预应力一次张拉,限制落地竖向结构侧移;采取预应力微调主动控制设计方法,减小托柱柱根弯矩;设计专门铸钢节点,实现索-钢结构-混凝土结构应力转移。比钢桁架结构节省结构成本 8 千万元人民币,获业主好评。

项目获 2012 年第七届全国优秀建筑结构设计一等奖、获 2014 年国际混凝土协会 FIB 特别贡献大奖、获美国建筑师学会建筑荣誉奖,中国土木工程詹天佑奖等,学术论文获美国土木工程协会结构工程学报(ASCE Journal of Structural Engineering)2012 年论文下载率最高之一。

图 3 万科中心项目效果图

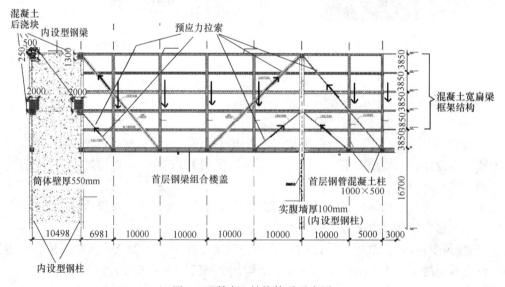

图 4 万科中心结构体系示意图

3. 深圳北站[5]

深圳北站位于深圳市龙华中心区，为京广港铁路重要交通枢纽。由站房建筑及两侧的无柱站台雨棚组成。站房 2 层，局部设夹层，屋盖结构"上平下曲"。两侧雨棚呈波浪形。总建筑面积 18 万 m²。

该项目为一复杂的交通枢纽项目，结构具有以下特点：

（1）复杂交通枢纽：轻轨高架穿越，需控制列车振动对结构的影响；多条交通线路下沉穿越，基础结构需特殊处理。

（2）结构超长：站房楼盖结构 339.06m×201.50m，屋盖结构 407.316m×203m，不设变形缝。

（3）大跨度、大悬挑：楼盖标准柱距 43m；屋盖结构柱网 86m×81m，悬挑 62.5m。

（4）创新：提出并实现四边形环索弦支网格梁创新结构。将国外引进的圆形弦支网壳结构改进为四边形环索弦支网格梁结构，提高环索效率 4～5 倍，获业主好评和国家发明专利，推广应用于福州体育中心。

该工程由中铁第四勘察设计院集团有限公司和深圳大学建筑设计研究院联合设计，2008 年 12 月通过铁道部鉴定中心结构设计专项审查，2011 年 3 月结构竣工验收，2011 年年底通车运行。该项目已获 2013 年中国土木工程詹天佑奖、2014 年第八届全国优秀建筑结构设计一等奖等。

图 5 深圳北站效果图

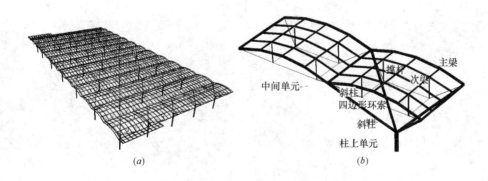

图 6 雨棚结构计算模型三维图及四边形环索弦支结构基本单元示意图
(a) 计算模型；(b) 四边形环索弦支结构基本单元

4. 济南奥林匹克中心体育场[6]

2009年第十一届全国运动会主体育场——济南奥林匹克中心体育场，建筑面积154,323m²，可容纳观众约6万人。该工程平面近椭圆形，南北长约360m，东西宽约310m，如图1所示。下部为看台及各功能用房，采用钢筋混凝土框架-剪力墙结构体系；上部钢结构分为东、西两个独立的钢结构悬挑罩棚，采用折板型悬挑空间桁架结构体系，由64榀径向主桁架和9榀环向次桁架组成，落地墙面结构为屋面折板结构的延伸，屋面罩棚的最前端为平板结构。中部最大悬挑长度约53 m，根部桁架高度7m；中间高、两边低，高差14m，最高点离地面约52m。上部钢结构采用内、外支座支承于下部混凝土结构上，外支座采用外包混凝土的圆钢管组合倒三角支承；内支座为四根圆钢管组合的交叉V形柱汇交于下部型钢混凝土柱。

罩棚钢结构总用钢量5153t，单片罩棚理论用钢量2562.9t，按其覆盖面积19000m²计134.9kg/m²；按屋面墙面展开面积32000m²计80kg/m²，约为目前同规模同标准国内体育场用钢量的1/2~1/3，由江苏沪宁钢机制作安装。钢结构东西罩棚临时支撑拆除于2007年10月。山东大学健康监测及第三方实测数据与设计理论模拟计算结果十分吻合。

在该项目提出整体结构总装分析设计方法，并进一步考虑基础土体塑性、混凝土徐变及钢结构节点刚度退化，提出刚度退化多模型包络设计方法，推广应用于大跨空间结构；提出超长结构温差收缩效应分析与控制设计新方法，引入地基基础有限约束刚度和混凝土徐变收缩时效特性，模拟结构生成全过程及施工月最不利温差，揭示超长结构实际受到的温差收缩效应，并进一步结合设置后浇带、低温合拢等方便易行技术措施，减小混凝土收缩、降温效应，既利于整体建筑使用，降低造价，又利于提高结构整体性及其抗震性能，突破规范伸缩缝间距规定，推广应用于超长无缝结构。

2006年8月通过初步设计审查，2006年10月通过了全国抗震超限审查，2007年12月主体结构验收，中国建筑设计研究院进行了总装结构振动台试验研究和节点试验研究验证。为十一届全运会成功举行作出重大贡献。获2010年中国土木工程詹天佑奖，2009年全国建筑结构设计一等奖等。

图7　济南奥体中心项目效果图

图8　济南奥体中心钢结构施工照片

5. 平安国际金融中心[7]

平安国际金融中心位于深圳市福田中心区，东边相邻的益田路是福田区的其中一条主干道路；南北分别是福华路与福华三路。

本项目是一幢以甲级写字楼为主的综合性大型超高层建筑，为目前国内在建第一高楼，其他功能包括商业、观光娱乐、会议中心和交易等五大功能区域，总用地面积为18931m^2，总建筑面积460665m^2，建筑基底面积为12305m^2。本项目包括一栋地上115层的塔楼，含塔尖高度为660m，结构上屋面高度549.1m。还包括一个11层高的商业裙楼，用来作为奢侈品零售，办公，餐饮和大堂等。地面以下为五层地下室，设计用作零售、泊车等功能。

塔楼主体采用巨型空间斜撑框架-劲性钢筋混凝土核心筒－外伸臂结构体系，合理配置内筒外框结构及其连接构件，形成多重抗侧力空间结构体系。

外框结构设置空间带状桁架、巨型钢斜撑和V形撑等，提高外框结构刚度，增加多余约束，形成较为可靠的二道防线，有利于增强整体结构稳定性，提高整体结构抗震、抗连续倒塌能力。

（1）沿塔楼高度均匀设置六道空间双桁架、一道单桁架及七道单角桁架。空间双桁架及单角桁架连接巨柱，塔楼的外围形成巨型框架，承担相当大部分由侧向力引起的倾覆力矩。

（2）在每两个相邻的周边桁架间布置一道巨型斜撑，形成外围"巨型支撑框架"结构作为抗侧力体系的第二道防线。该斜撑连接相邻两根巨柱，在每个斜撑始于下部周边桁架的上弦杆，止于上部周边桁架的下弦杆。该巨形斜撑体系进一步提高了结构抗侧力的安全富余度。

（3）首创的V形撑主要承担建筑角部重力荷载，控制角部区域楼板竖向振动，同时巧妙结合了建筑立面造型，体现了建筑与结构和谐。

（4）巨柱在平面上为基本上为长方形，但为了与建筑平面协调，其中在一角部有调整。巨柱底部的尺寸约为6.5×3.2m，在顶部逐渐减小至3.1×1.4m。巨柱内埋组合型钢由上而下连续变化，厚度75mm变化至25mm，内埋型钢均匀分布。巨柱含钢率由底部的8%至顶部的4%。通过整体结构的线性屈曲稳定分析，确定巨柱的屈曲模态，合理反推巨柱计算长度，保证巨柱的安全性。

（5）沿塔楼全高设置四道外伸臂，控制和减小结构层间位移。外伸臂将核心筒与巨柱有效的连接在一起，改善结构的性能，增加结构抗侧刚度。

内筒采有劲性钢筋混凝土核心筒，其边长约32m的正方形，底部外墙厚1.5m，内墙厚0.8m，混凝土采用C60。核心筒墙体厚度随高度增加逐渐减小，顶部外墙减为0.5m，内墙减为0.4m。核心筒提供了结构抗侧刚度以及抗剪承载力，承担大部分的基底剪力。考虑建筑功能的要求，核心筒的角部在第六区以上被切去，同时在112层观光层以上，南部及北部的墙体将部分切除，改为设置钢柱支承上部结构。核心筒角部及相交处内埋型钢柱以增加核心筒的延性及刚度。核心筒全高设置800mm高的连梁。大约六分之一的连梁需要内埋型钢梁加强。同时，在办公室楼层的需要设置部分双连梁，拟允许机电设备管道在双连梁之间穿过，使楼层有效利用高度增加。在底部加强区的墙体采用组合钢板剪力墙的形式，提高墙体抗弯及抗剪承载力。

提出重力荷载长期效应分析与控制方法，模拟结构生成及使用的全过程，引入混凝土收缩徐变时效特性并考虑含钢率影响，揭示超高结构长期变形规律及其对结构内力重分布影响，给出层高预调、构件长度预留的全世界首份施工图，以保持建筑层高和楼面平整，利于非结构构件、电梯运行和超高层建筑正常使用。

图 9　平安金融中心项目效果图

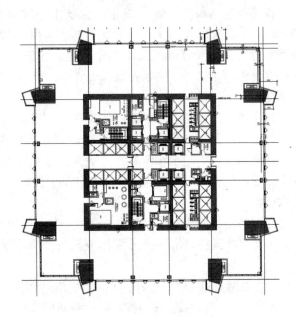

图 10　平安金融中心标准层平面布置图

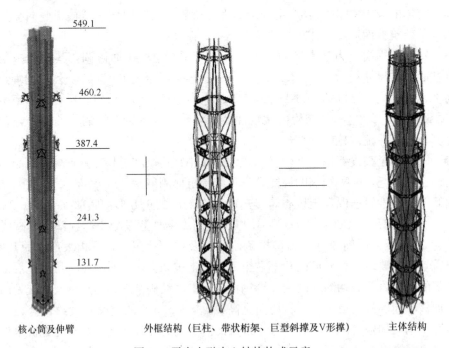

核心筒及伸臂　　　　外框结构（巨柱、带状桁架、巨型斜撑及V形撑）　　　　主体结构

图 11　平安金融中心结构构成示意

6. 天津响螺湾中钢广场[8]

中钢天津响螺湾项目占地 26666.7m²，总建筑面积 395181m²，包括 T1、T2 两座塔楼、裙房及扩大地下室。T1 高层酒店建筑，地面以上高度 102.9m，共 24 层，建筑面积约 65130m²；T2 超高层办公、酒店建筑，地面以上高 358m，共 82 层，标准层平面 53m×53m 建筑面积约 225370m²。两栋塔楼之间连接 3 层商业裙房，并且与塔楼之间不设永久缝，高度约 16.0m，建筑面积约 11070 m²。4 层扩大地下室，上部建筑主体结构沟通落地，扩大地下室柱网 8.5m×8.5m，建筑面积约 93611m²，主要为车库、机电用房。

T2 采用筒中筒结构体系。内筒为型钢钢筋混凝土筒体，墙体洞边角部受力较大处理设型钢柱。外框筒中下部楼层采用钢管混凝土＋矩形钢管六边形网格，上部楼层采用矩形钢管菱形网格，中部楼层网格过渡，采用矩形钢管梁。外筒设置楼面钢管环梁，工字钢楼面梁与之铰接。为满足建筑不规则开洞要求，底层外框筒1～4层采用钢管混凝土框架柱＋斜柱＋框架梁结构。

 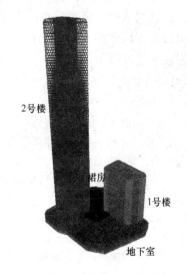

图 12　天津响螺湾中钢广场项目效果图　　图 13　天津响螺湾中钢广场计算模型

在深入研究六边形网格结构工作性能的基础上，提出采用以下结构创新技术：

（1）提出适当选取六边形网格横梁合理刚度，保证结构在竖向荷载作用下正常工作，有利于横梁在大震下首先进入弯曲屈服，为结构提供更好的延性，有利于整体结构抗震，同时具有良好的经济效益。

（2）适当调整扶直六边形斜柱，利于减小斜柱几何长度，有利于减小有利于减小竖向荷载作用下的水平力臂及斜柱杆端弯矩，从而减小斜柱弯曲变形及其对总竖向变形的贡献，提高结构竖向刚度。

（3）采用施工措施改善六边形网格结构重力荷载下受力性能；施工措施 1：主体结构施工期间，释放连接下部六边形角部斜柱的楼面水平钢梁的杆端弯矩及轴力，就可切断六边形网格中部结构自重向角部斜柱转移的路径，下部六边形网格区域结构自重自承担，不向角部转移，主体结构生成后，将此横梁与节点外伸短梁焊接连接。施工措施 2：由于六边形网格角部斜柱截面大于中部斜柱截面，上部30层菱形交叉网格结构中部自重仍将被部分转移至下部六边形网格角部斜柱。主体结构施工期间，过渡区顶层每个角部8根斜柱不连接，上部交叉网格结构自重只能向中部六边形斜柱传递，可进一步减轻角部斜柱负担，主体结构生成后，将此斜柱与节点外伸短管焊接连接。

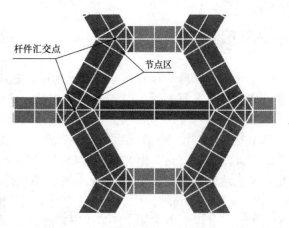

图14 节点刚域示意图

（4）杆元模型节点刚域合理选取：工程节点域占杆件长度比例较大，节点刚域影响较大。下图中红色区域为六边形斜柱与梁汇交节点区，在该重合区域，梁柱变形都会受到约束，节点刚域合理取值十分重要。采用ANSYS全壳元有限元分析得到本工程六边形网格结构节点刚域系数取0.6，较为合理。将其引入杆系模型，对比壳元模型计算结果表明：两个模型构件内力及整体刚度计算结果接近，误差小于5%。杆元模型节点刚域合理取值可较好地反映实际结构工作状态，同时便于构件截面设计。

参考文献

[1] 傅学怡. 实用高层建筑结构设计(第二版). 北京：中国建筑工业出版社，2011.
[2] 傅学怡，孙璨，吴兵. 高层及超高层钢筋混凝土结构的徐变影响分析. 深圳大学学报理工版，2006，23(4)：283-290.
[3] 傅学怡，吴兵，陈贤川，等. 卡塔尔某超高层建筑结构设计研究综述. 建筑结构学报. 2008, 29(1)：1-9，15.
[4] 傅学怡，高颖，肖从真，等. 深圳大梅沙万科总部上部结构设计综述. 建筑结构. 2009, 39(5)：90-96，79.
[5] 傅学怡，吴兵，陈朝晖，等. 深圳火车北站结构设计. 建筑结构学报. 2011.32(12)：98-107.
[6] 傅学怡，杨向兵，高颖，等. 济南奥体中心体育场结构设计. 空间结构. 2009.15(1)：11-16.
[7] 傅学怡，余卫江，孙璨，等. 深圳平安金融中心重力荷载作用下长期变形分析与控制. 建筑结构学报. 2014.35(1)：41-47.
[8] 傅学怡，高颖，周颖等. 天津响螺湾超高层结构设计. 土木工程学报. 2012.45(12)：1-8.

复杂多塔隔震结构设计

郁银泉,邓 烜,曾德民,高晓明,肖 明

(中国建筑标准设计研究院,北京 100048)

摘 要:本文对基底隔震多塔结构和层间隔震多塔结构进行了理想化质点系模型的分析,揭示了复杂多塔隔震结构地震响应的相关特点。同时针对两个分别为基础隔震和层间隔震的大底盘多塔结构实际工程案例的设计方法进行了探讨。通过实际工程案例表明,复杂多塔隔震结构设计时应该根据各塔楼的特性采用多种计算模型综合比较进行设计,对于基础隔震应尽量减小隔震层刚度,选择最优隔震层屈服力;对于层间隔震方案应根据具体的设计目标来选择合适的隔震层刚度和屈服力配置。

关键词:多塔结构;大底盘;隔震设计;复杂结构;层间隔震

DESIGN METHOD OF ISOLATED MULTI-TOWER STRUCTURES

YU Yinquan, DENG Xuan, ZENG Demin, GAO Xiaoming, Xiao Ming

(China Institute of Building Standard Design & Research, Beijing 100048, China)

Abstract: This paper presents analysis of base isolated and story isolated multi-tower structure including a three mass points model and two actual projects already constructed. By the analysis of the three mass points model, some basic rules of the seismic response of isolated multi-tower structures could be known. In practical design, several analysis models should be taken into account according to structural features of the projects. For base isolated structures, a smaller stiffness and an optimized yield force of isolated story will lead to better reduction of seismic reaction. For story isolated structures, the stiffness and yield force should be decided by the aim of design.

Key words: multi-tower structure; enlarged base; isolated structure design; complex structure; story isolated structure

1. 引言

2008 年 5.12 汶川地震以来,建筑的抗震性能越来越受到人们的关注,隔震建筑的数量也逐年增多。2013 年 4.20 雅安地震发生时,采用了隔震技术的芦山县人民医院在地震中的优异表现更加促进了隔震建筑在我国的发展[1]。目前,我国的大多数隔震建筑的结构形式仍然属于中低层且布置相对简单的结构,这种隔震结构可以近似为一个刚体放置于具有耗能能力的弹簧之上,地震时建筑的变形都集中在隔震层发生,因此地震对上部结构的损伤就非常之小。但是，随着我国隔震技术应用的不断发展以及人们对高抗震性能建筑的需求不断增加，一些结构形式相对比较复杂的建筑也开始采用隔震技术，这也给传统的结构隔震设计提出了新问题和挑战。

近年，对于大底盘多塔建筑的隔震已有若干工程案例[2-4]，但其计算分析和设计方法仍然较为简略，仍有很多需要注意和改进的地方。大底盘多塔结构本身因为各塔楼动力特性的差异，各塔楼之间的地震响应存在相互影响，如何正确评价上部塔楼的地震响应是多塔结构设计的重点之一。此外，上部塔楼对底盘的层剪力所产生的影响及其影响的范围将是底盘部分设计时需要给予注意的地方。为此本文将结合两个工程实例，针对大底盘多塔结构分别采用层间隔震和基础隔震时所遇到的相关问题，提出相应的设计方法。

2. 多塔结构隔震简述

2.1 多塔结构的隔震方式

大底盘多塔隔震结构可以采用如图1所示的两种隔震方式。具体方式的选用首先应根据建筑的功能用途和需要，当底盘和上部塔楼都是有提高抗震性能的需求的时候，通常采用基础隔震方式；当主要为了提高塔楼抗震性能，并适当减小底盘地震作用时候可采用层间隔震方式。

为明确以上两种多塔隔震结构的一些基本动力特性，假定在采用隔震技术后，底盘与上部塔楼结构本身在地震作用下均处于弹性状态，所有的地震能量均由隔震层的塑性变形以及结构的整体阻尼来消耗。因此，可将图1所示的两种多塔隔震结构简化为如图2所示的三质点系体系来简化模拟。

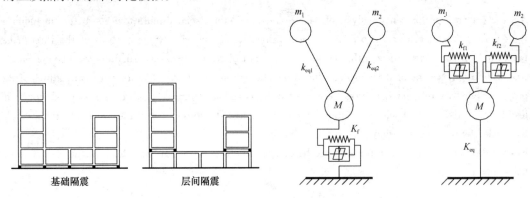

图1 多塔结构隔震方式　　　　图2 多塔质点系模型

图2中M代表大底盘的质量，m_1、m_2代表上部两个塔楼的质量，K_{eq}代表底盘部分的等效刚度，k_{eq1}、k_{eq2}代表上部塔楼的等效刚度。隔震层通过刚度为k_f、k_{f1}、k_{f2}的弹簧和弹塑性耗能单元组成。在隔震层刚度相对上部结构小很多的情况下，对于设置隔震层的部位将其上部结构按刚体考虑，不再考虑其等效刚度。

2.2 简化多塔质点系的地震响应特点

为便于对前述三质点系进行分析，现假定该质点系三个质点的总质量为22万吨，上部结构原始刚度=1×10⁶ kN/m，下部结构原始刚度=2×10⁷ kN/m，通过对三质点系在不同质量分布及不同隔震方式的情况的地震响应进行分析，可以得出多塔隔震结构地震反应的一些特点。本文针对表1所示的五种质量分布方案分别进行了基础隔震和层间隔震的试算，考虑不同的隔震层层刚度 K_f 以及不同的隔震层屈服力 Q_y 对隔震结构地震响应的影响。因隔震层初始布置的时候一般根据竖向承载力来按比例确定，试算中隔震层的刚度及屈服力也同样根据质量的比例关系进行设定。试算中采用EL-Centro（NS）进行时程分析，为便于各工况的比较，时程分析中不考虑结构等效黏滞阻尼的影响。

试算方案　　　　　　　　　　　　　　　　　　　　　　表1

计算方案	M（万吨）	m_1（万吨）	m_2（万吨）
方案Ⅰ	20	1	1
方案Ⅱ	18	2	2
方案Ⅲ	12	5	5
方案Ⅳ	18	1	3
方案Ⅴ	12	2	8

以上5种方案当假定承受每1万吨质量的隔震层屈服后刚度为30000kN/m时，在不同隔震层屈服力的情况下基底剪力的变化如图3所示。

通过比较方案Ⅱ、方案Ⅳ以及方案Ⅲ、方案Ⅴ可以得出，在结构的总质量不变的情况下，不管是基础隔震还是层间隔震，结构的基底剪力与塔楼的质量分布的关系不大。

从图3和图4可以看出，对于基础隔震，在隔震层刚度一定的条件下，随着隔震层屈服力的增加，其基底剪力呈先下降后上升的趋势，存在一个最优的隔震层屈服力。对于层间隔震，则是随着隔震层屈服力的增加，塔楼底部地震力基本不变，基底剪力逐渐减小，但其变化幅度很小。因此对于层间隔震建筑的设计，应尽量采用较小的隔震层屈服刚度以降低上部塔楼的地震反应。

此外，上部塔楼的塔底地震剪力系数试算值见图4，可以看出，对于顶部塔楼的隔震效果，层间隔震要明显优于基础隔震层。

当设定模型隔震层屈服力固定为每一万吨质量对应2500kN时，基础隔震和层间隔震的基底剪力与隔震层屈服后刚度的关系如

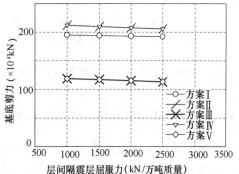

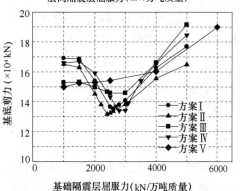

图3　地震剪力-隔震层屈服力的关系

图5所示。可以看出，基础隔震基底剪力将随着隔震层刚度的增加而增大，但是对于层间隔震方案的基底剪力则存在一个最优刚度值。因此当进行基础隔震或者层间隔震的设计目标是降低上部塔楼地震反应时，应尽可能地减小隔震层刚度；当层间隔震的设计目标是减

小隔震层下部结构地震作用时，则可能在合理的隔震层刚度范围内存在一个隔震层的最优刚度，设计中需要给予适当的研究。

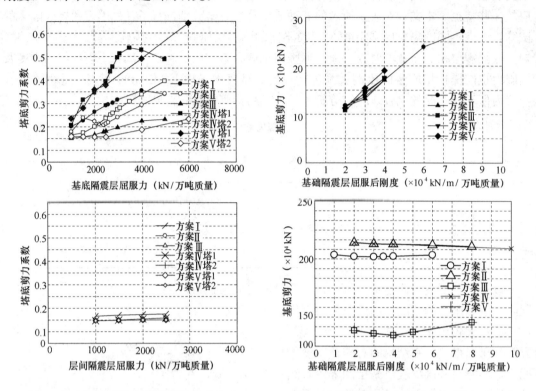

图 4 塔底剪力系数-隔震层屈服力的关系　　图 5 基底剪力-隔震层屈服后刚度的关系

2.3 基于能量法的隔震层最优屈服力校验

根据在某一水准地震作用下，结构单位质量输入的能量与结构的强度和刚度分布无关，而是与结构基本周期相关的一个大致的固定值的基本原理。对于基础隔震结构，可以推导出隔震层剪力响应与隔震层屈服力的关系式 1-式 3[5]，本节将探讨对于多塔基础隔震结构，该式的适用性。

$$基底剪力系数 = (-a + \sqrt{(a^2+1)})\alpha_0 + \alpha_s \tag{1}$$

$$a = 8\left(\frac{\alpha_s}{\alpha_0}\right) \tag{2}$$

$$a_0 = \frac{2\pi V_E}{T \cdot g} \tag{3}$$

其中 α_s 为铅芯阻尼产生的基底剪力系数，即隔震层的屈服力与总质量的比值，α_0 为当全部为无铅芯支座时体系的基底剪力系数，V_E 为地震输入总能量的等效速度，T 为仅考虑橡胶支座水平刚度而得到的基本周期。

根据上式，当隔震支座大小确定之后，可以求解得出隔震层的最优阻尼。若对于前述总质量为 22 万吨的单质点模型，采用 EL-Centro（NS）波对无铅芯阻尼模型进行分析后得到 α_0 为 0.1104。由此可求得隔震层的最优屈服力为 2760kN/每万吨质量。参照图 3 可以得出，多塔结构的最优隔震层屈服力与单质点系结果接近。

3. 某基础隔震大底盘多塔结构

3.1 工程概况

本工程为大型商业综合体（图6），总用地面积22604m^2（33.92亩），总建筑面积24.08万m^2，地上建筑面积17.08万m^2，地下部分4层，地上包括了6层裙房和4栋塔楼，4栋塔楼高度分别为119m（1号楼）、99m（4号楼）、75m（3号楼）、50m（2号楼），裙房总高度为32.7m。业主为了提高整个建筑的抗震安全性能，采用基础隔震技术。

本建筑所在地抗震设防烈度为8度，根据地震安全性评价报告结论，场地的特征周期T_g按0.5s计算。各塔楼采用钢筋混凝土框架－核心筒及框架剪力墙结构，首层平面布置如图7所示。

建筑的剖面示意如图8所示，隔震层设于地下室顶板与上部结构之间。为尽量减小各塔楼的实际高宽比，各塔楼与裙房之间不设置结构缝，连成整体，增加较高塔楼的抗倾覆能力。从裙房顶部标高开始计算，本工程中最高的1号楼其高宽比为2.84，适宜隔震技术的应用。

图6 某基础隔震大底盘多塔建筑

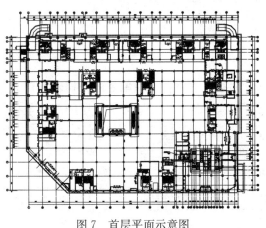

图7 首层平面示意图

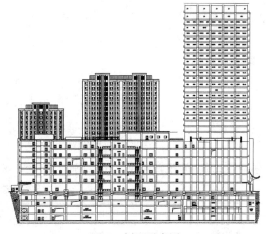

图8 剖面示意图

3.2 隔震结构设计

通过在隔震层合理布置铅芯叠层橡胶支座（图9），可以使隔震结构具备较大的竖向承载力、可变的水平刚度、水平弹性恢复力、足够的阻尼力，满足减小地震作用与抗风等要求。本工程首先根据各部位的竖向荷载共设置隔震支座388个，并大致确定各隔震支座的直径大小，其中纯橡胶支座的总刚度为808186kN/m，在纯橡胶支座的布置下按单质点系模型考虑，RG波作用下对无铅芯阻尼模型进行分析后得到α_0为0.1289，隔震层的最

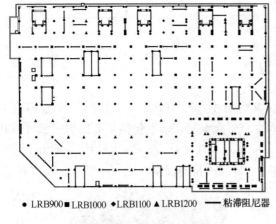

● LRB900 ■ LRB1000 ◆ LRB1100 ▲ LRB1200 ── 粘滞阻尼器

图 9 隔震层布置图

优屈服力为 93456kN。经过计算，对于所有支座采用铅芯阻尼支座（阻尼部分屈服力为 91028kN），其中 LRB900 隔震支座 131 个，LRB1000 隔震支座 113 个，LRB1100 隔震支座 98 个，LRB1200 隔震支座 46 个，并控制各个隔震支座的长期面压均在 12MPa 以内，隔震支座的参数如表 2 所示。为减小隔震层位移并控制结构的平面扭转变形，在隔震层设置了 70 个 150t 的黏滞阻尼器用于吸收地震能量，黏滞阻尼器的阻尼系数为 1500kN/（m/s）。

隔震层的偏心率是隔震层设计的一个重要指标，对于解决大底盘多塔结构通常存在的塔楼偏置与平面扭转不规则等不利问题具有明显的作用。适宜的隔震支座刚度布置，可以有效地缓解上部结构的偏心扭转作用，在本工程的设计中，按式 4～式 9 的方法进行了隔震层偏心率的验算，保证其偏心率不超过 3%。

$$X_g = \frac{\sum N_{l,i} \cdot X_i}{\sum N_{l,i}}, Y_g = \frac{\sum N_{l,i} \cdot Y_i}{\sum N_{l,i}} \quad (4)$$

$$X_k = \frac{\sum K_{ey,i} \cdot X_i}{\sum K_{ey,i}}, Y_k = \frac{\sum K_{ex,i} \cdot Y_i}{\sum K_{ex,i}} \quad (5)$$

$$e_x = |Y_g - Y_k|, e_y = |X_g - X_k| \quad (6)$$

$$K_t = \sum [K_{ex,i}(Y_i - Y_k)^2 + K_{ey,i}(X_i - X_k)^2] \quad (7)$$

$$R_x = \sqrt{\frac{K_t}{\sum K_{ex,i}}}, R_y = \sqrt{\frac{K_t}{\sum K_{ey,i}}} \quad (8)$$

$$\rho_x = \frac{e_y}{R_x}, \rho_y = \frac{e_x}{R_y} \quad (9)$$

公式中 X_g，Y_g 为上部结构的重心坐标，$N_{l,i}$ 为第 i 个隔震支座承受的长期轴压荷载，X_i，Y_i 为第 i 个隔震支座中心位置 X 方向和 Y 方向坐标；X_k，Y_k 为隔震层的刚心坐标；$K_{ex,i}$，$K_{ey,i}$ 为第 i 个隔震支座在隔震层发生位移 δ 时，X 方向和 Y 方向的等效刚度；e_x 为偏心距；K_t 为扭转刚度；R_x 为弹力半径；ρ_x 为偏心率。

隔震支座参数   表 2

型 号	LRB900	LRB1000	LRB1100	LRB1200
橡胶剪切模量（N/mm²）	0.392	0.55	0.55	0.55
有效直径（mm）	900	1000	1100	1200
铅芯直径（mm）	180	200	220	240
支座高度（mm）	360.5	392.0	389.0	389.0
竖向刚度（kN/m）	4415	5659	6459	8086
等效水平刚度（kN/m）	2521	3469	4094	4873
屈服后刚度（kN/m）	1394	2204	2602	3096
屈服力（kN）	202.9	250.4	303.0	360.6

隔震层必须满足风荷载和微震动的要求，将铅芯橡胶支座和叠层橡胶支座刚度简化成双线性，隔震层的水平恢复力特性由铅芯橡胶支座组成，图10给出了本工程隔震层的水平恢复力特性。

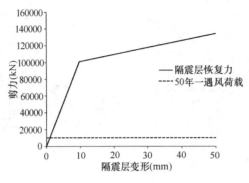

图10 隔震层的恢复力特性

3.3 计算分析方法的确定

为解决多塔结构可能存在各塔楼相互影响以及裙房面积过大，其层剪力不适宜简单评价的问题，在本工程的隔震设计中，采用了多种模型与多种计算方法并行的分析手段，然后结合各塔楼自身的结构特性，选取适宜的设计手法。本工程裙房部分的质量约17万吨，四个塔楼的质量分别为6.8万吨（1号楼）、2.6万吨（4号楼）、1.7万吨（3号楼）、0.4万吨（2号楼），大底盘裙房质量较大，其动力特性相对独立，2、3、4号楼质量很小，对大底盘裙房的动力特性影响较小。为此，本工程主要的计算分析模型有如下几种（图11）：

（a）模型Ⅰ：不考虑上部塔楼的独立裙房模型，将上部塔楼的质量直接施加于裙房顶部，该模型用于裙房部分的隔震分析和结构设计。

（b）模型Ⅱ：因1号楼的质量相对较大，且高达119m，其地震动力响应将相对独立，因此对于1号楼采用连带两跨裙房的单塔模型进行隔震分析和结构设计。

（c）模型Ⅲ：为综合的考虑各种因素，进行本工程的整体地震反应验证，采用实际的多塔整体模型进行隔震分析和验证。

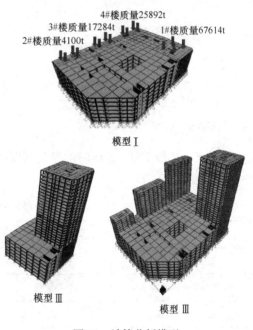

图11 计算分析模型

以上模型均分别进行了隔震模型和原结构模型的分析和比对。采用SAP2000V14.1对上述三个模型进行了上部结构为弹性的计算分析，采用ABAQUS6.11对模型Ⅲ的隔震模型进行了同时考虑上部结构弹塑性与隔震支座非线性行为的罕遇地震分析。计算中，隔震支座水平向采用双线性平行四边形滞回模型,竖向采用拉压刚度比为 1/10 的线弹性模型,阻尼器采用速度比例型的黏滞阻尼单元,用于分析的地震波包含了两条天然波和一条人工波,各波的主方向波形及其反应谱如图 12 所示。

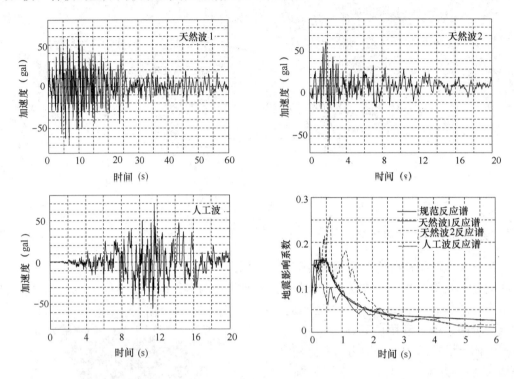

图 12 分析用地震波

3.4 地震反应分析

隔震支座按照 100% 变形时的等效刚度进行模态分析时,各模型隔震前后的自振周期特性如表 3 所示。

模态分析结果　　　　　　　　　　　　　　表 3

模 型	T1	T2	T3
模型Ⅰ—隔震	3.162	3.090	2.721
模型Ⅰ—原结构	0.843	0.759	0.603
模型Ⅱ—隔震	3.24	3.16	2.59
模型Ⅱ—原结构	2.04	1.91	1.30
模型Ⅲ—隔震	3.67	3.56	2.80
模型Ⅲ—原结构	2.222	1.891	1.741

为验证隔震结构对多塔结构扭转效应的缓解作用,模型Ⅲ小震时程分析中裙房各层的层间位移比如表 4 所示,最大层位移比为 1.185,结构变形以平动为主。

模型Ⅲ裙房部分的各层位移比 表4

楼层	天然波1 X向	天然波2 X向	人工波 X向	天然波1 Y向	天然波2 Y向	人工波 Y向
1	1.167	1.043	1.123	1.163	1.143	1.133
2	1.059	1.035	1.177	1.157	1.132	1.125
3	1.059	1.027	1.139	1.155	1.122	1.118
4	1.058	1.021	1.110	1.155	1.116	1.113
5	1.063	1.015	1.185	1.161	1.117	1.114
6	1.072	1.012	1.170	1.156	1.102	1.105

在8度设防地震水准下，上部结构按照弹性考虑时，模型Ⅰ与模型Ⅱ隔震结构与原结构对应的层剪力分布与层间位移角分布分别如图13、图14所示。模型Ⅰ求得的减震系数为0.18，模型Ⅱ求得的减震系数为0.40。

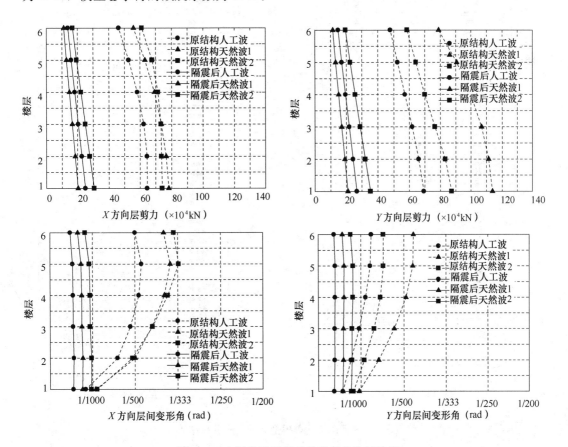

图13 8度设防地震模型Ⅰ的时程分析结果

整体模型Ⅲ中，隔震结构与原结构最高的两个塔楼（1号楼与4号楼）的时程分析结果如图15所示。

在对隔震层进行罕遇地震分析时，对模型Ⅰ、Ⅱ按上部结构弹性进行了计算，模型Ⅰ的隔震层最大位移为335mm，黏滞阻尼器的最大出力为1345kN。对于模型Ⅲ分别采用了

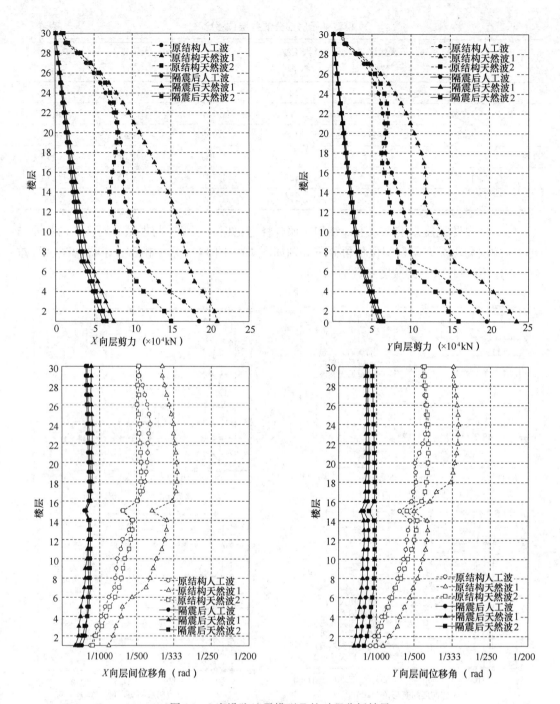

图14 8度设防地震模型Ⅱ的时程分析结果

SAP2000 和 ABAQUS 针对上部结构弹性和弹塑性的模型进行分析。当上部结构按照弹性计算时,隔震层的最大位移为290mm,当考虑上部结构塑性变形时,隔震层的最大位移为175mm。可以看出,隔震结构在无法保证上部结构大震弹性的情况下,简单的采用弹性计算分析方法会对隔震层位移过大评价。考虑上部结构弹塑性时1号楼及4号楼在罕遇地震作用下的层间位移角见图16。

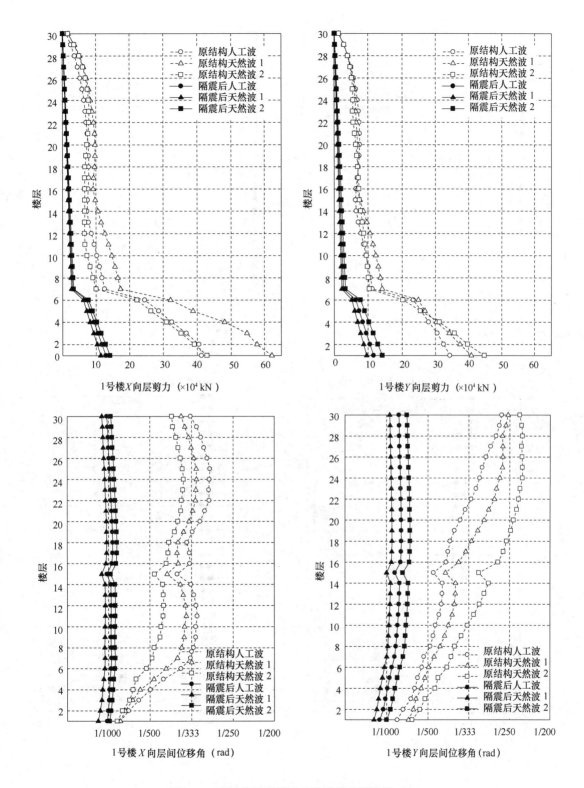

图 15 8 度设防地震模型Ⅲ的时程分析结果(一)

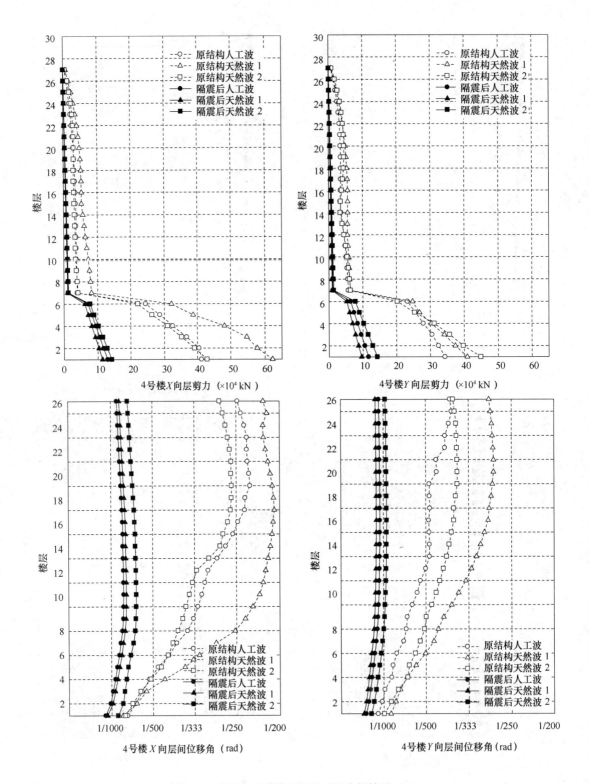

图15 8度设防地震模型Ⅲ的时程分析结果(二)

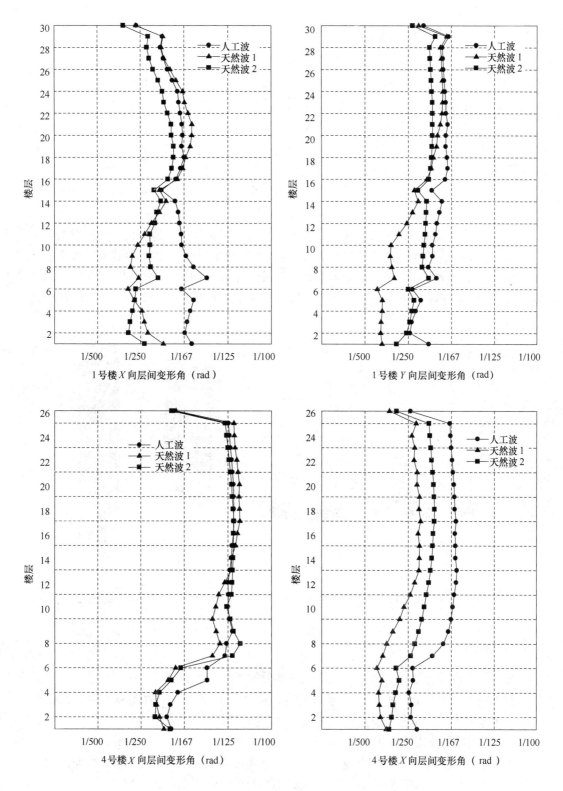

图16 8度罕遇地震模型Ⅲ的弹塑性时程分析

3.5 局部支座拉应力的对策

本工程 1 号楼为框架核心筒结构，高度为 119m、4 号楼为框架剪力墙结构，高度为 99m。此类结构形式，剪力墙及核心筒是主要的抗侧力构件，通常在首层承担的地震剪力和倾覆力矩均超过 80%，但其在建筑平面中负担的竖向荷载却相对较小，因此剪力墙和筒体的角部在罕遇地震作用下都不可避免地要承受一定的拉力。为避免罕遇地震中局部墙下支座可能产生的较大拉应力而引起的支座损伤，本工程对局部剪力墙及筒体下部的隔震支座采用了对竖向拉力进行释放的构造措施，隔震支座的具体连接方式如图 17 所示。类似的释放竖向拉力型隔震支座，日本学者山崎慎介[6]等有过一定研究并在一些高宽比较大的建筑中有实际的应用。

为进一步验证局部隔震支座释放后建筑的隔震性能及其对上部结构损伤的影响，在进行罕遇地震整体多塔模型弹塑性分析的时候，真实的考虑了竖向拉力释放型隔震支座的边界条件，经计算局部剪力墙下罕遇地震作用时的支座的竖向提离高度如图 18 所示，其最大值为 49mm，该提离高度对于整体结构的变形倾角影响极小，对整体结构的塑性损伤影响不大。

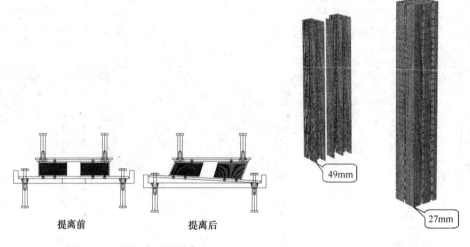

图 17　竖向释放型隔震支座　　　　图 18　局部墙下的支座提离高度

3.6 结构设计方法

在完成隔震设计之后，针对地上结构各部分的不同特点，需要采用不同的结构设计方法。对于底部裙房部分，可以采用考虑减震系数之后模型Ⅰ的非隔震模型进行构件的设计。对于 1 号楼则采用考虑减震系数之后模型Ⅱ的非隔震模型进行构件设计。对于 2、3、4 号楼，因其质量相对裙房来说很小，则分别针对这三个塔楼裙房以上部分，采用输入裙房顶楼面谱的方式进行设计。楼面谱的地震影响系数最大值按式 10 计算，楼面谱的特征周期取模型Ⅰ隔震结构的基本周期。4 号楼的设计楼面谱如图 19 所示。

$$\alpha_{\max 楼面} = \frac{PGA_{楼面}}{PGA_{基底}} \alpha_{\max} \tag{10}$$

其中，α_{max} 为规范反应谱中的地震影响系数最大值，$\alpha_{max地面}$ 为楼面谱的地震影响系数最大值，$PGA_{基底}$ 为输入地震波的峰值加速度，$PGA_{楼面}$ 为楼面相应部位的加速反应的峰值。

此外，各塔楼在满足前述计算结果的前提下，还需要满足我国《建筑抗震设计规范》GB 50011—2010中相应设防地震水准对于最小剪重比要求。通过对结构最小设计剪力的控制，

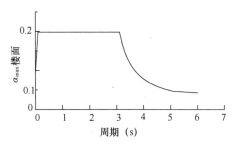

图19 4号楼的设计楼面谱

可以进一步提高上部结构抗震能力，减小隔震建筑上部结构在罕遇地震下的结构损伤。

对于多塔结构的地下室及基础部分，需要充分考虑上部结构在设防地震及罕遇地震作用下通过隔震层传递的外力，保证地下部分的构件达到中震抗弯弹性及大震抗剪弹性的性能目标。

4. 某层间隔震大底盘多塔结构

4.1 工程概况

本工程为某地铁车辆段上盖开发的项目，在大跨纯框架结构之上建设10栋8层的公租房（图20）。为降低上部住宅结构的结构设计难度并减小上部住宅对下部停车库的影响，各塔楼采用层间隔震技术，隔震层设置于大底盘顶板之上。本工程所在区域抗震设防烈度7度，场地特征周期0.45s。

首层大底盘通过伸缩缝分割成平面尺寸约70m×70m的结构单元，每个结构单元上有两栋住宅，结构剖面布置如图21所示。首层框架结构层高为9.3m，东西向柱距为10.8m、12.7m、14.4m，南北向柱距为5.7m、7.5m，住宅楼下部框架柱尺寸为1500mm×1500mm，其他框架柱为900mm×900mm。上部住宅也采用框架结构，层高2.8m，平面尺寸约15m×51m，塔楼高宽比约为1.9，框架柱尺寸为700mm×600mm～500mm×500mm。

图20 某层间隔震大底盘多塔建筑

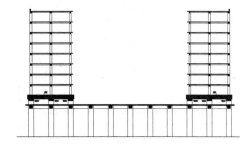

图21 结构剖面图

4.2 隔震结构设计及地震反应分析

隔震层的布置如图22所示，在结构的边柱及角柱部位布置铅芯叠层橡胶隔震支座提

供一定的刚度及阻尼耗能能力，其型号为LRB800、LRB1000，在结构的中柱部位布置无铅芯叠层橡胶隔震支座LNR1000保证隔震层获得适宜的刚度。

为充分考虑多塔结构的复杂性，在进行结构分析和设计时候，采用以下两种模型进行计算（图23）：

模型Ⅰ：上部单塔与部分底盘组合的模型

模型Ⅱ：上部双塔与整个底盘组合的模型

按隔震支座100%变形时的等效刚度计算，隔震后的结构周期见表5，相比原结构有明显的周期延长效果。

模态分析结果　　　　　　表5

模　　型	T1	T2	T3
模型Ⅰ—隔震	2.88	2.87	2.45
模型Ⅰ—原结构	1.10	1.04	0.99
模型Ⅱ—隔震	3.12	3.10	2.77
模型Ⅱ—原结构	1.89	1.79	1.68

模型Ⅰ隔震层以上结构质量为8547t，隔震层以下结构质量为7919t。模型Ⅱ隔震层以上结构质量为18762t，隔震层以下结构质量为15838t。时程分析时候采用两条天然波与一条人工波，各地震波及其加速度反应谱见图24。

采用设防地震水准进行时程分析之后，两个模型的层剪力及层间位移角见图25、图26。从图中可以看出，隔震对于减小上部及下部结构地震作用有明显的效果，并且采用模型Ⅰ进行计算分析将更加偏于安全。

在罕遇地震水准下，隔震层的最大位移为200mm，满足LRB800隔震支座的最

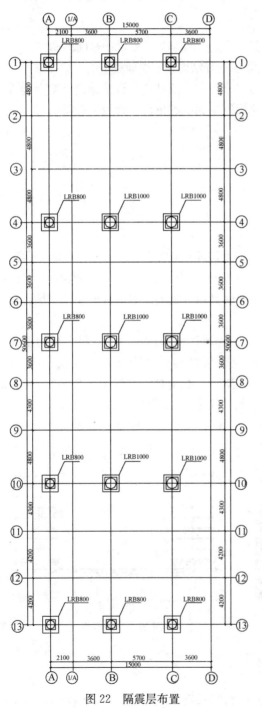

图22　隔震层布置

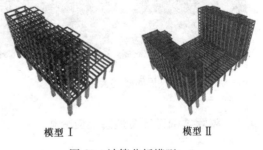

图23　计算分析模型

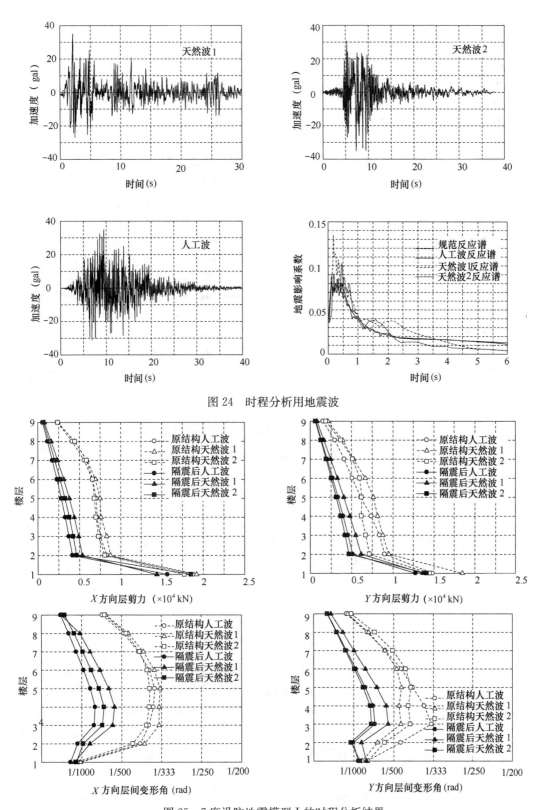

图 24 时程分析用地震波

图 25 7 度设防地震模型 Ⅰ 的时程分析结果

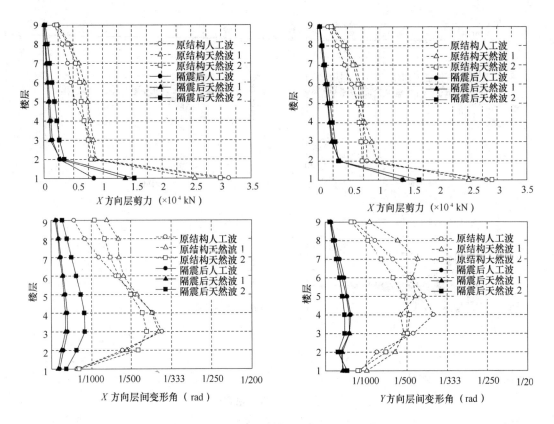

图 26　7 度设防地震模型 II 的时程分析结果

大位移限值要求，支座的最小面压为 0.52MPa，不存在支座受拉。

4.3　结构设计方法

对于层间隔震结构，结构设计中最重要的就是要保证隔震层以下结构的安全。为了保证在各水准地震作用下隔震层以上结构仍能够发挥良好的性能，需要对在塔楼平面范围内的底盘部分进行中震抗弯弹性和大震抗剪弹性设计。

5.　结论

本文通过对一个简化多塔隔震 3 质点系模型进行理论分析，并结合两个工程设计实例，对基础隔震多塔结构和层间隔震多塔结构各自的适用性及其地震反应的特点进行了分析，对两种形式隔震结构的设计方法进行了比对，得到以下结论：

（1）基础隔震多塔结构为获得更好的隔震效果，应尽可能减小隔震层刚度，并需要对最优隔震层屈服力进行分析和选择。

（2）经过优化布置的隔震层刚度分布，可以有效地缓解大底盘多塔基础隔震结构的平面扭转不规则效应。

（3）通过能量法基本原理对隔震层阻尼的最优屈服力进行预测对基础隔震设计具有较好的指导意义，能有效地配置铅芯阻尼的数量从而达到最优隔震效果。

(4)对于规模较大的大底盘多塔结构,应根据各塔楼自身的结构特点采用多种计算分析模型与设计方法进行综合评价。对质量较大且高度较高的塔楼,宜采用单塔模型进行计算,对于质量较小的塔楼,宜按楼面谱进行设计。

(5)对于高度较高的高层建筑结构,在进行罕遇地震分析时,应考虑上部结构弹塑性对隔震层地震响应的影响。

(6)对于层间隔震多塔结构,当主要目的为降低上部结构地震作用时,应尽可能减小隔震层刚度和屈服力。当需要考虑减小对下部结构影响时,则应选择一个适宜的隔震层刚度。

(7)层间隔震多塔结构设计时应按单塔、多塔模型分别计算,综合判断其最不利的情况,同时应保证隔震层以下结构具有足够的刚度和强度。

参考文献

[1] 周云,吴从晓,张崇凌,等. 芦山县人民医院门诊综合楼隔震结构分析与设计[J]. 建筑结构,2013(024):23-27.
Zhou Yun. Analysis and design of seismic isolation structure in outpatient building of the Lushan County People's Hospital [J]. Building Structure, 2013 (024): 23-27. (in Chinese)

[2] 赵楠,马凯,李婷,等. 大底盘多塔高层隔震结构的地震响应[J]. 土木工程学报,2010,43(1):254-258.
Zhao Nan. Seismic response of multi-tower isolated structure with an enlarged base [J]. China Civil Engineering Journal, 2010, 43(1): 254-258. (in Chinese)

[3] 吴曼林,谭平,唐述桥,等. 大底盘多塔楼结构的隔震减震策略研究[J]. 广州大学学报:自然科学版,2010,9(2):83-89.
WU Man-lin. Seismic isolation strategies for multi-tower structures with a large podium [J]. Journal of Guangzhou University(Natural Science Edition), 2010, 9(2): 83-89. (in Chinese)

[4] 谭平,周福霖. 大平台多塔楼结构的隔震减震控制[J]. 广州大学学报:自然科学版,2008,6(5):77-82.
TAN Ping. Seismic isolation and response control of multi-tower structure on a large platform[J]. Journal of Guangzhou University(Natural Science Edition), 2008, 6(5): 77-82.

[5] 日本建筑学会. 隔震结构设计[M]. 北京,地震出版社,2006.
Architectural Institute of Japan. Recommendation for the Design of Base Isolated Buildings[M]. Beijing, Earthquake Press, 2006.

[6] 山崎慎介,等. ストッパーピンと軸受を用いた積層ゴム支承の引張対応機構の開発 [J]. 学術講演梗概集. B-2,構造Ⅱ,振動,原子力プラント,2011,2011:511-512.
YAMAZAKI Shinsuke. The Development of Rubber Bearings Tensile Counter Measures System Using Stopper Pins and Bearings [J]. Transactions of AIJ. B-2, Structure II, 2011: 511-512.

屈曲约束支撑减震技术的研究与应用*

李国强[1]，胡大柱[2,3]，郭小康[3]，孙飞飞[1]

(1. 同济大学 土木工程防灾国家重点实验室，上海 200092；
2. 上海应用技术学院 城市建设与安全工程学院，上海 201418；
3. 上海蓝科建筑减震科技股份有限公司，上海 200433)

摘 要：简要介绍了屈曲约束支撑的组成，构件的各项力学性能以及对主体结构性能的影响。总结了屈曲约束支撑构件以及结构体系的研究现状。根据性能和构造的不同，将屈曲约束支撑进行了分类以及不同类型构件的试验要求。对屈曲约束支撑的施工技术进行详细介绍，并给出了具体施工步骤。给出了屈曲约束支撑的设计要求和工程应用方法，并通过四个典型工程案例介绍了屈曲约束支撑的应用效果。

关键词：屈曲约束支撑；性能分类；验收要求；应用方法；工程案例

中图分类号：TP391

STUDY AND APPLICATION OF BUCKLING-RESTRAINED BRACE ON SEISMIC REDUCTION TECHNOLOGY

G. Q. Li[1], D. Z. Hu[2,3], X. K. Guo[2], F. F. Sun[1]

(1. State Key Laboratory for Disaster Reduction in Civil Engineering, Tongji University, Shanghai 200092, China; 2. College of Urban Construction and Safety Engineering, Shanghai Institute of Technolgy, Shanghai 201418, China; 3. Shanghai Lanke Building Damping Technology Co., Ltd. Shanghai 200433, China)

Abstract: The primary configuration of buckling-restrained brace (BRB) is briefly introduced. The mechanical properties of BRBs and their influence on the main structure are presented. Research and developments of BRBs and their application in various types of structure systems are summarized. According to the performance and configuration, the classification and testing standards of BRBs are proposed. Construction technology of BRBs and the procedure are explicitly introduced. The design requirements and methods of BRBs for practical applications are presented, which is illuminated by four actual engineering cases to show the benefits of the BRB technology.

Key words: Buckling restrained brace; performance classification; testing standards; application method; engineering cases

---

* 基金项目：( )

第一作者：李国强(1963—)，男，博士，教授，博导，主要从事钢结构方面的研究，E-mail：gqli@tongji.cn.

1. 引言

进入 21 世纪后，特别是 2004 年发生在印尼由里氏 8.9 级地震引起的强烈海啸以后，地球进入到地震多发阶段，2008 年汶川 8.0 级地震、2010 年 1 月～4 月相继发生的海地 7.3 级地震、智利 8.8 级地震、玉树 7.1 级地震、2011 年日本 9.0 级地震、2012 墨西哥 7.4 级地震、2013 年雅安 7.0 级地震以及最近发生在新疆和田的 7.3 级地震等都是强度较大的地震。

中国地处于世界两大地震带—环太平洋地震带与欧亚地震带的交汇部位，一直是地震多发国家。统计数字表明，中国的陆地面积占全球陆地面积百分之六左右；中国的人口占全球人口百分之二十左右，然而中国的陆地地震占全球陆地地震百分之三十三左右，而地震造成死亡的人数达全球地震死亡人数的 50% 以上。

地震人员伤亡主要由建筑结构在地震下破坏和倒塌引起的[1]，提高建筑结构抗震能力有两种方法[2,3]，一种是提高结构的承载能力，另一种是提高结构的延性和耗能能力。其中第一种方法通常称为抗震，第二种方法通常称为减震。传统抗震设计方法成熟，各种分析手段和设计软件齐全，是以结构的承载力抵抗地震所产生的构件内力。在计算地震作用的时候是以既定的"设防烈度"作为设计依据，但当发生突发性超烈度地震时，房屋可能会严重破坏，并且由于地震的随机性，建筑结构的破损程度及倒塌可能性难以控制，故安全性难以保证。减震设计是近 30 年发展起来的新方法，它是通过附加减震设备耗散地震能量，主要梁柱构件较少的承担地震作用。但是减震设备不能提高结构的承载能力，结构整体造价增加。并且结构工程师一般在减震设计方面工程经验有限，而且减震设备大多数为专利产品，限制了使用。寻求一种结构构件，既能提高结构承载力又能作为耗能减震构件，已经成为广大建筑抗震工作者的追寻目标。屈曲约束支撑正是这样一种将承载构件和耗能减震构件合二为一的高效、经济、新技术型的结构构件。

屈曲约束支撑（Bucling-retrained-brace）又称无粘结支撑，是一种新型钢结构支撑，也是一种耗能支撑[4-6]。尽管目前屈曲约束支撑形式多样，但原理基本相似，都是利用刚度较大的外套筒拟制中心芯板的屈曲。一般而言，屈曲约束支撑由 3 部分所组成，即核心单元、约束单元和滑动机制。支撑的中心是芯材（Steel Core），是构件中的主要受力原件，由特定强度的钢板制成（一般是低屈服点的钢材），在轴向力作用下允许有较大的塑性变形，通过这种变形可以达到耗能的目的，常见形状有十字形和一字形。为避免芯材受压时整体屈曲，即在受拉和受压时都能达到屈服，芯材被置于一个钢套管内（Steel Tube），然后在套管内灌筑混凝土或砂浆等高强度填充物。构件的组成示意如图 1（a）所示。为减小或消除芯材受轴力时传给砂浆或混凝土的力，而且，由于泊松效应，芯材在受压情况下会膨胀，因此，在芯材和砂浆之间设有一层无粘结材料或非常狭小的空气层（Gap）。因而，此支撑在受压时亦能达到完全屈服，使支撑受压承载力与受拉承载力相当，克服了传统支撑受压屈曲的缺点，改善了支撑的承载能力，使支撑的滞回曲线饱满，如图 1（b）所示，提高了结构的抗震能力。

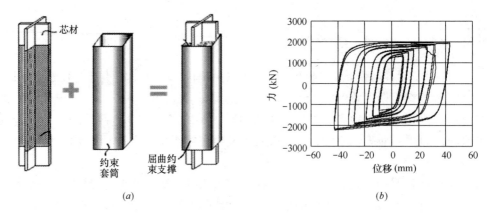

图 1 屈曲约束支撑原理图
(*a*) 构造图；(*b*) 滞回性能

2. 屈曲约束支撑优点

屈曲约束支撑作为一种改进的钢支撑，力学性能、滞回性能等都各项指标都有所提高。在工程应用中能够提供更优的结构性能。

2.1 承载力性能

与普通支撑相比，在相同轴向刚度条件下，屈曲约束支撑承载力高：
抗震设计中，普通支撑的轴向承载力设计值为：

$$N_b = \frac{\varphi A f}{1 + 0.35 \lambda_n} \tag{1}$$

式中：φ——轴心受压构件的稳定系数；
A——支撑的截面面积；
f——支撑材料强度设计值；
λ_n——支撑的正则化长细比，$\lambda_n = (\lambda/\pi)\sqrt{f_{ay}/E}$；
λ——支撑长细比；
f_{ay}——钢材屈服强度；
E——钢材弹性模量。
抗震设计中，屈曲约束支撑的轴向承载力设计值为：

$$N_b = A_1 f = 0.9 A_1 f_y \tag{2}$$

式中：A_1——约束屈服段的钢材截面面积；
f_y——芯板钢材的屈服强度标准值。
由式（1）和式（2）的对比可以看出，如果屈曲约束支撑的截面面积 A 与普通钢支撑相等，屈曲约束支撑的承载力与普通钢支撑承载力比值为：

$$\zeta = \frac{1+0.35\lambda_n}{\varphi} \tag{3}$$

以某长细比为100，材料Q235的钢支撑为例，由《钢结构设计规范》可得到稳定系数 $\varphi=0.463$，$\lambda_n=1.08\lambda_n=(\lambda/\pi)\sqrt{f_{ay}/E}$，因而，屈曲约束支撑的承载力为普通钢支撑的2.97倍。

2.2 滞回性能

屈曲约束支撑在弹性阶段工作时，就如同普通支撑可为结构提供很大的抗侧刚度，可用于抵抗小震以及风荷载的作用。屈曲约束支撑在弹塑性阶段工作时，变形能力强、滞回性能好，就如同一个性能优良的耗能阻尼器，可用于结构抵御强烈地震作用。

屈曲约束支撑区别于普通支撑的最大特点是解决了支撑杆件的受压屈曲问题，使得核心单元无论受拉还是受压都能达到全截面屈服。国外研究者利用试验方法对此做过分析对比[7]，如图2所示。

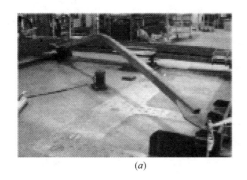
(a)

(b)

图 2 国外某对比试验图片
(a) 普通支撑；(b) 屈曲约束支撑

图3为普通支撑与屈曲约束支撑的轴向力－变形滞回曲线对比。在屈曲约束支撑的设计时，不必考虑失稳，只计算其强度，因为支撑受拉和受压性能基本相同，在受压时不会发生屈曲。

滞回性能是反映一个结构或者构件抗震性能的重要指标，曲线饱满，各种受力状态下均匀对称是一个结构或构件抗震性能好的表现。从屈曲约束支撑与普通钢支撑的滞回曲线对比可以看出，屈曲约束支撑符合这项要求，而普通钢支撑在受压状态下屈曲失稳，拉压不对称，因而不是一种好的耗能构件。

普通钢支撑受压弯曲，在由受压状态转向受拉状态时，构件的轴向刚度无法发挥作

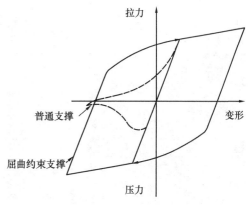

图 3 屈曲约束支撑与普通支撑性能对比

用，所能给结构提供的刚度较小，而当构件被拉直的瞬间，其刚度突然增大，在动荷载作用下，将会对相邻构件具有冲击的作用，这也是普通钢支撑地震作用下滞回性能较差的一个影响因素。

2.3 对主体结构性能的影响

根据建筑抗震设计规范中的规定，在强震作用下，如果人字形或V形支撑受压斜杆受压屈曲（如图4所示），则受拉斜杆内力将大于受压屈曲斜杆内力，这两个力在横梁交会点的合力的竖向分量使横梁产生较大的竖向变形，人字形支撑使梁下陷，V形支撑使梁上鼓。这可能引起横梁破坏，并在节点两侧的梁端产生塑性铰，同时体系的抗剪能力发生较大的退化，为了防止支撑斜杆受压屈曲，其组合的内力设计值应乘以增大系数1.5。同时为了防止横梁破坏，对普通钢支撑框架，应按下述方法进行横梁的设计：

人字形支撑和V形支撑的横梁在支撑连接处应保持连续，该横梁应承受支撑斜杆传来的内力，并应按不计入支撑支点作用的简支梁验算重力荷载和受压支撑屈曲后产生的不平衡力作用下的承载力，其中不平衡力值可取受拉支撑的竖向分量减去受压支撑屈曲压力竖向分量的30%。

在上述支撑布置方法中，如果采用屈曲约束支撑，由于支撑不会发生受压屈曲的现象，受拉和受压承载力相同，在支撑与梁的交点处就不会出现该不平衡力。

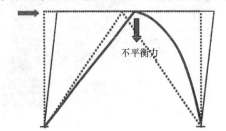 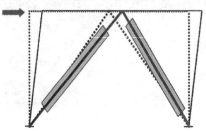

图4 梁不平衡力对比

2.4 对支撑节点设计的影响

节点作为框架中连接传力的枢纽，对整个结构性能的充分发挥具有十分重要的意义。支撑与主体结构连接常采用螺栓或者焊接连接形式。普通支撑受压屈曲，受拉与受压承载力差异很大，支撑的内力由支撑受压承载力控制，而与支撑相邻构件的内力由支撑受拉承载力控制。因而支撑节点是由构件的受拉承载力控制。支撑节点如果仅按照支撑受压最大内力设计，在往复荷载作用下，支撑由受压状态变为受拉状态，将会导致节点先于构件破坏。如图5所示的支撑破坏形式，即为螺栓连接节点处由于构件净截面强度不足而导致的破坏形式。

图5 支撑连接节点破坏

3 国内外研究现状

3.1 屈曲约束支撑构件研究

屈曲约束支撑起源于日本,在国外以及台湾地区研究起步较早。1976 年,Kimura[8]等首先提出了屈曲约束支撑这一概念,其将普通钢支撑置于钢管混凝土中,用于提高普通支撑的稳定承载力。由于屈曲约束支撑构造的特殊性,这种构件的研究以试验研究为主。对约束套筒形式的研究内容较为广泛,包括约束钢管的弹性屈曲强度与内核支撑屈服强度比值研究[9],双钢管约束形式[10,11],预制组合式屈曲约束支撑[12]。无粘结材料是屈曲约束支撑重要组成部分,它对构件滞回性能的影响也是重要研究方向[12]。除套筒形式、无粘结材料外,陈正诚对用低屈服点钢屈曲约束支撑恢复力特性进行了研究[13]。

我国大陆地区屈曲约束支撑研究时间较短,工程应用较晚。2006 年,欧进萍等[14]对 7 个"一"字形内芯外包钢管的屈曲约束支撑进行了静力反复试验和子结构拟动力试验。2006 年,李国强等[15-17]利用国产低碳钢和低屈服点钢研制了一种纯钢型的屈曲约束支撑构件,该支撑芯板为"一"字形,通过加劲钢套筒措施实现了对芯板的约束作用。该种屈曲约束支撑已在 2010 年上海世界博览会主场馆"世博中心"中应用 108 根,成为首个国产屈曲约束支撑大规模应用的案例。2008 年,吴斌等研发了三种屈曲约束支撑,"十"字形核心纯钢型屈曲约束支撑[18],角钢组合式屈曲约束支撑[18]与组合钢管式屈曲约束支撑[19,20],并进行了少量构件与子系统试验。2010 年,郭彦林等提出了两种组装式屈曲约束支撑,分别称为型钢组合装配式屈曲约束支撑[21]与双矩管带肋约束型屈曲约束支撑[22],并对其基本性能和设计方法进行了研究,提出了有效约束比的确定方法、外围约束构件螺栓强度验算方法和间距控制原则、内核与外围之间空隙的控制要求。2009 年至 2011 年,广州大学周云教授等提出了"核心单元局部削弱相当于其他部分加强"的新型防屈曲耗能支撑设计思想,给出"开孔式"和"开槽式"两种新型防屈曲耗能支撑设计方案[23-25]。

从屈曲约束支撑研究成果可以归纳出其基本的设计要求:
(1) 约束钢管的弹性屈曲强度与内核支撑屈服强度比值应大于 1.5;
(2) 无粘结材料压缩性不宜过大;
(3) 低屈服点钢材制作的屈曲约束支撑滞回性能更为优越。

3.2 屈曲约束支撑钢框架结构体系研究

在钢框架结构中,支撑是一种经济的抗侧力构件,可使钢框架具备更高的抗侧刚度,传统的带支撑框架有中心支撑框架 CBF(Concentrically Braced Frame)和偏心支撑框架 EBF(Eccentrically Braced Frame)。中震和强震时,CBF 中的支撑会受压屈曲和受拉屈服,而受压屈曲极大地限制了支撑作为抗侧力构件的耗散能力,因而大多数抗震规范都对支撑的抗震承载力进行了调低。EBF 通过偏心梁段的屈服,限制支撑的屈曲,可是结构具有较好的耗能性能,但是由于偏心梁段屈服,地震后结构修复较为困难。而采用屈曲约束支撑作钢框架的抗侧力构件,形成屈曲约束支撑框架 BRBF（Buckling Restrained Braced Frame）。

屈曲约束支撑钢框架体系的研究包括试验研究和理论研究两大内容。

试验研究包括大比例屈曲约束支撑钢框架的静力试验[26],拟动力试验[28],振动台试验[29,30]。这些试验结果都表明,屈曲约束支撑钢框架结构体系比传统中心支撑框架具有更为优越的抗震性能。

在理论研究方面,包括屈曲约束支撑钢结构的地震反应与耗能能力分析[31,32],附加阻尼分析[33]。近年来,对屈曲约束支撑钢框架的设计方法进行了广泛研究,形成了弹性设计方法[34]、基于等效阻尼比的弹塑性位移简化计算方法等[35]。

大量的试验和理论研究表明,屈曲约束支撑框架具有优越的抗震性能,在弹性阶段,屈曲约束支撑能起到抗侧力构件的作用,而在罕遇地震下,它成为主要的耗能构件。

3.3 屈曲约束支撑混凝土框架结构体系研究

混凝土框架结构由于良好的经济性,在我国的多层建筑中具有广泛应用。国内学者对屈曲约束支撑-混凝土框架结构相关性能进行了研究,并取得了丰富的研究成果,可概括为两大部分：一、屈曲约束支撑与混凝土框架连接节点的研究；二、抗震性能的研究。

目前屈曲约束支撑与混凝土框架的连接节点的研究主要有三种类型[36,37]：预埋钢筋式节点、预埋钢板式节点以及预埋型钢式节点。预埋钢筋式节点采用与传统的预埋件节点形式类似的构造,施工便捷,如能保证锚筋的锚固长度以及锚固板的抗弯刚度,则能获得较大的节点刚度以及一定的承载力,目前还缺乏综合考虑钢筋、节点连接板等因素共同作用的力学模型和设计方法,有待进一步研究。预埋型钢式节点分为两种,一种是将与支撑相连接的梁、柱设计为型钢混凝土构件,这种节点形式其设计方法及节点性能与型钢混凝土-钢支撑结构的节点性能接近。第二种节点形式在与支撑相连的混凝土梁、柱中埋入一段型钢,在保证混凝土、型钢荷载传递长度满足要求的前提下,其性能较为可靠。但是这两种节点形式都存在价格高、施工复杂的问题。预埋钢板式节点是一种将钢板埋入混凝土梁柱内,在钢板两侧焊接栓钉传递荷载的节点形式。这种连接节点施工便捷程度介于上两种节点形式之间,但存在承载力低、用钢量大的缺点,并且梁、柱截面高度较小的框架内该构造不易实现。

对屈曲约束支撑-混凝土框架结构的抗震性能研究方法包括有限元分析以及试验研究[38-49]。国内北京工业大学、中国建筑科学研究院、哈尔滨工业大学、北京建筑大学、上海建筑科学研究院分别进行了单榀平面框架的拟静力试验,得到了屈曲约束支撑进入屈服状态时的层间位移,整体结构不适宜继续承载的最大层间位移、结构达到最大层间位移时的损伤和破坏、延性系数等。这些静力试验结果都表明在混凝土框架结构中设置屈曲约束支撑后,结构的滞回曲线形状饱满,与普通支撑框架相比,屈曲约束支撑既可提高刚度和承载力,还可以大幅提高结构的耗能能力。

除试验研究外,兰州理工大学[50-52]、西南交通大学[53]、东南大学[53]对屈曲约束支撑-混凝土框架结构的抗震性能进行了系统的理论分析,以平面框架、规则空间框架以及不规则空间框架等为研究对象,从结构的动力特性、地震反应、耗能机理与特征等分析了屈曲约束支撑-混凝土结构的特点，从理论分析结果中可以看出屈曲约束支撑对改善混凝土结构的抗震性能具有显著贡献。

4 TJ型屈曲约束支撑

4.1 TJ型屈曲约束支撑的构造分类

TJ型屈曲约束支撑分为两大类：(1) 钢管混凝土型支撑；(2) 纯钢型屈曲约束支撑。

TJ钢管混凝土型屈曲约束支撑，以钢管混凝土为约束单元，其基本构造为："十"字形、"一"字形或"工"字形耗能核心外包钢管混凝土；耗能核心采用低碳钢或低屈服点钢；约束非屈服段与无约束非屈服段采用高强钢；屈服耗能核心与非屈服节点区采用焊接连接，用于减小支撑外观；核心耗能段与内填高强混凝土之间采用无粘结脱层材料隔离；耗能核心中部采用限位装置，防止支撑外套筒产生滑移，其具体构造示意见图10。

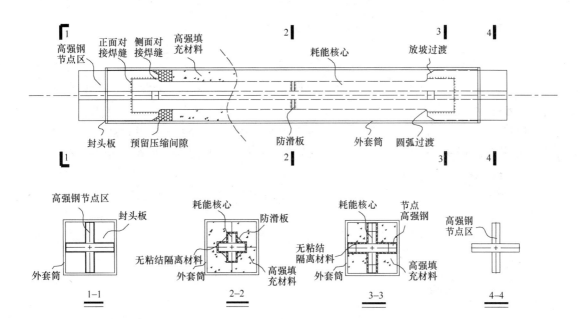

图6 TJII型屈曲约束支撑构造特点

TJ纯钢型屈曲约束支撑，其基本构造为：采用在"H"型或者"口"字型耗能核外部套箱型钢管的措施，钢管内部不填充防屈曲材料，通过限制内部耗能核心板件宽厚比的措施，使其不发生局部屈曲破坏，提高了经济性，加工简单，经济性高，适合用于吨位较大的支撑，其具体构造形式见图7。

4.2 屈曲约束支撑的性能分类

根据屈曲约束支撑性能和使用要求的不同，又可将其分为"承载型屈曲约束支撑"、

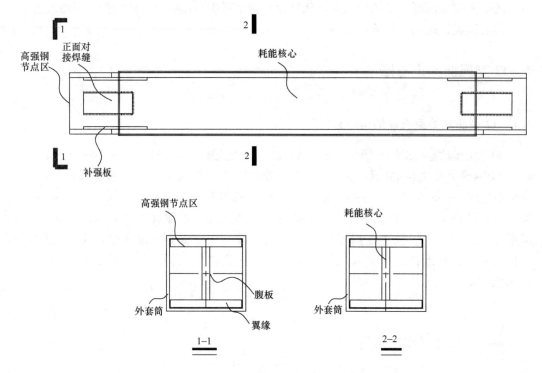

图 7 TJH 型屈曲约束支撑构造特点

"耗能型屈曲约束支撑"和"屈曲约束支撑型阻尼器"三种类型。

（1）作为承载构件使用，指通过引入屈曲约束机制来提高支撑构件的设计承载力，保证支撑在屈服前不会发生失稳破坏，从而充分发挥钢材强度，称之为"承载型屈曲约束支撑"；

（2）作为耗能构件使用，指在弹性阶段利用屈曲约束的原理来提高支撑的设计承载力，在弹塑性阶段利用芯板钢材的拉压屈服滞回来耗能的消能减震结构构件，称为"耗能型屈曲约束支撑"；

（3）作为拉压屈服型软钢阻尼器使用，一般控制在小震屈服，称为"屈曲约束支撑型阻尼器"。

4.3 试验研究与总结

通过对约 150 根足尺 TJ 型屈曲约束支撑的试验研究，充分验证了其构造的合理性，并积累了大量原始试验数据，对国产屈曲约束支撑的应用与推广奠定了坚实的基础。这些构件试验中包括"承载型屈曲约束支撑"、"耗能型屈曲约束支撑"和"屈曲约束支撑型阻尼器"。

按照不低于国家《建筑抗震设计规范》的要求，对上述试件进行了试验，承载型屈曲约束支撑典型滞回曲线见图 10，耗能型屈曲约束支撑见图 11，屈曲约束支撑型阻尼器见图 12 所示。

从上述构件滞回曲线可以看出，TJ 型系列屈曲约束支撑滞回曲线可以看出，构件拉压性能对称。耗能型屈曲约束支撑和屈曲约束支撑阻尼器低周疲劳性能满足规范要求，无退化和衰减现象发生。

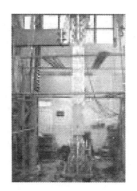

图 8 试验照片

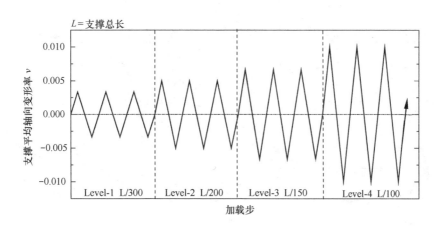

图 9 《建筑抗震设计规范》标准加载

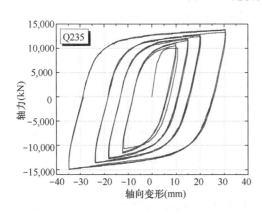
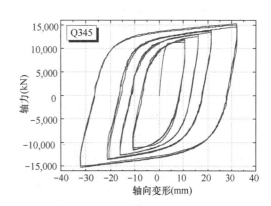

图 10 TJ 承载型屈曲约束支撑试验滞回曲线

通常采用一些重要的指标，应变强化系数 ω、材料超强系数 μ 与支撑承载力拉压比 β 等，评价屈曲约束支撑的性能。

应变强化系数 ω 为支撑的极限承载力与其初始屈服力的比值，ω 由下式确定：

$$\omega = \frac{F_u}{F'_y} \tag{4}$$

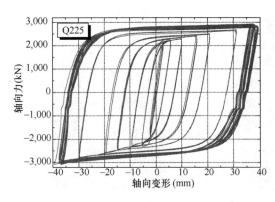
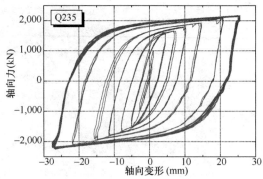

图 11 TJ 耗能型屈曲约束支撑试验滞回曲线

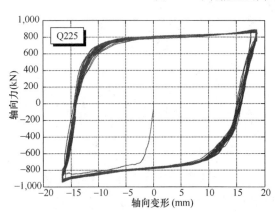
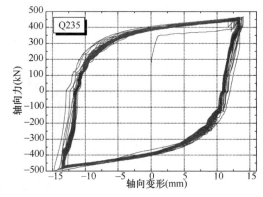

图 12 TJ 屈曲约束支撑阻尼器试验滞回曲线

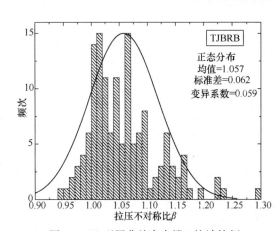

图 13 TJ 型屈曲约束支撑 β 统计特征

式中 F_u 为支撑的极限承载力；F'_y 为支撑的初始屈服力。

材料超强系数 μ 由下式确定

$$\mu = \frac{F'_y}{F_y} \quad (5)$$

式中 F_y 为支撑屈服力标准值，由下式确定

$$F_y = A f_y \quad (6)$$

式中 A 为支撑核心截面面积；f_y 为支撑核心钢材屈服强度标准值，按照中国钢结构设计规范取值。

参数 β 反应屈曲约束支撑拉压承载能力的差异，由下式确定

$$\beta = \left| F^-_{umax} / F^+_{umax} \right| \quad (7)$$

式中 F^-_{umax} 与 F^+_{umax} 分别为支撑最大的抗压与抗拉承载能力。

对 TJ 型屈曲约束支撑上述参数进行概率统计，结果如下：

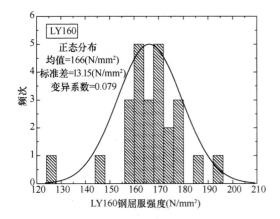
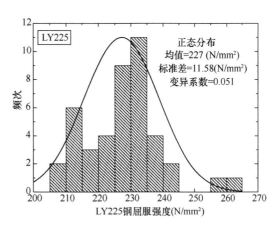

图 14　LY160 钢材料超强系数 μ 统计特征　　　　图 15　LY225 钢材料超强系数 μ 统计特征

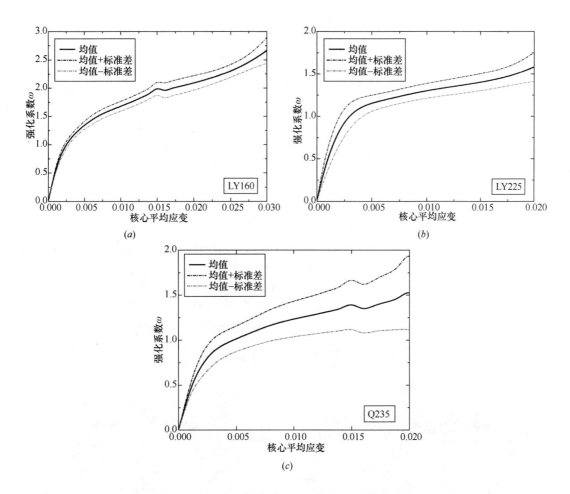

图 16　芯材应变强化系数 ω 均值及 ±1 倍标准差值
(a) LY160；(b) LY225；(c) Q235

4.4 屈曲约束支撑施工技术

对屈曲约束支撑构件设计、性能指标要求等，我国规范都有相应的原则性规定，但是目前我国并没有相应的屈曲约束支撑施工、验收的规范或规定，从而导致这项技术不能按照国家统一标准进行施工以及验收。本文系统提出屈曲约束支撑的就位方法，解决屈曲约束支撑现场运输、吊装、就位、固定、与非结构构件的连接等全套技术。

4.4.1 屈曲约束支撑就位方法

屈曲约束支撑的就位，就是将其从堆放位置运输到吊装位置，再将其吊装到安装位置，并临时固定于安装位置。对于轻型屈曲约束支撑与重型屈曲约束支撑的安装，其就位、水平运输及垂直吊装的方法也不同。

4.4.1.1 就位流程

屈曲约束支撑水平运输及垂直运输——架子搭设——现场尺寸复核——屈曲约束支撑节点板修正——屈曲约束支撑吊装——屈曲约束支撑临时固定及校正

4.4.1.2 屈曲约束支撑就位

（1）屈曲约束支撑水平运输及垂直运输

对于轻型屈曲约束支撑，其自重较小、长度较短，一般布置在层高不高、跨度不大的柱间，屈曲约束支撑通常采用自制小推车、钢滚轮小车水平运输，见图17。

对于屈曲约束支撑，其自重较大、长度较长，一般布置在层高高、较大跨度的空间结构中，屈曲约束支撑通常采用大型塔吊、汽车吊等完成水平运输，见图18。

（2）架子搭设

屈曲约束支撑水平运输到安装位置处后，根据现场情况及规范要求搭设活动架，做好安全防范，为后面的工作提供操作平台。

（3）现场尺寸复核

现场尺寸复核包括节点板间支撑的安装长度，节点板节点与施工图的偏位，以及节点板在安装过程中出现的出平面偏移，见图19。出现的偏差的原因一方面由于主体结构在施工中存在偏差，致使屈曲约束支撑节点板在安装中顺延了原有的偏差；另一方面节点板在安装过程中因定位不准出现偏差。

图17 水平运输用工具

图18 垂直运输用工具

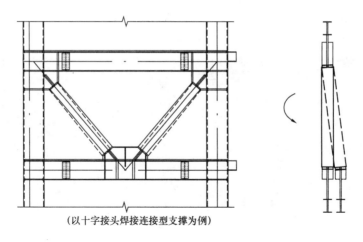

(以十字接头焊接连接型支撑为例)

图 19 安装偏差

（4）屈曲约束支撑节点板修正

现场尺寸出现偏差超出设计要求后，应采取相应的措施予以纠偏、矫正，保证支撑的安装满足设计要求。采取措施如下：

① 节点板切割

复核现场节点板间的净距小于支撑长度时，可通过现场火焰切割对节点板进行修正，保证支撑的安装长度，见图 20。

② 火焰校正

火焰校正的方法通常是在出平面偏移量不大，及板厚较小的情况下采用。

③ 安装一块底座钢板

安装一块底座钢板的方法是在出平面偏移量较大，及板厚较大的情况下以及节点板间的净距大于支撑长度时采用。采用此方法时节点处须切割与底座钢板厚度相同的长度，且此钢板须要检测 Z 向性能，见图 21。

图 20 修正节点板

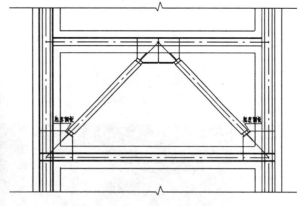

图 21 底座钢板

（5）屈曲约束支撑吊装

对于轻型屈曲约束支撑的吊装，通常采用手拉葫芦倒链来完成。成品支撑构件焊有专用的吊耳（沿支撑长度有两道），可直接穿入吊索进行吊装，支撑有吊耳的面朝上，起吊为两端不等高起吊,首先牵拉支撑下端达到安装部位,再牵拉支撑上端就位,见图22。

图 22 吊装

对于重型屈曲约束支撑的吊装,通常采用大型塔吊或者汽车吊。吊装时要遵循塔吊或者汽车吊的吊装规则要求,做好安全防范措施。吊装过程中,同样也是先保证下端就位,然后做临时固定措施,再调整上端就位,见图23。

图 23 重型支撑

(6)屈曲约束支撑临时固定及校正

支撑牵拉到位后,采用措施进行临时固定,再次校正后就位完成,见图24。

图 24 固定

5 屈曲约束支撑设计要求

5.1 设计准则

屈曲约束支撑有三种承载力,即设计承载力、屈服承载力与极限承载力,在结构设计中适用于不同阶段的计算。

5.1.1 设计承载力

屈曲约束支撑的设计承载力是按下式计算得到的:

$$N_b = Af \tag{8}$$

式中:A——屈曲约束支撑芯材截面面积;

f——芯材强度设计值,按照表1确定。

芯板钢材强度设计值　　　　　　　　表1

芯材型号	f(MPa)
BLY160	125
BLY225	180
Q195	175
Q235	215

5.1.2 屈服承载力

屈服承载力用于结构的弹塑性分析,为屈曲约束支撑首次进入屈服的轴向力,是按下式计算得到的:

$$N_{by} = Af_y \tag{9}$$

式中:A——屈曲约束支撑芯材截面面积;

f_y——芯材屈服强度,按照表2确定。

芯板钢材的屈服强度　　　　　　　　表2

材料型号	f_y(MPa)
BLY160	140
BLY225	205
Q195	195
Q235	235

5.1.3 极限承载力

国家规范中规定的钢材强度为下限,计算屈曲约束支撑极限承载力时应考虑钢材的超强系数,且屈曲约束支撑的芯材在地震作用下拉压屈服会产生应变强化效应,考虑应变强化后,支撑的最大承载力为极限承载力,可按下式计算:

$$N_{bu} = R_y \omega N_{by} \tag{10}$$

式中:R_y——芯板钢材的超强系数,根据表3确定;

ω——应变强化调整系数,根据表3确定;

N_{by} —— 屈曲约束支撑屈服承载力。

芯板钢材的超强系数和应变强化调整系数　　表3

材料型号	R_y	ω
BLY160	1.14	2.4
BLY225	1.10	1.6
Q195	1.15	1.6
Q235	1.15	1.6

5.1.4　外套筒抗弯刚度要求

为保证承载型屈曲约束支撑在轴力作用下不发生整体失稳，其套筒抗弯刚度应满足下式要求：

$$\frac{\pi^2 EI}{l^2} \geqslant 1.8 N_{by} \tag{11}$$

或：

$$I \geqslant \frac{1.8 N_{by} l^2}{\pi^2 E} \tag{12}$$

式中：I —— 屈曲约束支撑套筒的弱轴惯性矩；

E —— 套筒钢材弹性模量；

l —— 支撑长度；

N_{by} —— 承载型屈曲约束支撑的屈服承载力。

为保证耗能型屈曲约束支撑在大震作用下不发生整体失稳，其套筒抗弯刚度应满足下式要求：

$$\frac{\pi^2 EI}{l^2} \geqslant 1.8 N_{bu} \tag{13}$$

或：

$$I \geqslant \frac{1.8 N_{bu} l^2}{\pi^2 E} \tag{14}$$

式中：I —— 屈曲约束支撑套筒的弱轴惯性矩；

E —— 套筒钢材弹性模量；

l —— 支撑长度；

N_{bu} —— 耗能型屈曲约束支撑的极限承载力。

屈曲约束支撑型阻尼器的外套筒抗弯刚度要求可参考耗能型屈曲约束支撑。

5.1.5　小震验算

对承载型屈曲约束支撑和耗能型屈曲约束支撑，其设计承载力应满足下式的要求：

$$N \leqslant N_b \tag{15}$$

式中　N —— 屈曲约束支撑在风载或小震作用下与其他静力荷载的基本组合的最大轴力值（受拉或者受压）；

N_b —— 屈曲约束支撑的设计承载力。

对屈曲约束支撑型阻尼器，其计算可参照位移相关型消能器。

5.1.6 中震和大震验算

承载型屈曲约束支撑应根据抗震性能目标的不同进行中震和大震的验算，但在任何情况下，应能够满足芯材屈服前不发生整体屈曲的要求。

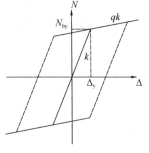

图 25 屈曲约束支撑
双线性恢复力模型

耗能型屈曲约束支撑在中震和大震下的验算应采用弹塑性分析方法。其滞回模型可选用图 25 所示的双线性恢复力模型。

对屈曲约束支撑型阻尼器，在中震和大震下的验算也应采用弹塑性分析方法。

5.1.7 连接节点设计

对于高强螺栓型连接节点，应保证与屈曲约束支撑相连节点在地震作用下不发生滑移，其连接高强度摩擦型螺栓的数量 n 可由下式确定：

$$n \geqslant \frac{1.2N_c}{0.9n_f \mu P} \tag{16}$$

式中：n_f——传力摩擦面数目；

μ——摩擦面的抗滑移系数；

P——每个高强螺栓的预拉力；

N_c——连接节点设计用屈曲约束支撑承载力代表值，对承载型屈曲约束支撑为屈服承载力 N_{by}；对耗能型屈曲约束支撑和屈曲约束支撑型阻尼器为极限承载力 N_{bu}。

对于焊接型连接可采用对接焊缝，焊接连接的承载力 N_f 应满足下式要求：

对承载型：

$$N_f \geqslant 1.2N_{by} \tag{17}$$

对于耗能型和阻尼器：

$$N_f \geqslant 1.2N_{bu} \tag{18}$$

5.2 性能标准

屈曲约束支撑的性能标准主要涵盖以下几个方面的内容：

5.2.1 芯板钢材

三种类型的屈曲约束支撑芯板钢材应满足表 3 的规定。

5.2.2 焊缝

屈曲约束支撑内部芯材的剖口焊缝应满足一级探伤要求，并应能够满足一级探伤的要求。

屈曲约束支撑芯板屈服段钢材性能指标　　表 4

类　型	屈强比	伸长率	冲击功韧性	屈服强度波动范围
承载型屈曲约束支撑	≤0.8	≥20%	≥27J（常温）	按一般钢材供货要求
耗能型屈曲约束支撑	≤0.8	≥30%	≥27J（常温）	±20MPa（低屈服点钢） Q235（235 MPa～295 MPa）
屈曲约束支撑型阻尼器	≤0.8	≥40%	≥27J（0℃）	±20MPa（低屈服点钢）

5.2.3 防火

屈曲约束支撑采用纯钢约束体系,则需要做防火保护,且其防火等级等同于相应的普通钢支撑;若采用填充材料约束体系,则自由伸缩段和封头板必须做防火保护,此时套筒内填充材料为具有隔热性能的不燃烧材料时套筒可不做防火保护,否则套筒也应采取防火保护。

5.3 产品验收

因为屈曲约束支撑不得当作一般的钢结构构件来设计制作,必须由专业厂家作为产品来生产和供货,其产品性能应通过验收后方可在工程中使用。屈曲约束支撑产品生产厂家的屈曲约束支撑生产相关的管理应通过 ISO 质量管理体系认证。

5.3.1 质量验收

质量验收分为产品检验批和施工检验批。产品验收又分为进场验收和竣工验收两阶段。屈曲约束支撑进场验收资料应包含:产品合格证、产品质保书。而屈曲约束支撑竣工验收资料应包含:产品型式检验报告、芯材质保书及复检报告、设计院对型式检验的确认。

屈曲约束支撑的施工验收应符合以下规定:

(1)焊缝外形尺寸应符合现行国家标准《钢结构焊缝外形尺寸》GB 10854 的规定。

(2)焊接接头内部缺陷分级应符合现行国家标准《钢焊缝手工超声波探伤方法和探伤结果分级》GB 11345 的规定,焊缝质量等级及缺陷分级应符合国家标准《建筑钢结构焊接技术规程》JGJ 81 的规定。

(3)其他验收项目应符合《钢结程施工质量验收规范》GB 50205 的规定。

产品检验批控制项目包括主控项目和一般项目,其中主控项目包括:产品合格证、产品质保书、产品型式检验报告、芯材质保书及复检报告。而一般项目包括:产品外观、尺寸检测、产品漆膜厚度检测(一般项目检测抽检率不少于 20%)。

屈曲约束支撑各项尺寸偏差应满足表 5 要求。

屈曲约束支撑尺寸偏差要求　　　　　表 5

项次	支撑长度	宽	高	外径	侧弯矢量	扭曲
偏差要求	±3mm	±2mm	±2mm	±2mm	$L/1000$,且≤10mm	$h(d)/250$,且≤5mm

注:L——支撑长度;h——支撑高度;d——支撑外径。3 产品漆膜厚度检测根据设计要求。

5.3.2 产品型式检验要求

屈曲约束支撑应按照同一工程中支撑的构造形式、芯材和屈服承载力分类进行抽样试验检验,构造形式和芯材相同且屈服承载力在 50% 至 150% 范围内的屈曲约束支撑划分为同一类别。每种类别抽样比例为 2%,且不少于 1 根。

对耗能型屈曲约束支撑,试验时依次在 1/300,1/200,1/150,1/100 支撑长度的拉伸和压缩往复各 3 次变形。试验得到的滞回曲线应稳定、饱满,具有正的增量刚度,且最后一级变形第 3 次循环的承载力不低于历经最大承载力的 85%,历经最大承载力不高于屈曲约束支撑极限承载力计算值的 1.1 倍。然后在 1/150 支撑长度的位移幅值下往复循环

30圈后,屈曲约束支撑的主要设计指标误差和衰减量不应超过15%,且不应有明显的低周疲劳现象。

对屈曲约束支撑型阻尼器,试验时在 n 倍(n 应不小于10)的阻尼器屈服位移下往复循环30圈后,屈曲约束支撑阻尼器最大承载力的衰减量不应超过15%,且不应有明显的低周疲劳现象。

对承载型屈曲约束支撑,试验时支撑在受拉和受压两种状态下均应能达到屈服承载力,且支撑不可出现屈曲的现象。对于有抗震要求的承载型屈曲约束支撑还应在1/100支撑长度的拉伸和压缩往复各一次变形,支撑不得出现屈曲失稳现象。

为降低工程成本,且鼓励生产制造厂家提高产品质量,对通过ISO质量管理体系认证的厂家生产的同一芯材及同一构造形式的屈曲约束支撑,在工程实践中若能够提供屈服承载力不小于该工程支撑最大屈服承载力的实际工程实验报告,并获得该工程设计单位认可的,可以作为工程验收依据,以减少抽检数量或免于抽检。

6 屈曲约束支撑应用方法

结构所受的地震作用与结构刚度有关,质量和结构布置接近的结构,刚度越大,所受的地震作用越大。通常作为抗侧力的构件有:支撑、剪力墙、抗弯框架等。其中剪力墙结构刚度最大,所受到的地震作用也最大。抗弯框架刚度最小,但是抗弯框架是通过加大梁柱截面的形式来达到抗侧力要求,经济性较差。支撑是一种较为经济的抗侧力构件,现常用的支撑有普通钢结构支撑、混凝土支撑、型钢混凝土支撑以及屈曲约束支撑。为提高普通钢结构支撑的稳定性能,在设计中需要加大其截面,导致结构刚度增大。混凝土支撑和型钢混凝土支撑由于混凝土开裂,在强地震作用下其耗能能力和延性性能降低。屈曲约束支撑解决了普通支撑的稳定性能,使钢结构支撑在受拉和受压时候性能一致,因而可以减小支撑的截面面积,降低结构的刚度,减低结构地震作用,且使结构延性性能好耗能能力增强。

6.1 钢框架屈曲约束支撑结构体系

钢框架屈曲约束支撑结构体系中抗侧力构件由两部分组成,分别是屈曲约束支撑和钢框架。可分解如下:

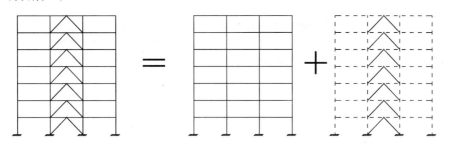

图26 钢框架屈曲约束支撑框架体系

结构设计中,框架部分的抗侧刚度主要由梁、柱的抗弯刚度来实现,增大梁柱的抗弯刚度则主要通过加大截面面积和缩小柱距,而柱距、梁柱截面又受建筑功能限制,无法无限制增大。因而支撑的刚度对建筑物的刚度影响较大，进而影响到整个结构的地震作用。

结构的强度、刚度和延性是一个建筑物抵抗地震作用的关键三个指标，较小的刚度、较大的强度和延性是建筑抗震设计的目标。屈曲约束支撑与普通钢支撑相比，最大的特点即是其通过较小的刚度获得更大的强度和延性。

屈曲约束支撑由于没有受压稳定问题，其在风载与小震下构件承载能力比普通支撑提高 2－10 倍，支撑构件越长其承载能力提高越多。相同承载力条件下与普通支撑相比，其截面可大大减小，所以结构的抗侧刚度变柔，周期相应加大。

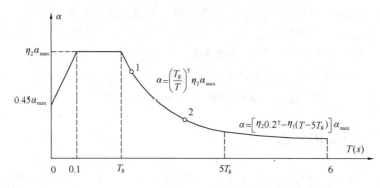

图 27 反应谱曲线

根据图 27 的反应谱曲线可以看出，结构周期加长，其地震反应就减小，例如，假设普通支撑方案结构周期在 1 点，那么采用屈曲约束支撑方案后，结构周期到达 2 点，地震反应加速度有很大的降低；当结构的地震作用以第一振型为主时候，地震反应将大大减小，减小幅度一般为 10～25％。如果结构由地震工况控制，地震作用减小后，所有构件截面都可以减小，一般可降低结构整体造价 10～30％。

6.2 混凝土框架屈曲约束支撑结构体系

钢筋混凝土框架结构应用广泛，在我国 6 层以下建筑结构中是主要的抗侧力体系。但是当楼层增高或者高烈度地震区，与钢框架相同，由于其抗侧刚度有限，通常需要设置抗侧力构件。常用剪力墙方案，形成框架－剪力墙结构体系。在我国 2010 版本的建筑抗震设计规范中，新增加了钢支撑-混凝土框架结构体系。这种结构体系中，对丙类建筑其主要的抗震指标如表 6 所示。

钢支撑-混凝土框架结构体系主要抗震要求　　　　表 6

项　　次		规　　定						
设防烈度		6		7		8（0.2g）		8（0.3g）
最大适用高度		95		85		70		57.5
抗震等级	高度	≤24	>24	≤24	>24	≤24	>24	≤24	>24
	钢支撑框架部分	三	二	二	一	一	一	一	一
	混凝土框架部分	四	三	三	二	二	二	二	一
结构布置	水平方向	应在两个主轴方向同时布置						
	竖向布置	上下连续布置，当无法连续时候，宜在邻跨连续布置						

续表

项 次	规 定
支撑间距	满足抗震墙的间距要求
地震作用分摊率要求	底层钢支撑框架按刚度分配的地震倾覆力矩应大于结构总地震倾覆力矩的50%
阻尼比	不应大于0.045,也可按混凝土框架部分和钢支撑部分在结构总变形能所占的比例折算为等效阻尼比
混凝土框架部分的地震作用	按纯框架结构和支撑框架结构分别进行计算,并取较大值
位移限值	1/652
连接部位	应符合连接不先于支撑破坏的要求

在表5中,其各项规定以及要求主要针对普通钢支撑混凝土框架结构体系,但部分要求也可用于混凝土框架屈曲约束支撑结构体系。

在表5中,支撑框架的抗震等级比混凝土框架的抗震等级高一级,这主要是为了确保支撑框架在地震作用下发挥主要作用,这一条也适合于钢筋混凝土框架屈曲约束支撑结构体系,为确保屈曲约束支撑发挥作用,需要与支撑直接相连的框架具有足够的安全储备。

屈曲约束支撑作为主要抗侧力构件应用的时候,其布置要求应与普通钢支撑相同。然而当屈曲约束支撑作为主要耗能构件应用时,其布置可更为灵活,但在结构设计中,要确保支撑产生的阻尼力能有效传递至结构的基础。

屈曲约束支撑的间距设置要求应与抗震墙、普通钢支撑的要求相同。

在抗震规范中,要求底层普通钢支撑框架按刚度分配的地震倾覆力矩应大于结构总地震倾覆力矩的50%。在纯框架结构中,层间位移最大的楼层一般都出现在底层。钢支撑混凝土框架结构中对底层地震倾覆力矩的规定和框架——剪力墙结构类似,都是要求主要抗侧力构件(钢支撑或者剪力墙)在底层所占地震倾覆力矩比值大于50%。但在实际工程中,对框架——剪力墙结构体系,当剪力墙设置的数量不足,导致在基本振型地震作用下,框架部分承受的地震倾覆力矩大于结构总地震倾覆力矩的50%,即剪力墙承担的地震倾覆力矩比值不超过50%时候,其框架部分的抗震等级按框架结构采用。对于钢支撑混凝土框架结构,框架部分的抗震等级与纯框架结构相比,并未降低。因而,对于混凝土框架屈曲约束支撑结构体系,当支撑框架承担的结构总地震倾覆力矩比值小于50%的时候,对混凝土框架部分,其抗震等级仍然按照框架结构,对于支撑框架,其抗震等级须提高一级,但建筑物的适用高度仍然应遵照框架结构的规定。

由于材料的特征,钢结构阻尼比小于混凝土结构,当钢支撑作为主要抗侧力构件时,其阻尼比应进行折算。但当屈曲约束支撑小震屈服时,应按照附加阻尼比后的总等效阻尼比进行计算。

在钢支撑混凝土框架结构中,由于普通钢支撑在地震作用下会失稳,从而丧失抗震能力,因而在计算结构地震作用的时候,要求按照有支撑框架和无支撑纯框架进行地震作用计算,从而进行配筋。当采用屈曲约束支撑后,由于屈曲约束支撑具有良好的延性和耗能能力,在强震作用下,也不会退出工作,仍将保持有承载能力,且会起到主要耗能构件的作用。因而,对于混凝土框架屈曲约束支撑结构体系,无需进行无框架的地震作用计算,仅仅需要计算带屈曲约束支撑的框架结构,并按照此内力进行配筋。

关于位移限值,钢支撑混凝土框架结构中按照框架结构和框架-抗震墙结构的层间位移限值进行插值决定。验算小震下结构的弹性层间位移是为了实现第一水准的抗震设防要求,即在频度高而强度低的地震作用下,要求建筑能完全履行其设计功能,即小震弹性的设计要求,确保结构构件不损坏。对于混凝土框架屈曲约束支撑结构体系,即使支撑屈服也不代表结构构件损坏,因此建议混凝土框架屈曲约束支撑结构层间位移角限值取值与混凝土框架相同,即1/550。

综合表6的各项设计指标要求,以及上述分析,对丙类建筑的混凝土框架屈曲约束支撑结构抗震要求提出以下规定:

混凝土框架屈曲约束支撑结构体系主要抗震要求 表7

项次		规 定						
设防烈度		6		7		8 (0.2g)		8 (0.3g)
最大适用高度		95		85		70		57.5
抗震等级	高度	≤24	>24	≤24	>24	≤24	>24	≤24	>24
	钢支撑框架部分	三	二	二	一	一	一	一	一
	混凝土框架部分	四	三	三	二	二	一	二	一
结构布置	水平方向	宜在两个主轴方向同时布置。							
	竖向布置	上下连续布置,当无法连续时候,宜在邻跨连续布置							
支撑间距		满足抗震墙的间距要求							
地震作用分摊率要求		可不作具体要求							
阻尼比		根据屈曲约束支撑是否耗能进行确定							
混凝土框架部分的地震作用		按照带屈曲约束支撑的框架进行结构计算							
位移限值		1/550							
连接部位		将屈曲约束支撑的极限承载力作为连接设计值							

注:1. 当采用屈曲约束支撑型阻尼器时,可仅在一个方向布置,但计算结构地震作用的时候需要分别计算两个主轴方向的阻尼比,对框架柱构件,需要按照两个方向较大地震作用的方向进行设计。

2. 当屈曲约束支撑的设计性能要求是小震弹性状态时候,阻尼比应按照普通钢支撑混凝土框架相同的要求。但当屈曲约束支撑作为小震屈服的阻尼器应用时候,应计算附加阻尼比的大小,然后叠加混凝土结构5%的阻尼比,得到总阻尼比。

6.3 平面不规则结构体系中的应用

历次大地震都对不规则结构造成了严重破坏,甚至倒塌,给人民生命财产带来不可估量的重大损失,因而提高不规则结构的抗震安全性能是结构设计的重要任务。随着建筑设计的发展,现阶段在结构设计过程中,设计师经常会遇到结构平面不规则的情况,例如,扭转位移比超过限值、扭转周期比与平动周期比的比值不满足规范要求等。为解决平面不规则的问题,通常会采用加大结构抗扭刚度的方法。增加结构抗扭刚度有效方法是在扭转效应较大的部位设置抗侧力构件,例如剪力墙、支撑或抗弯框架。

设置屈曲约束支撑解决平面不规则是一种简便易行的方法。

造成平面不规则结构的原因有多种,平面尺寸不规则、结构布置平面不规则等,计算中主要表现为扭转周期比超限、结构扭转位移比超限。

6.3.1 平面尺寸不规则情况

在平面不规则结构中,平面尺寸的凹角和凸角不规则是重要原因之一,在我国现行

《建筑抗震设计规范》中对平面凹角、凸角做了详细规定。针对这几种不规类型，分别给出几种常用的解决方案。

对如图 28 所示的 L 形平面结构体系，通常在 L 形的两个端部设置抗侧力构件，既能增加结构的抗侧刚度，又能增加结构的抗扭刚度。

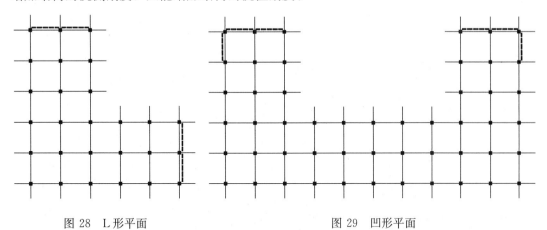

图 28　L 形平面　　　　　　　　　图 29　凹形平面

对如图 29 所示的凹形平面结构体系，结构位移通常在两个凸起部位，通过在这两个部位设置抗侧力构件后，能使整体结构振型合理，控制结构刚心和质心基本重合。同时也起到增加结构抗侧刚度的作用。

对如图 30 所示的 Y 形平面结构体系，在上海市 90 年代的高层住宅建筑和商住混合建筑中大量使用，这种结构体系在计算中有时候并不会出现数值上的超限（例如扭转位移比超限、扭转周期比超限），然而如果变换结构地震作用输入方向，或者进行扭联地震反应分析，结构扭转明显。因而需要在结构端部设置抗侧力构件，如图 30 所示，起到控制结构扭转位移比的作用。保证结构不会出现单肢反应的不利情况出现。

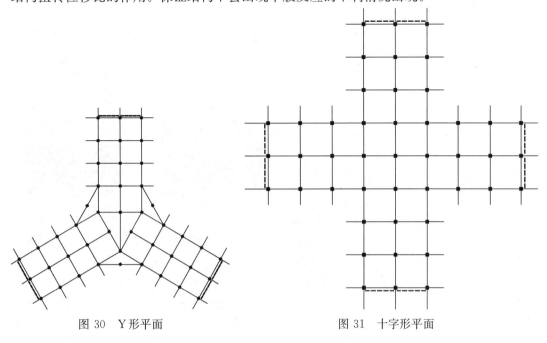

图 30　Y 形平面　　　　　　　　　图 31　十字形平面

73

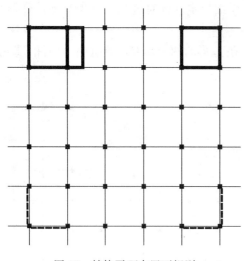

图 32 结构平面布置不规则

对如图 31 所示的十字形平面结构体系，在建筑比较密集的上海等南方大型城市中也经常采用。这种结构体系扭转明显，通过和 L 形结构体系相同的方法，在端部设置抗侧力构件，能够控制结构的扭转位移比和扭转周期比。

6.3.2 结构平面布置不规则情况

在现代建筑设计中，希望结构尺寸小且不影响建筑功能。为满足该要求，结构设计中通常在建筑的竖向通道（电梯、楼梯）部位设置剪力墙筒体。然而当建筑布置中竖向通道远离结构质心时，会导致结构扭转明显，如图 32 所示，由于为满足建筑要求，导致结构布置平面不规则，在计算中，结构扭转位移比超限。为了解决这个问题，通常在远离结构筒体部位设置抗侧力构件，如图 32 所示。

6.4 结构抗震加固中的应用

屈曲约束支撑有优良的滞回耗能性能和施工安装方便、经济、设计灵活且不影响建筑物美观等诸多优点，使其不仅成为新建结构抗震设计的较佳选择，也成为已有结构抗震加固和改造的重要手段。屈曲约束支撑对已有结构的抗震加固不仅用于钢结构体系，还可用于混凝土结构体系；不仅可用于房屋建筑结构，还可用于桥梁、通信铁塔等结构设施。在日本，每年有一半的屈曲约束支撑用于各类建筑抗震加固中，通过测算，屈曲约束支撑为抗震加固最为经济与便捷的加固手段。

日本竹中公司广岛分公司办公楼于 1970 年建成，由于日本规范的修订，使该建筑不满足抗震设防标准。因此，在 1998 年采取利用低屈服点钢制成的屈曲约束支撑进行了加固。屈曲约束支撑在 2~9 层的每一层的四角窗户垂直设置，共设计装有 32 个屈曲约束支撑，且支撑的部位只限于建筑物的四角，不会影响采光，由于布置在室内也不会影响建筑的外观（如图 33 所示）。

图 33 日本竹中公司加固

1999 年台湾集集地震以后，对台北县政府行政大楼重新进行了抗震验算，发现需要增加结构的抗侧刚度以及消能能力，以提高结构的抗震等级。因而共安装了 562 根屈曲约束支撑来提高结构的抗震等级。位于美国盐湖城的 Wallace F. Bennett 联邦大楼，建于 20 世纪 60 年代初，为钢筋混凝土结构，共六层，建筑面积 300000m²，是该市的标志性建筑。根据研究报告，该市可能会发生大地震，因此对其进行抗震加固。通过 10 多种方案的比较，最终选取屈曲约束支撑。屈曲约束支撑布置在结构外柱之间,采用跨层交叉的形式,该工程一共用了 344 根屈曲约束支撑,设计对比表明,在达到所需要的抗震能力的前提下,比使用中心支撑节省 250 万美元,并可节约 2 个月的工期。该工程获得 2002 年度美国犹他州优秀工程奖。

国内 2009 年开始,在校安工程的推动下,大批学校建筑采用屈曲约束支撑技术进行抗震加固。

7 典型工程实例

7.1 上海世博中心

上海世博中心(如图 34 所示)部位采用 508 根屈曲约束支撑构件产品,采用普通支撑结构的整体性能与屈曲约束支撑结构的对比见表 8。

表 8 中数值说明,普通支撑结构的刚度明显大于屈曲约束支撑,这是由于为了保证在设防烈度地震作用下支撑不失稳,普通支撑的截面尺寸明显加大(如 BRB 的截面面积为 $0.0045m^2$,普通支撑的截面面积增大为 $0.0384m^2$,为 BRB 截面面积的 8.5 倍)。结构刚度的增大使得地震作用下结构吸收的地

图 34 上海世博中心

震力也显著增大,表 8 中数值为多遇地震下结构的基底剪力,说明普通支撑结构的地震力比屈曲约束支撑结构增大 24%~30%。普通支撑结构的用钢量与屈曲约束支撑结构的比较见表 9。两者差别主要为抗侧力结构(框架柱和支撑)的钢材用量,与屈曲约束支撑相比,普通支撑的总用钢量多 1014t。

防屈曲耗能支撑结构与普通支撑结构的比较　　　　　　　表 8

结构体系		防屈曲耗能支撑	普通支撑
结构自振周期(s)	T1	2.23(X)	1.42(Y)
	T2	1.93(Y/T)	1.32(X)
	T3	1.75(Y/T)	1.24(T)
基底剪力	X	39785	52421
	Y	40850	50755

防屈曲耗能支撑结构与普通支撑结构(东区)用钢量比较(t)　　　　表 9

结构类型	支撑	框架梁	框架柱	桁架	主体结构总用钢量
屈曲约束支撑	431	2341	3063	1959	7794
普通支撑	1120	2458	3271	1959	8808

普通支撑结构由于地震力的增大,不仅增大结构用钢量,同时也增加地基基础、节点

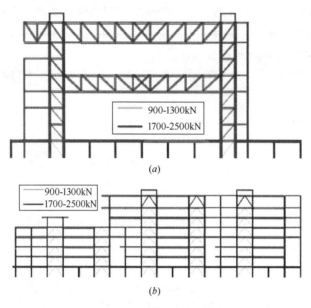

图 35 屈曲约束支撑布置
(a) 屈曲约束支撑横向布置；(b) 纵向屈曲约束支撑布置

连接的费用。此外罕遇地震作用下，普通支撑可能屈曲，抗震性能不易保证。

普通支撑结构和防屈曲耗能支撑结构的技术经济性比较可总结如下：

(1) 对普通支撑结构，大震作用下一旦支撑屈曲，结构刚度迅速退化，大震作用下的抗震性能不易保证。

(2) 由于普通支撑截面比屈曲约束支撑大，结构刚度明显增大，常遇地震下地震力也显著增大，X、Y 向分别增大 30% 和 24%。

(3) 核心筒处柱脚支座竖向反力增大约 1.5～2.0 倍，需要增加 20% 抗压桩，抗拔桩配筋增加，柱脚构造措施抗拔承载力需提高。

(4) 如采用普通支撑，东区和西区抗侧力结构（框架柱、梁及支撑）用钢量增加约 2000t，增加钢结构部分造价 2400 万。

7.2 天津 117

天津高银 117 大厦，外形设计独特，见图 36，总建筑面积约 37 万 m²，建筑高度约为 597m，共 117 层，结构高度 587m。塔楼首层至 93 层将用作甲级写字楼，94 层至顶层将用作六星级豪华商务酒店。

该项目外围框架采用巨型支撑+巨柱+环带桁架的结构形式，形成支撑外框筒，由地面往上依次分为 1-9 区（见图 37）。巨型支撑在 1 区为人字形布置，2-9 区为交叉支撑布置。其中人字支撑采用屈曲约束支撑，原因主要如下：1) 由于支撑不约束长度较大，而地震作用的不确定性较强，要满足大震不屈曲，稳定性上不容易满足要求；2) 由于支撑水平杆跨度达 44m，若采用普通支撑，该杆件为承担受压支撑屈曲后不平衡力需要截面很大，承载力难以满足要求且经济性较差。

该巨型屈曲约束支撑总长度达到 48m，截面面积 373,026mm²，屈服承载 3,730t，无论在设计上还是制造上都克服了诸多技术难题，由于支撑较长较重，在运输以及现场吊装都无法采用整根支撑一次成型的技术。根据我国公路运输长度要求、现场塔吊起吊重量，将整根支撑分三段，并采用三向定位、空中拼装方案，平均每段长度约 16m，均在工厂制作完毕运往现场。

根据设计要求，在区间 1 巨型屈曲约束与巨型框架之间的布置关系见图 38 所示。

整根巨型屈曲约束支撑分三段进行设计，见图 39，其中两端的两段芯材设计为屈服段，中间一段为弹性段。

图 36　天津高银 117 大厦效果图　　　　　图 37　外框筒

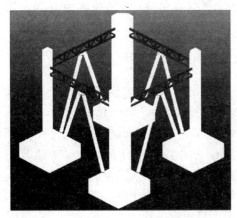

图 38　结构区间 1 巨型外围框架示意图　　图 39　巨型屈曲约束支撑分段数设计/mm

巨型屈曲约束支撑在天津高银 117 大厦中的应用刷新了该项技术世界应用历史，让我国屈曲约束支撑技术应用在某些指标上迈入世界领先水平，同时为今后超高层建筑结构巨型支撑的广泛应用提供参考。

安装过程照片如图 40 所示：

图 40　安装进度照片
(a) 第一段；(b) 第二段；(c) 第三段；(d) 整体成型

7.3　上海市防震减灾中心大楼

上海市防震减灾中心大楼（图）共10层，地下1层，地上9层，为混凝土框架-屈曲约束支撑结构体系。由于该楼为上海市震后抗震减灾指挥中心，其抗震设防等级高于上海市标准，按8度抗震设防。作为结构主要抗侧力体系和第一道抗震防线设置的TJ型屈曲约束支撑在小震下为结构提供抗侧刚度，在罕遇地震作用下率先进入屈服耗能，吸收地震输入结构的能量，保护混凝土框架梁、柱不受地震作用发生破坏。全楼共设置75根屈曲约束支撑，屈服承载力分别为800kN和1300kN。

图 41　上海市防震减灾中心

7.4　深圳航天国际中心

深圳航天国际中心，塔楼地上49层，标准层层高4.1m，结构总高225m；地下室4层，深17.1m。塔楼平面尺寸45m×39.6m，核心筒尺寸15m×26m。建筑在13层、26层、39层设置了三个设备层（兼作避难层），层高7.2m。综合建筑的功能和结构受力需求，主塔楼采用型钢混凝土框架柱＋楼面钢梁＋钢筋混凝土核心筒的混合结构体系，并结合建筑避难层设置了三个结构加强层。由于项目地质条件较差，场地大震卓越周期0.9s，地震作用较大。如果加强层采用普通支撑，会导致承载力和刚度突变，大震下加强层及相邻楼层损伤严重。为提高结构整体抗震性能，航天国际中心主塔楼加强层伸臂杆件采用屈曲约束支撑。结合建筑设备层，在分别在13层、26层、39层X向设置四道伸臂，加强层平面布置如图42所示。核心筒Y向尺寸较大，刚度足够，Y向不设置伸臂。

如采用普通支撑，由于构件屈曲之后将丧失承载力并导致结构刚度的突变，因此普通

 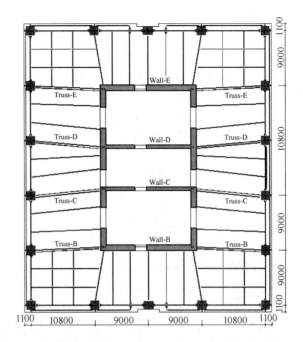

图 42 建筑示意及伸臂桁架设置

支撑伸臂性能目标设定为大震不屈曲，支撑基本上不参与耗能。如果采用 BRB，构件不管是受拉或受压均允许屈服，构件屈服后还会继续保持承载力，结构刚度只有轻微程度的下降，因此，BRB 支撑性能目标可设定为中震允许屈服，大震下 BRB 通过屈服耗能以降低地震的反应，减小地震作用对结构的破坏。本塔楼计算结果表明，如所有伸臂均采用 BRB，大震下它耗散的能量约为所有构件耗散能量的 10%。

8. 总结

通过对屈曲约束支撑特点、研究现状的总结，大量的构件试验以及施工技术探讨，屈曲约束支撑即是结构构件也是性能优越耗能构件。

（1）根据屈曲约束支撑性能和使用要求的不同，又可将其分为"承载型屈曲约束支撑"、"耗能型屈曲约束支撑"和"屈曲约束支撑型阻尼器"三种类型。

（2）通过 150 根屈曲约束支撑构件试验的归纳总结，应变率超过 0.01 后，普通低碳钢芯材应变强化系数变异性大于低屈服点钢芯材。

（3）屈曲约束支撑宜采用后装法，消除施工误差应采用修正节点板的做法，而不能对屈曲约束支撑进行现场修改。

（4）屈曲约束支撑即可应用于钢框架结构，也可用于混凝土框架结构。在设计中，弹性阶段应保证屈曲约束支撑的设计承载力大于构件内力，连接节点承载力应不小于 1.2 倍构件极限承载力。

（5）屈曲约束支撑采用纯钢约束体系，需要做防火保护；若采用填充材料约束体系，则自由伸缩段和封头板必须做防火保护，套筒可不做防火保护。

参考文献

[1] Roger Bilham. Lessons from the Haiti earthquake[J]. Nature, 2010, 463: 878-879.

[2] GB 50011—2010 建筑抗震设计规范[S]. 北京: 中国建筑工业出版社, 2010.

[3] JGJ—2013 建筑消能减震技术规程[S]. 北京: 中国建筑工业出版社, 2010.

[4] Iwata, M., Kato, T., Wada, A. Buckling Restrained Braces as Hysteretic Dampers[J]. Proceedings, STESSA, Quebec, Canada, 2000: 33-38.

[5] Iwata, M., Murai, M. Buckling-restrained brace using steel mortar planks: performance evaluation as a hysteretic damper[J]. Earthquake Engineering and Structural Dynamics, 2006, 35: 1807-1826.

[6] Uang Chia-ming, Nakashima Masayoshi, 陆烨. 屈曲约束支撑体系的应用与研究进展(I)[J], 建筑钢结构进展, 2005, Vol. 7(1): 1-12.

[7] 刘建彬. 防屈曲支撑及防屈曲支撑钢框架设计理论研究[D]. 清华大学硕士学位论文, 2005.

[8] Kimura K, Yoshioka K, Takeda T, Fukuya Z, Takemoto K. Tests on braces encased by mortar in-filled steel tubes. Summaries of technical papers of annual meeting, Architectural Institute of Japan 1976, 1041-2. [in Japanese]

[9] Watanbe A., Hitomi Y, and Saekie, et al. Properties of brace encased in buckling-restraining concrete and steel tube. In: 9th World conf. on Earthquake Engineering, Vol. IV, Japan 719-724, 1988.

[10] 蔡克铨, 翁崇兴. 双钢管型挫屈束制耗能支撑之耐震行为与应用研究. 台湾大学地震工程研究中心报告, NCREE R91-02, 2002.

[11] 蔡克铨, 黄彦智, 翁崇兴. 双管挫屈束制支撑之耐震行为与应用. 建筑钢结构进展. vol. 7(3), 2005. 2. 39.

[12] 魏志毓. 挫屈束制支撑局部挫屈与框架耐震效益分析研究[D], 台湾大学, 2006.

[13] 陈正诚. 韧性同心斜撑构架与韧性斜撑构材之耐震行为与设计. 结构工程, 第15卷, 第1期, 53-78, 2003.

[14] 李妍, 吴斌, 王倩颖, 欧进萍. 防屈曲钢支撑阻尼器的试验研究. 土木工程学报. Vol. 39(7): 9-14, 2006.

[15] 胡宝琳. 屈曲约束支撑框架抗震设计的理论和试验研究. 上海: 同济大学, 2008.

[16] Guo-qiang Li, Fei-fei Sun, Su-wen Chen, Xiao-kang Guo. Development of TJ-type Buckling-Restrained Braces and application [C]. //International Conference on Earthquake Engineering. The 1st Anniversary of Wenchuan Earthquake 10-12 May, 2009 Chengdu, P. R. China.

[17] 孙飞飞, 刘猛, 李国强, 胡宝琳, 郭小康. TJ-I 型屈曲约束支撑的试验研究和数值模拟[J]. 河北工程大学学报, Vol. 26(1): 5-9, 2009.

[18] 马宁, 吴斌, 赵俊贤, 等. 十字形内芯全钢防屈曲支撑构件及子系统足尺试验. 土木工程学报, Vol. 42(4): 1-4, 2010.

[19] 赵俊贤, 吴斌, 梅洋, 等. 防屈曲支撑的研究现状及关键理论问题. 防灾减灾工程学报, Vol. 30suppl. 93-101, 2010.

[20] 梅洋, 吴斌, 赵俊贤, 等. 组合钢管混凝土式防屈曲支撑. 灾害学, Vol. 25(s0), 99-104, 2010.

[21] 郭彦林, 江磊鑫. 型钢组合装配式防屈曲支撑性能及设计方法研究[J]. 建筑结构. Vol. 40(1): 30-37, 2010.

[22] 郭彦林, 江磊鑫. 双矩管带肋约束型装配式防屈曲支撑的设计方法[J]. 建筑科学与工程学报. Vol. 27(6): 67-74, 2010.

[23] 邓雪松, 邹征敏, 周云等. 开槽式三重钢管防屈曲耗能支撑试验研究[J]. 土木工程学报. Vol. 44(7): 37-44, 2011.

[24] 周云, 邓雪松, 钱洪涛等. 开孔式三重钢管防屈曲耗能支撑性能试验研究[J]. 土木工程学报.

Vol. 43(9): 77-87, 2010.

[25] 周云,钱洪涛,褚洪民等. 新型防屈曲耗能支撑设计原理与性能研究[J]. 土木工程学报. Vol. 42(4): 74-70, 2009.

[26] 吴从永,吴从晓. 周云. 新型开槽式防屈曲耗能支撑力学模型研究及应用[J]. 土木工程学报. Vol. 43: 297-402, 2010.

[27] Clark P., Aiken I, etc. Design procedures for buildings incorporating hysteretic devices proceedings [A]. 68th Annual Convention SEAOC, Sacramento CA, 1999.

[28] K. C. Tsai, Y. T. Weng, etc. Pseudo-dynamic test of a full-scale CFT/BRB frame: Part I-Performance based specimen design[A]. 13WCEE, Vancouer, Canada, 2004.

[29] 胡大柱,李国强,等. 屈曲约束支撑铰接框架足尺模型模拟地震振动台试验[J]. 土木工程学报,2010, Vol. 43: 5520-525.

[30] 程绍革,罗开海,等. 含有屈曲约束支撑框架的振动台试验研究[J]. 建筑结构,2010, Vol: 40: 11-14.

[31] Jinkoo Kim, Youngill Seo. Seismic design of steel structures with buckling-restrained knee braces [J]. Journal of Constructional Steel Research, 2003, Vol. 59: 1477-1497.

[32] Jinkoo Kim, Youngill Seo. Seismic design of low-rise steel frames with buckling-restrained braces [J]. Engineering Structures, 2004, Vol. 26: 543-551.

[33] 欧进萍,吴斌,等. 耗能减振结构的抗震设计方法[J]. 地震工程与工程振动,1998, Vol.18(2): 98-107.

[34] 贾明明,张素梅,等. 基于Benchmark模型的抑制屈曲支撑耗能减振作用分析[J]. 地震工程与工程振动,2009(3): 140-145.

[35] 李国强,胡大柱,等. 屈曲约束支撑半刚性连接框架弹塑性地震位移简化计算[J]. 地震工程与工程振动,2009(4): 33-40.

[36] 李国强,郭小康,孙飞飞,刘玉姝,陈琛. 屈曲约束支撑混凝土锚固节点力学性能试验研究[J]. 建筑结构学报,2012, Vol. 33(3): 89-95.

[37] 李帼昌,林丹,高成富. 静载下BRB与混凝土连接节点的有限元分析[J]. 沈阳建筑大学学吧(自然科学版),2012, Vol. 28(2): 249-256.

[38] Cale Ash, S. E., Stacy Bartoletti, S. E. Seismic rehabilitation of an existing braced frame hospital building by direct replacement with buckling-restrained-braces[A]. ATC & SEI 2009 Conference on Improving the Seismic Performance of Existing Buildings and other Structures.

[39] Samer El-Bahey, Michel Bruneau. Buckling restrained braces as structural fuses for the seismic retrofit of reinforced concrete bridge bents[J]. Engineering Structures, 2011, Vol. 33: 1052-1061.

[40] Ishii T, Mukai T, Kitamura H, Shimizu T, Fujisawa K, Ishida Y. Seismic retrofit for existing R/C building using energy dissipative braces. In: Proceedings of the 13th world conference on earthquake engineering. Vancouver (Canada): 2004.

[41] L. Di Sarno, G. Manfredi. Seismic retrofitting with buckling restrained braces: Application to an existing non-ductile RC framed building[J]. Soil Dynamics and Earthquake Engineering, 2010, Vol. 30(11): 1279-1297.

[42] Juan Andres Oviedo A., Mitsumasa Midorikawa, Tetsuhiro Asiari. Earthquake response of ten-story story-dirft-controlled reinforced concrete frames with hysteretic dampers[J]. Engineering Structures, 2010, Vol. 32: 1735-1746.

[43] 顾炉忠,高向宇,徐建伟,等. 防屈曲支撑混凝土框架结构抗震性能试验研究[J]. 建筑结构学报,2011, Vol. 32(7): 101-111.

[44] 孔祥雄, 罗开海, 程绍革. 含有屈曲约束支撑平面框架的抗震性能试验研究[J]. 建筑结构, 2010, Vol.40(10): 7-18.

[45] 薛彦涛, 金林飞, 韩雪, 程小燕, 佟道林. 钢筋混凝土框架屈曲约束支撑试验研究[J]. 2013, Vol.43(1): 1-4.

[46] 贾明明, 孙霖, 郭兰慧, 吕大刚. 防屈曲支撑非屈服段平面外屈曲对组合框架支撑结构性能的影响[J]. 建筑结构学报, 2013, Vol.34(S): 383-388.

[47] 吴微, 张国伟, 赵健, 张扬. 防屈曲支撑加固既有 RC 框架结构抗震性能研究[J]. 土木工程学报, 2013, Vol.46(7): 37-46.

[48] 曹炳政, 朱春明, 郑昊, 李向民. 屈曲约束支撑加固框架抗震性能试验研究[J]. 工程抗震与加固改造, 2012, Vol.34(2): 1-7.

[49] 郑昊, 郑乔文, 朱春明, 等. 双层双跨混凝土框架结构加固后抗震性能试验研究[J]. 工程抗震与加固改造, 2010, Vol.32(5): 57-62.

[50] 王秀丽, 李涛, 金建民, 莫庸. 高烈度区新型钢筋砼框架的抗震性能[J]. 兰州理工大学学报, 2009, Vol.35(2): 105-109.

[51] 张晓将. 新型钢筋混凝土框架-支撑减震体系设计方法研究[D]. 兰州理工大学硕士学位论文, 2009.

[52] 蔡建辉. 带 BRB 的不规则钢筋混凝土框架结构抗震性能研究[D]. 兰州理工大学硕士学位论文, 2010.

[53] 胡其强. BRB-钢筋混凝土框架结构抗震性能研究[D]. 西南交通大学硕士学位论文, 2008.

[54] 袁钰, 吴京. 屈曲约束支撑框架 Pushover 分析的加载模式研究[J]. 世界地震工程, 2010 (2): 207-211.

钢板剪力墙底层边界柱耐震设计、分析与试验

蔡克铨[1*]，李弘祺[2]，李昭贤[3]，苏磊[4]

摘　要：本研究探讨钢板剪力墙底层边界柱的耐震设计方法，首先介绍底层柱的容量设计方法，设计方法采用等效斜撑模型搭配简算公式，边界构件之设计需求以迭加法考虑钢板内拉与构架侧位移效应，并探讨不同底层边界柱塑性弯矩容量与其塑铰位置之关系。本研究设计目标为当钢板剪力墙系统承受最大考虑地震时，限制底层柱顶须保持弹性，检核其弯矩及剪力塑铰两种限制状态，但放宽柱底塑性铰至四分之一底层高。本研究已于2011年在国家地震工程研究中心完成三座实尺寸两层楼单跨钢板剪力墙试体的反复侧推试验，用以验证本研究建议之容量设计方法适用性，同时以有限元素模型分析试体反应。结果证实本研究建议之容量设计方法确实能控制底层柱之塑性行为，而底层柱顶产生些微或明显屈服的试体，其边界柱均有明显内拉变形以及侧向扭转变形，但底层柱顶保持弹性的试体，其迟滞循环较饱满且边界柱无侧向扭转变形，证实底层柱顶的塑性铰会导致软层机构或系统失稳。试体从弹性进入塑性后拉力场角度有明显改变，进行容量设计时建议设定底层钢板拉力场为40度而其余楼层为45度。

关键词：钢板剪力墙；容量设计；底层边界柱；拉力场角度；有限元素模型

SEISMIC DESIGN OF THE BOTTOM COLUMN IN STEEL PLATE SHEAR WALLS: CAPACITY DEISGN, ANALYTICAL AND EXPERIMENTAL RESULTS

Hung-Chi Lee, Chao-Hsien Li and Keh-Chyuan Tsai

Abstract: This research focuses on the design method and experimental study of the bottom column in steel plate shear walls (SPSWs). A capacity design method, using equivalent brace model and simplified analysis procedures, for the bottom column in SPSWs is proposed. The design target is to prevent either a flexural or a shear hinge from forming at top of bottom column when the SPSW is subjected to the maximum considered earthquake. However, the method allows the plastic zone to form approximately at the 1/4-height of the bottom column. In this paper, the design requirements are considered by superposition of the frame sway action and panel yielding force. Cyclic test of three two-story SPSWs were conducted in Taiwan National Center for Research on Earthquake Engineering in 2011. Test results confirm that the proposed ca-

---

* 通讯作者，国立台湾大学土木工程学系，10668 台北市维斯福路四段一号
[1] 国立台湾大学土木工程学系教授，E-mail：kctsai@ntu.edu.tw
[2] 国立台湾大学土木工程学系硕士，E-mail：r98521206@ntu.edu.tw
[3] 国家地震工程研究中心助理研究员，E-mail：chli@ncree.narl.org.tw
[4] 国立台湾大学土木工程学系硕士，E-mail：r00521217@ntu.edu.tw

pacity design method can satisfactorily prevent the plastic hinge from forming at the top of the bottom column. The bottom column of specimens having minor or major yielding occurred at the column top had observable lateral torsional buckling. However, the specimen in which the top of bottom column remained elastic still had a rather good load-carrying capacity, in spite of large range of yielding developed in the midheight. Test results indicate that the formation of the top-end plastic hinge in the SPSW bottom column would lead to a soft story mechanism or instability problem. Tests showed that the tension field angle changed as the boundary frame went from elastic to inelastic states. Analytical and experimental results suggest that the tension field angle can be assumed as 40 and 45 degrees for the capacity design of the bottom and other stories column, respectively.

Key words: steel plate shear walls; capacity design; bottom column; tension field angle; finite element model.

1. 前言

1.1 介绍

钢板剪力墙系统（steel plate shear walls，SPSWs）为一新型之钢结构耐震系统，在边界梁柱构件（boundary elements，BEs）所构成的平面内安装薄钢板即能显著地提升结构系统之侧向劲度与强度，不论是在新建结构或是结构补强，已逐渐受到采用。设计恰当之钢板剪力墙系统，地震输入的能量可借由钢板屈服后发展之拉力场消散，震后可进行钢板之拆换，而针对有众多管线穿通结构体之建筑还可设计不同的钢板开孔形式[1, 2]，因此有极广泛之应用性。由于薄钢板剪力墙系统的总用钢量相对于其他耐震系统更少，因此钢板剪力墙系统可视为是一具有经济效益之耐震系统。

台湾耐震设计规范并未对钢板剪力墙系统有明确规定，以致台湾工程界对此系统尚不熟悉，虽国外[3, 4]对于钢板剪力墙系统已有规范及建议之分析方法，然而其分析方法常须由工程师建构复杂之板条模型来检核钢板剪力墙边界构件，使得设计效率降低，且建议边界柱承受地震力时保持弹性或强制塑性铰只能发生于柱底，使边界柱抗弯能力的要求增加，往往容易设计出极不经济之柱断面，此两点因素对推广钢板剪力墙系统极不利。

近年台大土木系在国家地震中心完成的两个试验[5-10]显示，允许底层柱塑性铰产生于四分之一柱高之钢板剪力墙系统仍有相当良好之韧性行为，且边界柱并无失稳之疑虑，显示底层柱的塑性反应位置可大幅放宽，如此可有较经济之柱断面选择，然而对于放宽的设计方法仍须进一步研究。因此，若能提出简化且可靠的分析方法，搭配经济之设计准则，必能对于钢板剪力墙系统之应用有极大帮助。

本研究尝试以等效斜撑模型设计钢板剪力墙系统，并且以底层柱顶不产生塑性铰的情况下，对钢板剪力墙之底层柱构件弯矩及剪力需求研究更大幅度的放宽。本文首先讨论边界柱构件的受力情形，以适当参数描述内力分布，并提出简算公式预测系统承受最大考虑地震下底层柱顶弯矩及剪力强度需求。为验证设计方法可行性及钢板剪力墙受震行为，本研究设计三座实尺寸两层楼钢板剪力墙试体进行反复侧推试验，后面章节将介绍试验规划及试验结果、讨论重要的实验观察结果，并以有限元素数值模型模拟试体反应，提供钢板剪力墙设计与分析的建议。

1.2 容量设计概念

在弹性阶段时，钢板剪力墙系统主要是由钢板之剪应力来抵抗侧向力，根据应力转换剪应力可视为纯拉及纯压之主应力，当剪力持续增加，钢板主压应力超过其弯曲应力强度后，钢板会发生明显之面外弯曲，若设定钢板不得弯曲为极限状态，往往会设计出厚重之钢板。研究显示[11]，钢板虽然弯曲但主拉应力方向上强度并不会丧失，会发展成拉力场，剪力墙仍具有相当好之强度与韧性，因此各规范开始接受剪力墙弯曲之行为，以薄钢板剪力墙来当作抗侧力系统，极具经济性。

完全发展的拉力场行为就像无限多根的二力杆件，会增加边界构件的内力需求，因此边界构件的设计需考虑受力变形下所形成之机构恰当与否。不恰当的机构包含软层现象，或是因边界梁跨中塑性铰使得拉力场无法充分地拉伸，进而降低钢板剪力墙抵抗侧力的效率，这种考虑完全降伏之拉力场增加的载重，所进行边界构件的设计方法属于容量设计（capacity design）的范围。

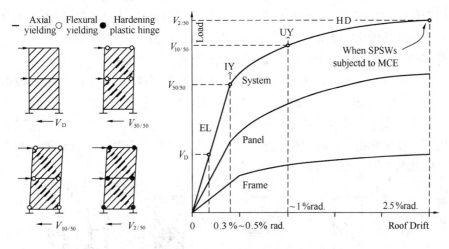

图 1　钢板剪力墙系统非线性侧推过程

以图1两层单跨的钢板剪力墙系统为例，经过适当的容量设计，符合均匀屈服机构[12]或本研究建议之放宽柱底塑性行为之设计，其非线性侧推（nonlinear pushover）受力变形过程可以细分如下：

（1）弹性阶段（Elastic Level，EL）：代表当系统承受设计地震力，顶层侧位移角约在0.3％弧度之前，系统保持弹性，但钢板有轻微的弹性变形。

（2）初始降伏阶段（Initial Yield，IY）：代表当系统受力变形至顶层侧位移角介于0.3％～0.5％弧度之间，相当系统经历50年50％超越概率之地震时，系统初始屈服，钢板有非弹性变形现象并由对角线最长距离处开始产生屈服拉力场，但边界梁柱构件仍保持弹性。

（3）全面屈服阶段（Uniform Yielding，UY）：当系统受侧力变形至顶层侧位移角介于1％～2％弧度之间，相当系统承受50年10％超越概率之地震时，系统进入全面屈服阶段，此时钢板拉力场发展完全，边界构件承受大量来自于钢板内拉分量及构架变形引致的内力，边界构件开始产生塑性铰，过去研究边界构件容量设计的需求是以此阶段的评估为主。

（4）应变硬化阶段（Hardening level，HD）：当钢板剪力墙系统受侧力顶层侧位移角超过1.5%弧度后，恰当的边界梁柱构件容量设计可确保整个系统在钢板及边界构件塑性铰经过应变硬化后，不会另有其他的塑性行为，不致发生系统失稳或丧失载重能力。过去研究结果显示经过恰当容量设计之钢板剪力墙系统承受最大考虑地震时，楼层侧位移角会到达2.5%弧度[13,14]，本研究以此侧向变形作为系统容量设计的目标，并估算此时底层柱内力，下文分别讨论系统全面屈服与应变硬化阶段底层柱之挠曲与剪力需求。

2. 底层柱设计容量设计方法

2.1 迭加法与柱弯矩参数 λ

本研究所建议的钢板剪力墙构架内力分析方法主要架构在迭加法（principle of superposition）上，迭加法首先被成功地利用于研究边界梁的容量设计方法[15]，近年来也成功地被利用于预测钢板剪力墙全面屈服阶段之底层柱弯矩分布趋势[6-9]，如图2（a）所示，钢板剪力墙构架的内力分布形式与传统之抗弯构架不同，所受侧力可视为两个分量：V_f为驱动构架变形之侧向力及V_p为驱动钢板变形之侧向力，钢板拉力场可分为垂直于边界构件的分量及平行于边界构件的分量，将两者造成的内力迭加即可得钢板剪力墙边界构件的内力，图2（b）与图2（c）分别为以迭加法推估钢板剪力墙边界构件受侧力时弯矩及剪力分布，除了在边界梁两端及受拉边界柱柱底产生塑性铰，由于弯矩与剪力在受压边界柱各层柱顶相加而非抵消，各层楼柱顶可能产生弯矩或剪力塑性铰，钢板剪力墙边界柱即是由受压柱此两限制状态（limit state）控制。

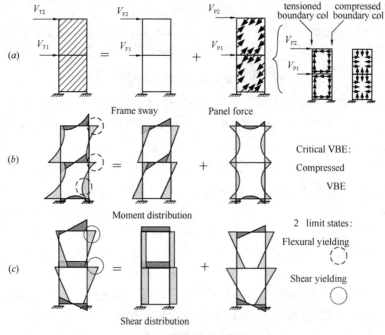

图2 钢板剪力墙构架
（a）受力分解；（b）弯矩分布；（c）剪力分布

当钢板产生拉力场后可视为所有的板条均匀屈服产生均布荷重，若考虑F_{yp}为钢板降伏应力，t_p为钢板厚度，α为拉力场角度，则钢板所产生的均布荷重可以分成四个分量[15]：

$$\begin{cases} \omega_{bvi} = F_{ypi}t_{pi}\cos^2\alpha_i \\ \omega_{bhi} = F_{ypi}t_{pi}\cos\alpha_i\sin\alpha_i \\ \omega_{chi} = F_{ypi}t_{pi}\cos\alpha_i\sin\alpha_i \\ \omega_{chi} = F_{ypi}t_{pi}\sin^2\alpha_i \end{cases} \quad (1)$$

其中ω_{bv}及ω_{bh}分别为边界梁所受之垂直及水平均布载重，ω_{cv}及ω_{ch}分别为边界柱所受之垂直及水平均布荷重，i为i楼层之钢板。

抗弯构架变形所引致之内力可将设计地震力施加在等效斜撑模型上[16]，如图3所示，由一般的结构分析软件即可得知弹性阶段下的边界柱内力分布，比起施加侧力于纯抗弯构架，等效斜撑模型能反映拉力场劲度对于边界柱内力分布之影响。由结果定义受压底层边界柱柱底与柱顶弯矩的比值为λ[6-9]，柱端弯矩比值λ反映边界柱在抗弯构架受侧力变形时弯矩分布情况，此与该楼层所承受之剪力以及柱上下两端点梁柱劲度比例有关：柱断面尺寸越大或劲度越高，柱底所分配到的弯矩会较高，即λ较大，反之若柱顶的梁劲度相对较高，λ就会变得较小。

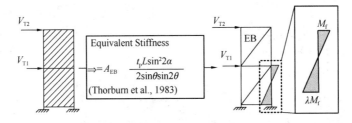

图3 以等效斜撑模型计算构架变形之边界柱弯矩

需注意柱端弯矩比值λ由线弹性分析之结果定义，虽无法完全反映构架产生塑性变形后之弯矩分布，然研究证实[6-9]假设λ从构架弹性进入塑性过程中保持定值，搭配简算公式来预估钢板剪力墙的底层柱塑性挠曲反应仍是相当准确的。

2.2 全面屈服阶段底层柱挠曲强度控制塑性铰位置

依习见钢结构设计规范所建议之容量设计原则，边界柱的屈服必须在钢板全面屈服后才发生，因此可假设底层柱的上下端为固接且承受均匀载重ω_{ch1}，以考虑钢板屈服后拉力场所造成之弯矩。抗弯构架受剪力侧向变形所造成之柱弯矩则为线性分布，若定义柱底的位置为$y=0$，柱高y坐标向上为正，此时受压柱弯矩沿柱高y分布如右侧所示：

$$M(y) = \underbrace{\lambda M_f\left(1 - \frac{1+\lambda}{\lambda h_1}y\right)}_{\text{frame sway action}} + \underbrace{\frac{\omega_{ch1}}{12}(-6y^2 + 6h_1y - h_1^2)}_{\text{panel force effect}} \quad (2)$$

其中M_f为底层柱顶受抗弯构架变形效应所造成之弯矩，h_1为底层楼高。此时$M(y)$所评估的是底层边界柱于系统全面屈服阶段之弯矩需求。假设最大弯矩发生位置于$y'=xh_1$之高度如图4所示，x为弯矩塑铰发生高度与底层柱高之比值，假设其范围介于0到0.5

之间，借由弯矩一次微分为零的条件，亦即：

$$M'(xh_1) = 0 \tag{3}$$

可求得柱内弯矩之最大值，并假设弯矩最大值发生位置与柱中弯矩塑性铰位置相同，借此可求得控制柱中塑性铰发生位置之弯矩需求为：

$$M_{\text{require}}(xh_1) = \left[\frac{\lambda(0.5-x)}{\lambda+1} + \frac{1}{2}x^2 - \frac{1}{12}\right]\omega_{\text{ch1}}h_1^2 \tag{4}$$

本研究依照过去实验研究结果之建议，将柱中弯矩塑性铰假设约在四分之一柱高即 x 为 0.25 作为试体设计之准则，这个假设反映了底层柱塑性弯矩容量较低时的塑性铰实际位置，亦即式（2）须满足：

$$M'\left(\frac{1}{4}h_1\right) = 0 \tag{5}$$

针对本研究底层柱容量设计方法设定之参数，x 值之设定若越大，边界柱的弯矩容量将越小，比值 λ 也将越小，因此可借由控制柱中弯矩塑性铰发生位置高低之条件达到设计经济边界柱断面之目的。

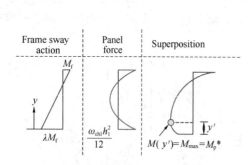

图 4　系统全面屈服时受压底层柱弯矩分布　　图 5　底层柱挠曲强度与塑铰位置

2.3　应变硬化阶段防止底层柱顶塑性铰之弯矩需求

本研究假设系统应变硬化阶段下之弯矩需求如图 5 所示，与全面屈服阶段相比，钢板应变硬化及构架超强会增加额外之弯矩需求，考虑增加之弯矩需求后，假设从全面屈服阶段进入应变硬化阶段柱中弯矩塑性铰发生位置不变，同样以迭加法计算底层边界柱柱顶会到达之弯矩需求，亦即：

$$M_{\text{d-HD}}^t = \Omega_{\text{Hf}}M_{\text{f-UY}}^t + \frac{\Omega_{\text{Hp}}\omega_{\text{ch1}}h_1^2}{12} = \left[\left(\frac{\Omega_{\text{Hf}}}{\lambda+1}\right)(0.5-x) + \frac{\Omega_{\text{Hp}}}{12}\right]\omega_{\text{ch1}}h_1^2 \tag{6}$$

其中上标 t 表示底层柱顶，$M_{\text{f-Uf}}^t$ 为全面屈服阶段由构架变形引致的底层柱顶弯矩，Ω_{Hf} 与 Ω_{Hp} 分别为构架超强因子及钢板硬化因子。在本研究中定义构架超强因子（frame overstrength factor，Ω_{Hf}）为钢板剪力墙系统中边界构件极限强度与初始降伏强度之比值，另外定义钢板应变硬化因子 Ω_{Hp} 来量化应变硬化效应，以系统承受最大考虑地震时系统顶层侧位移角为 2.5% 弧度作为目标，假设对应于 45 度对角线钢板应变约为 1.25% 如图 6 所示，因此钢板应变硬化因子 Ω_{Hp} 为：

$$\Omega_{\text{Hp}} = \frac{\sigma_{1.25\%}}{F_{\text{yp}}} \tag{7}$$

本研究假设底层柱第一个塑性铰发生位置在四分之一底层柱高，即 $x=0.25$ 带入式（6）即可得到当系统承受最大考虑地震时，欲防止底层边界柱柱顶发生挠曲塑性铰而必须有之弯矩容量：

$$M_p^* \geqslant M_{d-HD}^t = \left[\left(\frac{0.25}{\lambda+1}\right)\Omega_{Hf} + \frac{\Omega_{Hp}}{12}\right]\omega_{ch1}h_1^2 = \eta_{HD}^t \omega_{ch1}h_1^2 \qquad (8)$$

其中 η 为底层柱挠曲需求系数，由于底层柱须承受倾倒力矩所造成的轴力，因此柱弯矩容量需考虑经过轴弯互制的影响，本研究建议采用 AISC 的规定：

$$M_p^* = 1.18 M_p (1 - P_u/P_y) \leqslant M_p \qquad (9)$$

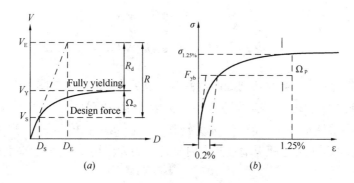

图 6 底层柱受力自由体图与柱剪力计算法

2.4 应变硬化阶段防止底层柱顶产生剪力塑性铰之设计要求

本研究假设系统应变硬化阶段下之剪力需求是由底层柱自由体中之力平衡所推导求得，如图 7 之自由体图所示，同样假设从全面屈服阶段进入应变硬化阶段柱中弯矩塑性铰位置不变，由自由体图中对柱中塑性铰取弯矩平衡可求得在应变硬化阶段柱顶剪力强度之需求为：

$$V_{d-HD}^t = \left\{\left[\frac{6(1-x)^2-1}{12(1-x)}\right]\Omega_{Hf} + \left[\frac{6(1-x)^2+1}{12(1-x)}\right]\Omega_{Hp}\right\}\omega_{ch1}h_1$$

$$+ \frac{\Omega_{Hp}\omega_{cv1}d_{c1}}{2} \qquad (10)$$

其中 d_{c1} 为底层柱深，若同样假设底层柱弯矩塑铰首先形成于四分之一柱高，代入式（10）即可得到当系统承受最大考虑地震时，底层边界柱柱顶之剪力需求为：

$$V_{d-HD}^t = \left(\frac{19}{72}\Omega_{Hf} + \frac{35}{72}\Omega_{Hp}\right)\omega_{ch1}h_1 + \frac{\Omega_{Hp}\omega_{cv1}d_{c1}}{2} \qquad (11)$$

而柱剪力容量则以 Von Mises 屈服准则考虑轴应力效应：

$$V_n = V_p\sqrt{1-(P_u/P_y)^2} \qquad (12)$$

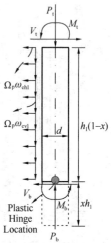

图 7 (a) 系统超强与 (b) 钢板应变硬化因子示意图

其中 $V_P = 0.6 f_y (d_{c1} - 2t_{fl})t_{w1}$，$t_{w1}$ 与 t_{fl} 分别为底层柱腹板厚与柱翼板厚。

2.5 轴力需求

前面已提及边界柱构件弯矩及剪力容量须考虑柱轴力而折减，柱轴力主要来自于重力及系统承受侧向力时所产生的倾倒力矩，考虑图8所示之自由体图来计算轴力，边界梁通常会采用梁翼切削（Reduced Beam Section，RBS）以减低其挠曲需求[15]，假设各层梁翼切削断面发展出塑性铰且钢板全面应变硬化来计算其梁端剪力，再加上钢板对于边界柱拉力的垂直分量与静活载，即可保守的估计底层柱轴力需求：

$$\begin{cases} V_{Ri}^{RBS} = \dfrac{2M_{pi}^{RBS}}{L_i^{RBS}} + \dfrac{\Omega_p \omega_{bvi} L_i^{RBS}}{2} \\ V_{Ri}^{cf} = V_{Ri}^{RBS} + \Omega_p \omega_{bvi} s_i \\ P_d^{h1} = \sum_{i=1}^{n} V_{Ri}^{cf} + \sum_{i=2}^{n} \Omega_p \omega_{cvi} h_i + \text{Gravity Load} \end{cases} \quad (13)$$

其中 M_{pi}^{RBS} 与 V_{Ri}^{RBS} 为 i 楼层梁翼切削处的塑性弯矩及剪力，R 指的是图中受压柱侧，L_i^{RBS} 为梁两端梁翼切削之中心距离，V_{Ri}^{cf} 为边界梁传递至受压柱的剪力，s_i 为梁翼切削中心至柱面距离，P_d^{h1} 为底层柱顶轴压力需求，n 为钢板剪力墙系统总楼层数，须注意对中间梁而言 ω_{bvi} 应扣除第 $i+1$ 楼层之钢板垂直向上内拉力，始为作用于中间梁之垂直分量。

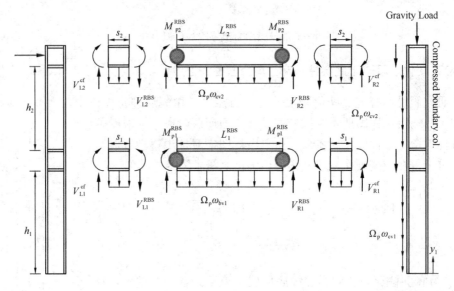

图8 系统应变硬化阶段下柱轴力需求

2.6 底层柱设计程序

本节建议之钢板剪力墙底层边界柱容量设计流程，系假设钢板厚度以及梁的尺寸已事先设计完成才进而对底层柱做设计，设计步骤如下：

步骤一：挑选底层柱断面。

步骤二：利用弹性结构分析方法建置等效斜撑构架模型，经由弹性侧推分析可得受压底层边界柱柱底与柱顶弯矩的比值 λ。

步骤三：以假设之 x 值与步骤二所得之 λ 值代入式（4）检核底层柱中弯矩塑性铰发生位置之强度是否满足需求，若不满足则回到步骤一重新挑选底层柱断面。

步骤四：检核应变硬化阶段底层受压柱顶之需求与容量比（demand to capacity ratio, DCR）是否小于 1.0，选取适合之 Ω_{Hf} 与 Ω_{Hp} 因子后，分别以式（6）与式（10）检核底层柱柱顶之弯矩强度以及剪力强度是否满足需求，若不满足则回到步骤一重新挑选底层柱断面。

3. 试体设计

为验证本研究所提出之容量设计方法之适用性，且探讨不同边界柱塑性弯矩容量对钢板剪力墙系统受力变形行为之影响，本研究利用国家地震工程研究中心之 15m 高反力墙与强力地板实验系统进行三座实尺寸构架实验，试体皆为两层楼单跨钢板剪力墙，以此探讨多层楼系统边界构件之耐震设计。三座试体规模均采用相同足尺寸如图 9 所示，构架跨距为 3420mm，两层楼楼高皆相同为 3820mm，楼层宽高比接近 0.9，上下层钢板厚度皆相同为 2.7mm，钢板材料采用"中国钢铁公司"赞助之低屈服点钢板，屈服强度约为 220MPa，每层墙体由六片钢板组成。除钢板外其余构件与加劲板皆使用 A572 GR50，试体之设计原则为：根据钢板厚度，以防止塑性铰发生于梁跨中之容量设计选取相同的边界梁[15]，边界梁端点采用符合 FEMA 350 规定[20]之梁翼切削设计，以前述不同的边界柱容量设计准则来决定三种不同边界柱构件，底层与二楼使用相同柱断面，断面尺寸同列于表 1，所有边界构件皆为组合断面且符合 AISC 所规定之耐震结实断面[3] 以避免局部变形过早发生控制试体行为。

以下分别以表 1 材料的拉力试验结果及本文第二章所整理边界柱弯矩及剪力需求，以容量比（Demand-to-Capacity Ratio，DCR）来控制试体反应，拉力场角度以二楼为 45 度与一楼 40 度进行计算，其理由将于后文交代，DCR 计算结果如表 2 所示，其中 P_u/P_y 为边界柱轴力需求与轴力容量之比值。

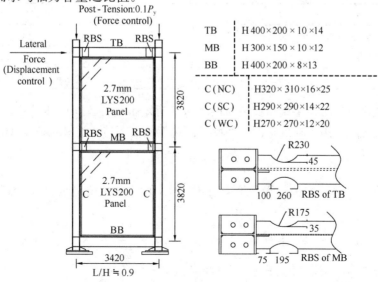

图 9 试体规模及其断面尺寸

拉力试片结果　　　　　　　　　　　　　　　　　　　表1

Location	Section	Plate	F_y (MPa)	F_u (MPa)
Top Beam	H400×200×10×14	flange	405	537
		Web	464	561
Middle Beam	H300×150×10×12	flange	372	506
		Web	464	561
Bottom Beam	H400×200×8×13	flange	390	502
		Web	430	509
Column (NC)	H320×310×16×25	flange	386	547
		Web	389	564
Column (SC)	H290×290×14×22	flange	375	518
		Web	405	537
Column (WC)	H270×270×12×20	flange	390	530
		Web	372	506

试体考虑材料试验结果之底层柱容量设计检核　　　　表2
The DCRs of the 1st story column considering coupon yielding stresses

Roof drift	1.0% rad. (in UY)		2.5% rad. (MCE)		
Specimen (λ)	$\dfrac{P_u}{P_y}$	$M_d^{h1/4}$	$\dfrac{P_u}{P_y}$	M_{d-HD}^t	V_{d-HD}^t
NC (23.5)	0.45	0.86	0.36	0.54	0.81
SC (9.53)	0.57	1.40	0.46	0.96	1.01
WC (6.37)	0.67	2.14	0.54	1.50	1.45

试体考虑材料试验结果之二楼柱顶容量设计检核　　　　表3
The DCRs at top of the 2nd story column considering coupon stresses

Specimen	M_{R2}^{cf}	V_{d-HD}^{t2}
NC	0.58	0.92
SC	0.89	1.15
WC	1.13	1.59

以系统全面屈服时顶梁梁翼切削断面之塑性弯矩推估传递至受压柱柱面之弯矩 M_{R2}^{cf}，以此作为二楼柱顶弯矩需求的来源，而此时由构架侧变形造成之柱顶弯矩 M_{f2} 为：

$$M_{f2} = M_{R2}^{cf} - \frac{\omega_{ch2} h_2^2}{12} \tag{14}$$

以二楼柱端点弯矩比值 λ 及钢板应变硬化因子考虑系统应变硬化后二楼柱顶剪力需求：

$$V_{d-HD}^{t2} = V_{f2} + \Omega_p V_{panel2} = \frac{1+\lambda_2}{h_2} M_{f2} + \left(\frac{\omega_{ch2} h_2}{2} + \frac{\omega_{cv2} d_2}{2}\right) \Omega_p \tag{15}$$

检核二楼柱顶的结果列于表3。以下将各试体的特色依边界柱由强到弱分别介绍，读者可配合图10的试体行为预测图了解试体设计考虑的差异。

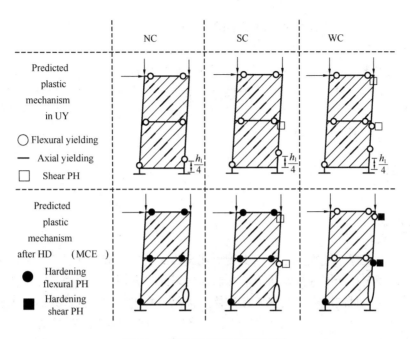

图 10 三座试体预测之塑性行为

3.1 试体 NC（Normal Column）

试体 NC 为三座试体中边界柱尺寸最大者，此试体之设计目标为系统经应变硬化后，一楼边界柱之柱顶仍不会有任何的塑性行为发生，但不强制塑性铰发生于柱底，因此除要求塑性铰须发生于柱底之弯矩需求 M_a^b 检核未通过外，其余弯矩需求与容量比皆小于 1.0，显示塑性铰位置大约介于柱底与四分之一底层柱高之间，而底层柱顶并无弯矩屈服，在底层柱顶剪力需求方面，以本研究建议之需求 V_{d-UY}^t 及 V_{d-HD}^t 作为标准检核，其需求与容量比小于 1.0，因此底层柱顶应无剪力屈服。

3.2 试体 SC（Small Column）

试体 SC 是以试体 NC 为标准，但缩小其边界柱的尺寸，试体之设计目标为使其在顶层侧位移 2.5% 弧度后，底层边界柱柱顶有发生塑性铰之可能，因此控制弯矩 M_{d-HD}^t 及剪力 V_{d-HD}^t 之需求与容量比分别为 0.96 及 1.01，使其于 2.5% 弧度顶层侧位移时发生弯矩及剪力塑性铰。但钢板全面屈服时需保持弹性，因此控制全面屈服阶段底层柱顶产生挠曲塑性铰之弯矩需求 M_{d-UY}^t 其需求与容量比检核要小于 1.0，但以 V_{d-UY}^t 检核之容量需求比为 0.91，所以预测会有轻微剪力屈服的机会。而检核弯矩需求 $V_d^{h1/4}$ 为 1.11，可知柱塑性铰的位置将略高于四分之一底层柱高。二楼边界柱部分，试体 SC 由于需求 M_{R2}^c 与容量比检核小于 1.0，但 V_{d-HD}^2 的检核为 1.15，因此预测试体 SC 在全面屈服阶段下二楼柱顶无弯矩屈服，但在系统应变硬化后产生剪力屈服。

3.3 试体 WC（Weak Column）

与试体 NC 及试体 SC 相比，试体 WC 采更小的边界柱尺寸，试体之设计目标为使其在钢板全面屈服阶段时底层柱顶即发生屈服，因此设计断面使其 $M^{\text{t}}_{\text{d-UY}}$ 及 $V^{\text{t}}_{\text{d-UY}}$ 的需求与容量比分别为 0.92 及 1.29，由于 $M^{\text{t}}_{\text{d-UY}}$ 的 DCR 接近 1，故底层柱顶之弯矩塑铰与柱跨中之弯矩塑性铰出现时机接近。而因 $M^{\text{t}}_{\text{d-HD}}$ 及 $V^{\text{t}}_{\text{d-HD}}$ 的检核分别为 1.50 及 1.45，故系统经应变硬化后，应有明显之弯矩与剪力塑性铰产生于底层柱顶。二楼边界柱考虑钢板应变硬化因子的剪力检核 $V^{\text{t2}}_{\text{d-UD}}$ 为 1.59 与弯矩检核 M^{t}_{R2} 为 1.13 均大于 1，故预测二楼柱顶在系统应变硬化后会有剪力与弯矩塑性铰。

4. 试验规划

4.1 试验配置

试体的侧向支撑系统主要是由四支侧撑杆及四支斜撑构成，如图 11（a）所示，四支侧撑柱与试体固定在相同的锚定板上，在试体之东面，分别以斜撑接合在试体二楼柱顶及二楼柱半高位置，且同时以四支横梁在试体一楼及二楼接合侧撑柱，并透过顶层连接梁连结试体两面侧撑柱以提供二楼之面外劲度，而试体之西面以两支斜撑接合在一楼柱顶位置，侧撑柱翼板在试体之一楼与二楼高开洞以安置含滚棒之挡板，以便顶住试体之梁柱接头侧撑组件，提供侧向抵抗且降低试体与侧撑系统摩擦力，其位置可参考图 11（b）所示。

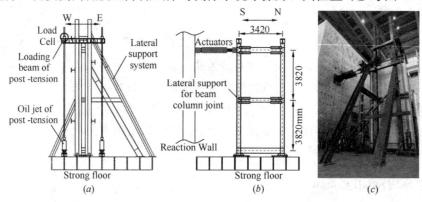

图 11　试验配置图

本试验之施力方式采单侧顶层侧推配置，采用两支轴力容量约为 980kN 的 MTS243.70 油压致动器，两支致动器采相同的位移控制，以降低梁翼切削处塑性铰发生后可能的严重扭转，柱顶侧位移历时分别于 0.1％、0.25％、0.5％、0.75％、1％、1.5％、2％、2.5％、3％、4％与 4.5％弧度各进行两循环的反复载重试验，最大顶层侧位移角 4.5％弧度为本试验配置之限制，因更大的位移将使施加垂直力的传力梁与侧撑柱相碰撞。为模拟静活载重对柱构件之轴力，本次试验透过预力螺杆来加载垂直力，预力螺杆透过传力梁将垂直力施加于柱顶上，两边界柱顶各施加约 10％柱轴力容量之垂直力，螺杆上方配有荷重计可读取每支螺杆之预力，预力螺杆与可自动修正预力损失或增加的油压千斤顶相接，以确保试验过程中预力的恒定，而图 11（c）为所有试验系统完成配置的景象。

4.2 量测计划

本次量测计划最重要的就是要记录试体受拉与受压边界柱的详细反应，因此在试体南北侧各安装一支参考柱，可参考图 12（a），将拉线式位移计沿边界柱高度等分的配置，而南侧二楼柱顶位置因油压致动器与传力梁连结而无法架设拉线式位移计，故此点之位移以千斤顶内建之位移计读值取代之。本次实验供使用 26 个角度计，量测范围有 5 度及 14.5 度两种规格，14.5 度配置位置主要是以可能产生塑性铰的边界构件腹板中心位置为主，如图 12（b）所示，在边界柱柱底以约一倍柱深之间距配置角度计至底层柱半高，以便量测柱底附近大范围之塑性变形转角，此外在边界梁的梁翼切削中心、梁柱交会区及边界柱的端点均有安装角度计。前述量测仪器所传出的讯号由可接收 50 组数据线的 TML-Switching Box SHW-50D 汇集，本次试验所使用的频道总数为 190，因此同时使用四台 Switching Box 接收讯号，再将数据使用 TML-THS-100 高速数据收集记录器记录，并配合国家地震中心新开发的实验程序 SFQSST（Software Framework for Quasi-Static Structural Testing）[22]，同时控制油压致动器速率与收集讯号之频率，且 SFQSST 配有实时运算连网的功能，可将预先想观察的实验数据频道输入，实验进行中或事后可透过网络播放观察试体之反应。

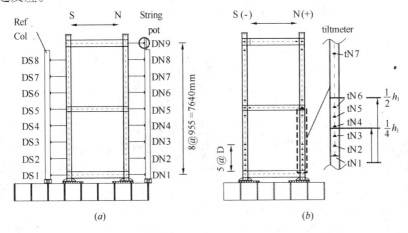

图 12 （a）拉线式位移计及（b）角度计配置位置图

5. 试验与有限元素模型分析结果

图 13 分别为三座试体之顶楼、一楼与二楼之受力及侧移角迟滞循环与 ABAQUS 有限元素模型非线性侧推模拟比较结果。由于试体并无安装楼板，边界梁受钢板大量内拉轴力而有缩短的现象，两支边界柱的侧变形并不一样，因此图 13 中楼层变位是以两支柱子的位移平均而得，图 13 比较显示有限元素模型的初始劲度虽未能完全与试验之反应相同，但当试体屈服后有限元素模型对于每一循环之尖峰强度均模拟良好，试体的最大侧向强度分别为 NC 为 1596kN、SC 为 1406kN 以及 WC 为 1327kN，试体的试验详细反应记录可参考 5.1 节。

很多研究均已证实设计与制造恰当的薄钢板剪力墙能有效发展拉力场效应来抵抗侧向

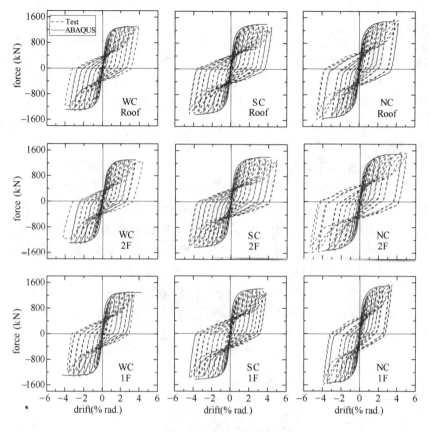

图 13 试体试验迟滞循环

外力，侧向位移增加至够大时，钢板可在拉力场方向发生屈服与应变硬化，钢板在拉力场的正交向则会发生波浪状的高模态变形，侧力移除后结构系统会有残余侧向位移。此时若侧向位移沿反方向施加，则先前在拉力场方向发生屈服之钢板会因没有显著抗压强度，而使系统在侧向反复施载下之迟滞反应循环有内缩之现象（pinching）。研究证实板条模型配合仅有拉力强度的双向桁架元素来模拟钢墙板，及习见之弹塑性梁柱元素来模拟边界构件，就能有效模拟前述系统在侧向反复施载下之内缩迟滞反应[13,14,21]。若采用非线性板壳元素的有限元素模型进行反复加载分析，更可精确地预测钢板剪力墙系统的反复受力反应[15,23,24]。本研究主要在探讨钢板剪力墙边界柱之耐震设计与分析方法，利用 ABAQUS 有限元素软件单向侧推分析，即可得到系统反应之包络曲线及试体反应。分析模型完全按

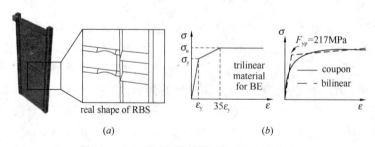

图 14 （a）ABAQUS 模型与其（b）材料定义
（a）ABAQUS Model；（b）material definition

照试体实际之尺寸与几何形状，如图 14（a）所示，并采用 S4R 的减积分薄壳元素，梁柱构件及钢板材料分别以三线性模型及双线性模型模拟材料拉力试验结果，如图 14（b），并给予适当的边界条件以模拟真实的试体配置，钢板输入前三个变形模态作为初始缺陷，使钢板墙能在分析中产生面外变形，并设定几何非线性侧推分析至试体顶层侧位移 4.0% 弧度。

5.1 试体反应观察

三座试体在顶层侧位移角达 1.0% 弧度时，均可以观察到系统全面屈服，钢板产生明显皱折，在顶层侧位移角超过 1.5% 弧度钢板应变硬化后，三座试体拉力场均呈波浪状均匀分布，惟三者拉力场角度略有差异，将于后面章节讨论，本节着重于讨论边界构件的实验反应。

5.1.1 试体 NC

试体的初始劲度为 43.5kN/mm，试体于顶层位移 0.1% 弧度及 0.25% 弧度阶段构架仍为弹性变形，无明显石膏漆脱落。试体于 1.0% 弧度及 1.5% 弧度顶层侧位移时底层柱底腹板与底梁腹板剪力石膏漆裂痕增加许多，从底层柱底至四分之一与二分之一柱高之间有较为明显翼板石膏漆裂痕产生，显示挠曲塑性铰已发生于此处，而顶梁切削处之上下翼板降伏之石膏漆条纹皆延伸至梁跨中。试体于顶层侧位移 2.5% 弧度时，南侧顶梁与中间梁切削处腹板向外凸出局部变形明显，顶梁南侧梁上翼板切削处发生剧烈扭转并已开裂。两边界柱底层柱底除轻微石膏漆脱落外均无任何明显变化，如图 15（a）所示，而底层柱腹板出现水平裂纹直至二分之一柱高，底层柱外侧翼板虽然石膏漆剥落严重但未超过二分之一底层高，显示底层柱有大范围的屈服。而直到试验结束，亦即顶层侧位移 4.5% 弧度试体仍无新增之石膏漆脱落反应方结束试验。

5.1.2 试体 SC

试体的初始劲度为 35.0kN/mm，试体于 0.75% 弧度顶层侧位移时两底层边界柱翼板之石膏漆折痕均匀分布于柱底至二分之一柱高之间，同时顶梁切削处也可观察到石膏漆裂痕，边界构件已发生初始屈服。试体于顶层侧位移 1.0% 弧度及 1.5% 弧度时底层柱翼板之石膏漆裂痕继续往上延伸到四分之三柱高。试体顶层侧位移达 2.5% 弧度时，底层边界柱的石膏漆折痕已非常明显，且集中于四分之一与二分之一底层高之间，显示塑性铰位于此处。而顶梁与中间梁切削处皆有轻微的腹板变形，且北侧底层柱顶腹板产生石膏漆剪力裂痕，应有剪力屈服产生，如图 15（b）所示。过了顶层侧位移 3.0% 弧度后边界柱有侧向扭转变形的现象，但试体强度并未明显降低因此执行到预定之顶层位移 4.5% 弧度后结束试验。

5.1.3 试体 WC

试体的初始劲度约为 33.4kN/mm，试体于顶层侧位移 0.75% 弧度时，底层边界柱石膏漆折痕已相当明显且集中在柱翼板四分之一至二分之一底层高，底层及二楼之两端边界柱的腹板均有剪力屈服之石膏漆裂缝。而试体于顶层侧位移 2.0% 弧度时，北侧底层柱顶有大范围之剪力屈服石膏漆裂痕，判断为剪力塑性铰产生，可参考图 15（c），于顶层侧位移 2.5% 弧度时两边界柱已有明显之侧向扭转现象，而顶梁切削处石膏漆脱落不明显只有轻微的面外变形，于顶层侧位移 4.0% 弧度时底层以及二楼柱顶都发生剪力塑性铰，且底层二分之一高以下柱翼板的内侧也有屈服，底层柱几乎整体都有石膏漆脱落之现象，且侧向扭转甚为严重，如图 16（a）所示，造成边界梁有外拱变形，参考图 16（b），由于担心试体边界柱与梁的侧向扭转会破坏侧向支撑系统，因此中止试验。

图 15　(a) 试体 NC、(b) 试体 SC 与 (c) 试体 WC 底层柱顶屈服情况

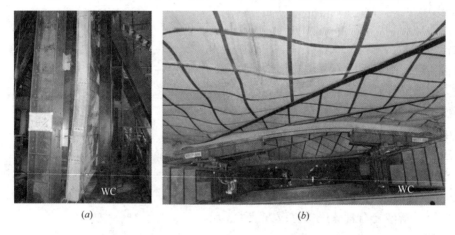

图 16　(a) 试体 WC 边界柱侧向扭转挫屈与 (b) 中间梁外拱之情况

5.1.4　疲劳破坏试验

在前述第一阶段反复载重试验结束后，若试体强度并未明显退化，则施加固定之反复顶层侧位移以及相同柱预力来进行疲劳试验，当每个循环中试体尖峰强度小于第一阶段试验试体之最大强度 70% 即停止疲劳试验。由于钢板之残余变形不会消除，卸除后需再度加载到达上回的最大应变量后，才会发展其劲度及强度，此即钢板剪力墙迟滞循环反应常具有之内缩特性，为观察钢板剪力墙系统极限强度之衰退，因此疲劳载重所设定之位移，以第一阶段试验之最大顶层位移 4.5% 弧度做为目标。由于 WC 试体在反复载重试验中即因边界柱严重侧向扭转变形而判定试体失稳，因此并未将 WC 试体纳入疲劳试验中。

试体 NC 与试体 SC 分别在经历五及六个位移循环后最大强度反应降至反复载重试验的最大强度 70%，两试体最大强度的逐圈衰退幅度接近，如图 17 (a) 所示，而试体 NC 因于第五个循环时南侧顶梁完全断开而强度明显降低，可参考图 17 (b)，而本次疲劳试验所设定之反复载重顶层位移 4.5% 弧度，高出前述容量设计目标承受最大考虑地震对应的 2.5% 弧度顶层侧位移[13,14]甚多，可证明本研究设计之钢板剪力墙系统卓越之耐震能力。图 18 为所有试体测试后并排之景象，由远观即可发现试体 NC 的边界柱无明显的受钢板内拉变形，而试体 SC 底层柱则有轻微内凹现象，试体 WC 则底层柱内凹最为严重，有漏斗形状产生。

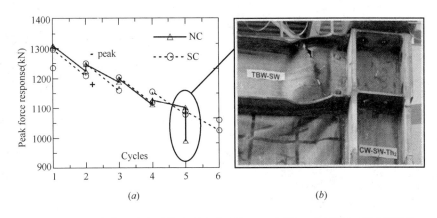

图 17 （a）疲劳试验中试体之强度衰退与（b）试体 NC 顶梁南侧断裂情形

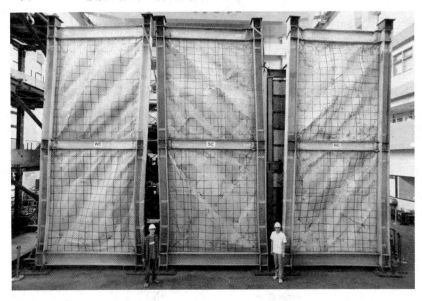

图 18　三座试体试验后并排景象

5.2　层间位移角分配比例

图 19（a）为试体之层间变位角分配比例与顶层侧位移之关系，层间变位角比例定义为每一受力循环中最大位移时一楼层间位移角除以二楼层间位移角，由南北侧两边界柱之位移读值平均制成，并各以多项式趋势线标记，若 θ_1/θ_2 的比例越高则代表试体变形越集中于一楼，反之亦然，可作为某楼层是否已有软层现象的指标。在顶层侧位移 0.75％弧度构架屈服前，三座试体的层间位移角分配比例相近，但构架屈服后则相差甚多，边界柱的强度越弱则一楼层间变位越大，而当 2.5％弧度顶层位移角后层间变位比例有减少的现象，主因为当底层柱塑性铰不发生于底部时，边界柱受钢板内拉作用会有内凹之现象，可以参考图 19（b）的受压柱变形示意，边界柱的内凹变形在 5.3 节有更定量的讨论，这造成了底层柱虽然变形严重但是楼层变位并未相同剧烈增加。

相同楼层侧位移下负侧位移角（亦即南侧）时一楼位移及力量反应均较大，主因是侧向力是由南侧的油压制动器输入，侧向力不是等分的由两边界柱顶来驱动试体之反应，

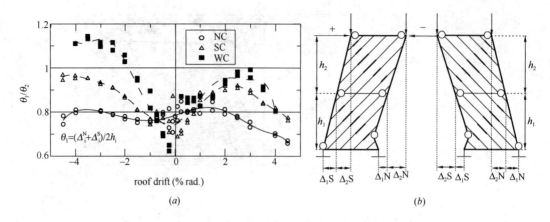

图19 楼层位移角分配比例

造成系统的反应并不对称。当油压制动器将试休推向正位移侧推动时，对角线的拉力场是透过北侧边界柱来拉伸钢板，如图20(a)，然而边界梁承受大量的钢板内拉造成之轴力，会有明显缩短的现象，可参考图20(b)所示，因此降低北侧边界柱之位移与钢板变形量。反之当油压制动器将试体柱顶往负位移角驱动时，对角线的拉力场是透过南侧边界柱来拉伸钢板，在传力路径上有最少之变形损失，因此钢板拉力场所发展的拉力也较大。中间梁因为这种传力路径的差别，试体在负顶楼位移角时会有较大之中间梁缩短现象，如图20(c)所示。而随边界柱强度越高，中间梁内缩越明显，主因是较强的边界柱不易受钢板内拉变形，拉力场发展效率较佳，因而梁轴力较高，在大侧位移下则颠倒，主因是边界柱的侧向扭转变形，造成中间梁外拱之变形。

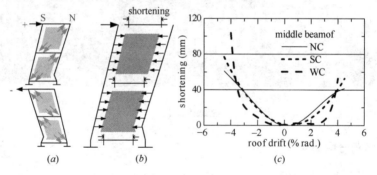

图20 (a)钢板传力路径 中间梁(b)内缩现象与(c)内缩量

5.3 边界柱塑性行为

角度计以约一倍柱深300mm等间距从一楼边界柱底分布至柱半高，因此试验时可以获得这些量测点的绝对旋转角，由材料力学梁挠曲变形理论可知，当固定的间距下相对旋转角或曲率越大，则该处弯矩越大，亦即：$d\theta \propto Mds$，将角度计读值两两相减获得的相对转角沿柱高分布，如图21，再将ABAQUS模型非线性侧推模拟所得旋转角度由浅至深上色，不同角度利用等高轮廓线区隔，若旋转角度变化较剧烈，则等转角轮廓线将会集中，此亦即塑性铰位置，如图22所示并配合试体顶层位移4.0%弧度时底层柱照片以验证，而图23为ABAQUS模拟所得试体屈服反应与试体在试验结束后的照片，模型中白色区域即为以屈服之元素。

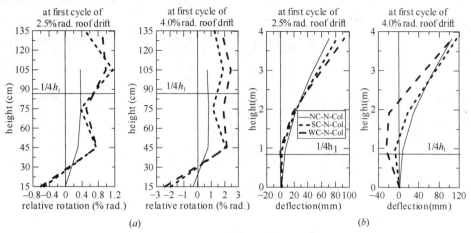

图 21 底层柱之 (a) 相对转角及 (b) 挠度

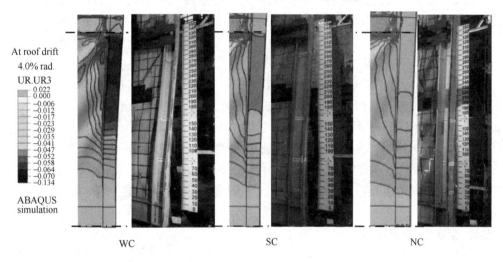

图 22 试体北侧底层边界柱角度轮廓线及变形状况

首先讨论系统顶层侧位移 2.5% 弧度时之反应，不论从角度计读值或是 ABAQUS 模拟，都显示试体 NC 柱底上方的旋转角度均匀分配，且转角不大，最大的相对转角发生靠近四分之一底层高下方，以控制塑性铰发生于四分之一底层楼高的弯矩需求 $M_d^{h1/4}$ 检核试体 NC 底层柱需求与容量比为 0.86，表示弯矩最大值位置会略低于四分之一底层高，与所观察的结果吻合，而试验过程中 tN6 角度计失效，因此并无该处高度之转角。试体 SC 从 ABAQUS 模拟可看到等旋转角的轮廓线集中于较高的位置，并非如试体 NC 均匀分布，最大相对旋转角位于四分之一底层高上方，以 $M_d^{h1/4}$ 比检核试体 SC 的 DCR 为 1.40，表示塑性铰位置会略高于四分之一底层高。而试体 WC 的 ABAQUS 模拟可看到等旋转角轮廓线集中在高于四分之一底层高甚多，最大相对转角在四分之一底层高上方约 20cm，且柱顶与柱跨中的挠曲屈服时机接近，图 23 显示系统应变硬化后试体 NC 底层柱顶仍保持弹性，试体 SC 有明显剪力屈服，而试体 WC 则有明显剪力塑性铰，以上底层柱顶与柱跨中反应与各需求与容量比验证结果，均证实本研究所提出的容量设计方法是可行的。

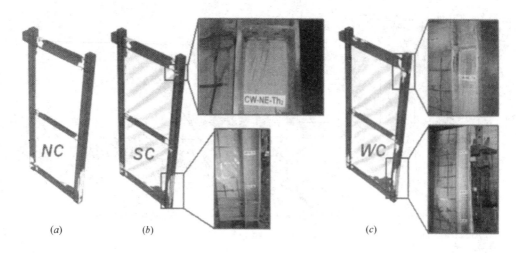

图 23 试体之有限元素屈服模拟

由图 21（a）系统顶层侧位移 4.0％弧度时之相对转角显示，三座试体旋转角度并不只集中于一处，高于 45cm 底层楼高的边界柱转角趋于相同，显示此区域边界柱也进入屈服，而三座试体皆可借由石膏漆脱落观察到受压底层边界柱大范围的屈服现象，比较图 21（b）试体 SC 与 WC 于 2.5％与 4.0％弧度顶层侧位移时底层柱挠度的差异，可发现 4.0％弧度时即因大范围屈服而边界柱有内凹之现象，因此读者需注意本研究虽能预测底层边界柱塑性铰位置，然在大侧位移下底层柱塑性区会有扩散之现象，可参考图 22 试验结束试体 SC 与 WC 的底层柱照片。

在试体设计时即对试体的二楼柱顶进行剪力屈服的检核，从图 23 同时可以观察到试体 SC 与 WC 二楼柱顶亦有剪力屈服或剪力塑性铰产生，而试体 NC 仍保持弹性，显示本研究以柱端弯矩比值 λ 及迭加法来估计弯矩及剪力容量同样也适用于其他楼层。

6. 设计与分析方法建议

6.1 底层柱设计原则

本文探讨多种设计弯矩及剪力的设计标准，其最终目的即为提供工程师设计钢板剪力墙上安全且经济的建议。试验结果显示三座试体行为的区别，主要在于系统应变硬化后底层柱顶是否会产生弯矩或剪力塑性铰：试体 NC 底层边界柱柱顶在试验结束前均没有任何的塑性行为，其边界柱也无侧向扭转变形；试体 SC 于全面屈服阶段底层柱顶轻微剪力屈服，系统应变硬化后底层柱顶同时有挠曲及剪力屈服，其边界柱则有轻微侧向扭转变形；试体 WC 的底层柱顶则在系统全面屈服阶段即有明显剪力屈服产生，系统应变硬化后更产生弯矩与剪力塑性铰，导致边界柱的侧向扭转变形最为明显，由图 24 可清楚看出三座试体试验结束后底层边界柱的变形状态。

底层柱顶与柱跨中同时产生塑性铰，可视为底层柱两端已从固接变为铰接，为不稳定的机构型式，在底层柱承受极大弯矩与轴力情况下极易产生侧向扭转变形[18]，另一原因为实际试体边界柱的侧向支撑不若数值分析模型完美，在最大侧位移时，边界柱已离侧撑系统柱构件太远，造成边界柱的侧向约束并不足够。综合以上所述，以系统应变硬化后底层柱顶仍保持弹性作为设计准则最为恰当。

图 24 三座试体（a）北侧及（b）南侧景象

6.2 拉力场角度

AISC 规范计算拉力场角度的公式是由系统弹性应变能推导而得，并未能反应当钢板与构架进入塑性时之应变能分布，本研究利用实验试体观测与有限元素分析来探讨钢板剪力墙在塑性变形下之拉力场角度变化。图 25 为三座试体试验后的拉力场角度观察，试验结束施力归零时试体还余有明显的残余侧向变位，因此照片上所标示之拉力场角度均已扣掉楼层残余变形角，以得到拉力场与边界柱之夹角。图 26 为试体 NC、SC 与 WC 以有限元素模型分析在顶层侧位移 2.5% 弧度时之拉力场角度，利用 ABAQUS CAE 读出有限元素模型中钢板面外变形量，并将面外变形等高线的轮廓两端相连，延伸联机与边界柱夹角视为有限元素拉力场角度。利用边界梁柱弹性应变能[13,14]计算所得拉力场角度列于表 4，以便与试验观察及有限元素模型分析结果比较。

拉力场角度计算值与实验及分析结果比较表　　　表 4

Story	NC	SC	WC
Computed tension field angle (elastic strain energy)			
2nd	40.4°	39.4°	38.5°
1st	40.3°	39.3°	38.5°
Experimental observation from test specimens (average of all the strips)			
2nd	44.3°	45.8°	44.2°
1st	40.3°	41.3°	40.5°
ABAQUS simulation (average of all the strips)			
2nd (+0.5% rad.)	41.0°	40.8°	39.5°
1st (+0.5% rad.)	40.3°	40.3°	39.7°
2nd (+2.5% rad.)	43.8°	43.2°	42.5°
1st (+2.5% rad.)	42.8°	42.3°	40.6°

图 25　试体试验后拉力场角度观察，由右至左为 NC、SC 与 WC

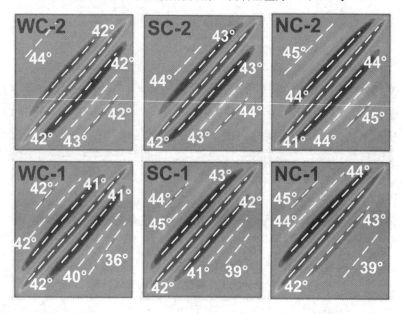

图 26　有限元素模拟之试体拉力场角度

钢板拉力场角度与钢板相连的边界构件劲度有关，当边界梁构件劲度相对边界柱较高时拉力场会偏向边界梁，亦即拉力场角度会变小，反之亦然，因此当梁柱构件劲度皆趋于无限大时，拉力场角度应趋近 45 度，而本研究试体边界柱尺寸大小由试体 NC 往 SC 与 WC 递减，因此由拉力场角度公式与 ABAQUS 分析的结果皆反映了拉力场角度由试体 NC 往 SC 与 WC 递减之现象，而观察试体拉力场角度则试体 SC 角度最大，主因为试体 SC 试验结束时停留在正顶层侧位移角，而试体 NC 与 WC 则停在负顶层侧位移角，如同 5.2 节所述，试体行为在正负侧位移并不对称，因此在边界柱强度差异不够大时，上面的讨论不适用。

从有限元素模型可观察到二楼拉力场角度由侧位移小到大会有 2 度到 3 度不等的增

加，反映了拉力场角度计算公式与试验观察之差距，因二楼的屈服区域集中在边界梁上，造成边界梁的劲度有软化，拉力场偏向边界柱，造成角度增加，根据观察三座试体的二楼非弹性拉力场角度不论是试体观察或是用有限元素仿型，皆接近45度。

由于三座试体均容许在底层柱跨内产生塑性铰，因此底层柱产生塑性铰后的劲度与其他楼层完全不允许塑性行为者不同，因此拉力场不若其余楼层接近45度，此外底层钢板对角线两侧的拉力场角度并非对称，观察试体于正顶层侧位移角时，钢板右下半部因塑性铰产生于底层柱跨中，边界柱明显有被钢板内拉而软化，因此拉力场会往没发生塑性铰的底梁改变，拉力场角度会变小，根据有限元素分析的观察，可发现右下半部的拉力场角度接近40度，而试体的边界柱行为还包含了侧向扭转变形，比起纯受挠曲之边界柱有更严重之软化现象，拉力场角度会更小，然系统设计过程中假设过小之拉力场角度可能会导致不保守的结果，因此本研究建议底层柱采用40度，其余楼层柱采用45度的拉力场角度来计算内力。本研究规划之试体宽高比为0.9，已具有代表性，虽本分析未能包含所有不同宽高比的钢板剪力墙系统，但宽高比接近1.0时，此结论可供参考使用。

6.3 SAP2000简化模型分析

本节利用工程实务习用之SAP2000程序进行单向侧推分析以探讨系统内各部构件之最大受力与对应之变形反应，以提供工程师建构钢板剪力墙模型设定建议。SAP2000模型规模以试体构件中心线距离为准，每面钢板墙以十根板条模型取代[24]，拉力场角度均设定为45度，模型中包含梁翼切削及梁柱交会区的设定如图27所示，同样也使用双线性模型来反映材料拉力试验结果。

SAP2000是在各杆件上配置非线性铰来模拟系统非线性反应，以钢板剪力墙系统而言，须在钢板板条模型中设置轴力P非线性铰，梁柱中设置轴弯互制P-M3及剪力V2非线性铰，各非线性铰的行为遵守各设定之屈服准则，例如：P-M3非线性铰以AISC所建议之轴弯互制式作为屈服曲线，而钢板塑性应变甚大，使用

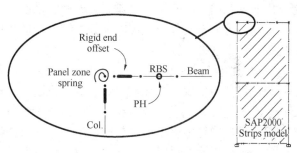

图27 试体SAP2000模型

内建的P非线性铰应重新调整可承受之应变及其强度，以免在大应变下钢板强度无法计算。由于本研究所规划的三座试体塑性区除钢板外主要集中于边界柱上，因此边界柱上非线性铰的设定直接关系到分析是否准确或是收敛，以下将以各试体非线性铰位置设定及其结果进行讨论。

试体NC受压柱于底层柱底、四分之一底层高、底层柱顶及顶层柱顶共设定四个P-M3非线性铰，并考虑几何非线性进行侧推分析，由于试体实际塑性铰位置略低于四分之一底层高，所以初始屈服强度SAP2000的分析结果会略高，但系统全面屈服后强度模拟仍良好，可参考图28，与试体NC相比，试体SC底层受压柱顶多设定了V2非线性剪力塑性铰，分析结果显示模型确实于顶层位移3.0%弧度时，底层柱顶产生剪力屈服，而强度反应也吻合试验结果。试体WC由于边界柱塑性变形较大，因此必须关闭几何非线

使能让分析收敛，其余模型设定均延用试体 SC 之设定，分析结果显示模型确实于顶层位移 1.0% 弧度时，底层柱顶与底层柱跨中同时产生塑性铰，其屈服强度的模拟也良好，但试体强度反应在顶层位移 2.5% 弧度后即逐圈衰退，而分析模型因为关闭几何非线性且无法模拟柱侧向扭转变形，模型力量反应仍有增加，未能反映实际试体的情况。本节分析证实了采用 SAP2000 建构钢板剪力墙的板条模型，对于分析底层柱大量屈服的钢板剪力墙系统，其结果足够精准可供工程上使用。

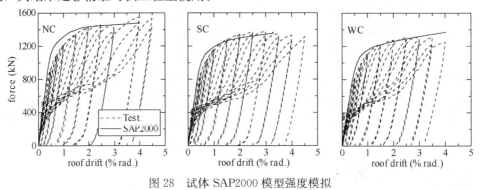

图 28 试体 SAP2000 模型强度模拟

7. 结论

1. 过去已有研究成功地以弹性模型来预估钢板剪力墙系统进入非线性后的内力分布；本研究讨论的底层柱内力需求是在大侧位移下，系统进入应变硬化阶段的内力分布，除了考虑应变硬化因子的迭加法以建立边界柱容量设计方法，尚利用实验与有限元素模型分析验证系统内力分布及行为，实验与分析结果均证实本研究所建议之容量设计方法对于控制底层边界柱塑性铰位置是可行的。

2. 本研究已讨论多种设计弯矩及剪力的设计标准，并根据不同的标准设计试体以验证行为，目的为提供安全且经济的钢板剪力墙设计建议。试验结果显示试体的行为区别主要在于系统应变硬化后是否会在底层柱顶产生弯矩或剪力塑性铰，柱顶产生屈服或塑性铰的底层边界柱均发生侧向扭转变形，反之柱顶保持弹性的试体行为良好，因此本研究建议考虑应变硬化后底层柱顶仍保持弹性作为耐震设计原则较为恰当。

3. 随底层边界柱强度容量越小，系统应变硬化后柱跨中之塑性区扩散越大，且层间变位越集中于底层，其拉力场发展效率也较差。

4. 钢板剪力墙系统由弹性进入塑性后，拉力场角度有明显的改变，本研究透过试体与 ABAQUS 有限元素模型分析证实这种现象，进行容量设计时建议底层拉力场角度可假定为 40 度，其余楼层可假设角度为 45 度。

5. 不论是利用 SAP2000 简化的板条模型或是采 S4R 薄壳元素的 ABAQUS 模型，采本文建议的有限元素模型设定，均可对试体的劲度与强度有效预测，使用 ABAQUS 模型分析更可了解试体细部反应。

致谢

本研究感谢国家科学委员会与国家地震工程研究中心给予的财务支持，以及中国钢铁股份有限公司致贈低屈服点钢板。实验期间作者获得国家地震工程研究中同仁、王孔君、莊勝智、陳家乾及林志翰先生的帮忙，数值分析获得蔡青宜小姐的协助，在此一并感谢。

參考文獻

[1] 林盈成、林志翰、蔡克銓，"束制型鋼板剪力牆之耐震試驗與分析"，結構工程，第二十四卷，第四期，第49-72頁（2009）。

[2] Vian, D., Bruneau, M., Tsai, K. C. and Lin, Y. C., "Special Perforated Steel Plate Shear Walls with Reduced Beam Section Anchor Beams. I: Experimental Investigation", *Journal of Structural Engineering*, ASCE, Vol. 135, No. 3 (2009).

[3] AISC, "Seismic Provisions for Structural Steel Buildings" (2005).

[4] CSA, "Limit States Design of Steel Structures", CAN/CSA-S16-01, Canadian Standards Association, Willowdale, Ontario, Canada (2001).

[5] 李昭賢、蔡克銓，「鋼板剪力牆系統之耐震設計研究」，國家地震工程研究中心研究報告，編號NCREE-08-019（2008）。

[6] 蔡克銓、李昭賢、林志翰、蔡青宜、遊宜哲、朱駿魁，"未束制與束制型鋼板剪力牆邊界柱構件之耐震設計(一)：數值分析研究"，結構工程，第二十五卷，第三期，第37-54頁（2010）。

[7] 李昭賢、蔡克銓、林志翰、陳沛清、朱駿魁，"未束制與束制型鋼板剪力牆邊界柱構件之耐震設計(二)：試驗研究"，結構工程，第二十五卷，第四期，第3－26頁（2010）。

[8] Tsai, K. C., Li, C. H., Lin, C. H., Tsai, C. Y. and Yu, Y. J. "Cyclic tests of four two-story narrow steel plate shear walls-Part 1: Analytical studies and specimen design.", *Earthquake Engineering & Structural Dynamics*, Vol. 39, Issue 7, pp. 775－799 (2010).

[9] Tsai, K. C., Li, C. H., Lin, C. H., Tsai, C. Y. and Yu, Y. J. "Cyclic tests of four two-story narrow steel plate shear walls-Part 2: Experimental results and design implications", *Earthquake Engineering & Structural Dynamics*, Vol. 39, Issue 7, pp. 801－826 (2010).

[10] 張景棠，「連梁式多樓層鋼板剪力牆耐震行為研究」，台灣大學土木工程學系結構組碩士論文，台北（2009）。

[11] Timler, P. A. and Kulak, G. L., "Experimental Study of Steel Plate Shear Walls", Structural Engineering Report No. 114, Department of Civil Engineering, University of Alberta, Edmonton, Alberta, Canada (1983).

[12] Berman, J., and Bruneau, M., "Plastic Analysis and Design of Steel Plate Shear Walls", *Journal of Structural Engineering*, ASCE, Vol. 129, No. 11, November 1, 2003, pp. 1448－1456 (2003).

[13] 蔡克銓、林志翰、林盈成、謝旺達、曲冰，"實尺寸兩層樓鋼板剪力牆子結構擬動態試驗"，國家地震工程研究中心技術報告，編號：NCREE－06－017，台北（2006）。

[14] C. H. Lin., Tsai K. C., Qu, B., Bruneau, M., "Sub-structural pseudo-dynamic performance of two full-scale two-story steel plate shear walls", *Journal of Constructional Steel Research*, Vol. 66, no. 12, pp. 1467－1482 (2010).

[15] Vian, D., "Steel plate shear walls for seismic design and retrofit of building structures", Ph.D. Dissertation, Department of Civil, Structural and Environmental Engineering, University at Buffalo, Buffalo, NY (2005).

[16] Thorburn, L. J., Kulak, G. L. and Montogomer, C. J., "Analysis of Steel Plate Shear Walls", Structural Engineering Report No. 107, Department of Civil Engineering, University of Alberta, Edmonton, Alberta, Canada (1983).

[17]　朱駿魁，「多樓層鋼板剪力牆結構耐震分析與設計之研究」，國立台灣大學土木工程學系結構組碩士論文，台北（2010）。

[18]　Qu, B., and Bruneau, M., "Behavior of Vertical Boundary Elements in Steel Plate Shear Walls", *Engineering Journal*, Vol. 47, no. 2, pp. 109－122,（2010）.

[19]　FEMA, "FEMA369 NEHPR Recommended Provisions for Seismic Regulations, for New Buildings and Other Structures"（2000）.

[20]　FEMA, "FEMA350 Recommended Seismic Design Criteria for New Steel Moment-Frame Building,"（2001）.

[21]　A. S. Lubell, H. G. L. Prion, C. E. Ventura, and Mahmoud Rezai, "Unstiffened Steel Plate Shear Wall Performance under Cyclic Loading", *Journal of Structural Engineering*, Vol. 126, No. 4, pp. 453－460 (2000).

[22]　Wang K. J., "An Integrated Environment for Structural Testing", Ph. D. Thesis, National Taiwan University, Taipei (2011).

[23]　M. R. Behbahanifard, G. Y. Grondin and A. E. Elwi, "Experimental and numerical investigation of steel plate shear walls", *Structural Engineering Report*, No. 254, Department of Civil Engineering, University of Alberta, Edmonton, Alberta, Canada (2003).

[24]　謝旺達，「鋼板剪力牆之有限元素分析與耐震設計研究」，國立台灣大學土木工程學系結構組碩士論文，蔡克銓教授指導，台北（2006）。

工程结构黏滞阻尼减振技术的研究与应用

李爱群[1]，陈 鑫[2,1]，张志强[1]，黄 镇[1]

（1. 东南大学 混凝土及预应力混凝土结构教育部重点实验室，南京 210096；
2. 苏州科技学院 江苏省结构工程重点实验室，苏州 215011）

摘 要：本文针对工程结构，围绕黏滞阻尼减振技术及其应用，首先介绍了对硅油进行改性的试验研究，总结了改性黏滞流体的力学特性，并建立了力学模型；随后，设计了6种黏滞流体减振装置，开展了理论和试验研究；在黏滞流体减振装置力学模型的基础上，建立了工程结构减振的分析模型和简化模型，提出了黏滞阻尼减振的智能优化方法和三阶段设计方法；进而，介绍了技术规程和减振设计图集的编制工作；最后，介绍了黏滞阻尼减振技术在北京奥林匹克国家会议中心、南京奥体中心观光塔、宿迁民丰农村合作银行等项目中的应用情况。本文的工作可为黏滞阻尼减振技术的进一步研究和推广提供参考和借鉴。

关键词：消能减振；阻尼器；黏滞流体
中图分类号：文献标识码：A
文章编号：

Investigation and Application of Viscous Damping Technology in Engineering Structures

Li Ai-qun[1], Chen Xin[2,1], Zhang Zhi-qiang[1], Huang Zhen[1]

(1. Key Laboratory of Concrete and Prestressed Concrete Structure
of Ministry of Education, Southeast University, Nanjing 210096, China;
2. Jiangsu Province Key Laboratory of Structure Engineering,
Suzhou University of Science and Technology, Suzhou 215011, China)

Abstract: In this paper, the viscous damping technology and its applications in engineering structures are studied. Firstly, the experimental study of modified silicone is introduced, and the mechanical characteristics of the modified viscous fluid are summarized, then the mechanical model is established. Secondly, six viscous damping devices are designed, and the theoretical and experimental study are conducted. Thirdly, on the basis of the study above, both the analysis model and simplified model of the controlled structures are established, then intelligent optimal method and three segment design method are proposed. Lastly, works of a technical specification and a vibration control design atlas are presented. And the applications of viscous damping technology in Olympic Park National Conference Center, Nanjing Olympic Center Sightseeing Tower and Suqian Minfeng Rural Cooperative Bank are introduced in detail. The study of this paper can provide an important reference for the investigation and application of viscous damping technology in engineering structures.

Key words：Vibration energy dissipation；Dampers；Viscous fluid；Application；Engineering structures

1. 前言

　　强震和飓风一直在威胁着人类的生存，带给人们灾难。随着经济发展和城市化进程加速，土木工程结构朝着高度更高、跨度更大、结构更加复杂的方向发展，随之而来的是一旦强震和飓风到来，将有可能带来更加重大的社会经济损失。因此，最大限度地减轻震灾和风灾所造成的损失，是人类必须解决的一个重要问题[1]。为了确保工程结构的安全性及与建筑、环境和使用要求的协调性，结构振动控制技术作为一种安全、经济、有效的措施便成了当前最佳选择之一[2,3]。结构振动控制的概念最早由华裔美国学者 J. T. P. Yao 首先提出，经过四十余年的发展，逐步形成了包含被动控制、主动控制、半主动控制和混合控制等方向，涉及控制论、计算机科学、振动力学、新材料科学等学科的新兴前沿研究领域[4-6]。近年来，该领域的研究取得了巨大的发展，其中被动控制的一些理论和技术已趋于成熟，逐渐走向应用[7]。在这之中，黏滞阻尼减振技术由于具有不提供附加刚度、受激励频率和温度影响较小等优点，成为被动控制技术中目前应用最为广泛的技术之一。

　　黏滞流体阻尼器（Viscous Fluid Damper，VFD）是指通过黏滞液体运动产生阻尼，消耗结构振动能量的一种被动速度相关型消能阻尼器。该类装置最早用于军事工业和航天工业等领域，进入 20 世纪 80 年代以来，人们一直尝试通过将军工、航空和重工业上的技术进行转换，从而将黏滞流体阻尼器运用到结构振动控制上。早在 20 世纪 80 年代末，美国纽约州立大学 Baffalo 分校就在美国国家自然科学基金会的资助下开展黏滞流体阻尼器研制和应用的相关研究工作[8,9]。目前，美国和日本两国在该领域的研究工作较为系统和成熟，而我国对黏滞阻尼减振技术的研究起步于 20 世纪 90 年代初[10-13]，但黏滞阻尼减振技术的核心——黏滞流体阻尼器的研究，则是从 90 年代末开始。在这一方面，国内的研究与国外相类似，均是工程应用早于（或同步于）科学研究[14]。在公开可见的文献中，1999 年中国建筑科学研究院利用法国 JARRET 公司生产的黏滞流体阻尼器对北京饭店进行了加固，这是我国在黏滞阻尼减振技术领域的第一例工程应用[15]；同年，该院又在北京火车站的加固工程中采用了美国泰勒公司（Taylor Devices Inc.）的 32 个黏滞流体阻尼器[16]。而国内关于黏滞流体阻尼器研制也是始于这一时期[17,18]，并在随后迅速发展，东南大学、哈尔滨工业大学、同济大学、武汉理工大学、上海材料研究所等众多科研单位对此开展了一系列的研究工作[14]。

　　笔者团队于 1999 年 7 月完成了单出杆和双出杆黏滞流体阻尼器的设计，并制作样品开展了大量的试验研究[18]。在此后的十余年内，团队从黏滞阻尼减振技术中的材料（改性流体）、构件（阻尼器）、结构（减振体系）三个层次，系统地开展了理论、方法和技术

项目基金：国家杰出青年基金项目（50725828），国家自然科学基金项目（51278104），江苏省高校自然科学研究面上项目（13KJB560012），东南大学混凝土及预应力混凝土结构教育部重点实验室开放课题
作者简介：李爱群（1962-），男，博士，教授，博士生导师，从事结构减振控制、结构健康监测研究（E-mail：aiqunli@seu.edu.cn）陈鑫（1983-），男，博士，讲师，从事结构振动控制及智能材料应用研究（E-mail：civil.chenxin@gmail.com）。

等方面的研究,并对该项技术的推广应用做了一定的努力。本文将对笔者团队近年来在黏滞阻尼减振技术方面开展的研究工作进行介绍。

2 黏滞流体阻尼材料研究

黏滞流体阻尼器通常是由黏滞液体通过阻尼孔的流动来耗散能量的,因此,其力学性能一方面要考虑阻尼器构造的影响,另一方面则要考虑阻尼介质的性能。在工程结构领域中应用的阻尼器,要求内部液体介质具有抗火、无毒以及温度稳定性,且性能不会随时间而退化,目前仅有硅基系列的液体可以满足这种要求。因此,随着阻尼器密封性能的解决,采用液体硅油作为黏滞流体阻尼器的主要材料已成为工程界的主要选择。在实际设计和生产过程中,由于硅油本构关系的不确定性和部分更高性能的需要,材性的改进和试验显得尤为重要,本节主要介绍笔者团队采用颗粒粉末对硅油进行改进的研究工作。

2.1 黏滞流体材料改性

硅油在剪切速率较低时表现为牛顿流体,考虑在硅油中加入颗粒直径极小(微米级)的颗粒粉末,由于微小颗粒的存在,阻碍介质间的相互运动,增加阻尼材料的黏度。又由于阻尼介质与颗粒之间的物理化学作用,形成某种松散的结构,随着剪切流动的进行,结构逐渐被破坏,使表观黏度随应变速度的增大而减小。此外,当介质静止时,细微颗粒在硅油中呈杂乱卷曲状态,随着流动的进行,它们沿流动方向排列起来,显然应变速度越大,定向排列整齐,流动阻尼就愈小,其表观黏度也就愈小。这样由于细微颗粒的存在,将原有硅油由牛顿流体变为非牛顿流体,增加阻尼材料的阻尼特性的同时,能更好地适应阻尼器对阻尼介质的要求。因此,笔者团队从多种可能的颗粒粉末添加物中选择了三种(表1)进行研究。

颗粒粉末添加物　　　　　　　　　　　　　表1

添加物	产地来源	粒径/μm
生产石膏的残渣物(简称 KS 粉)	日本,生产石膏的残渣物	50
硅石粉	玻璃制品的残渣物	5
硫酸钙晶须	沈阳东大富龙矿物材料研发有限公司	0.2—4

2.2 黏滞流体材性试验

为了研究颗粒粉末添加物对硅油特性的影响,以国产二甲硅油(标号 500#,硅油的标号表示硅油动力黏度的数量级(单位为 CST,即 mm^2/s))为基础,进行了对比试验研究。本试验采用的仪器是由成都仪器厂生产的 NXS-4C 型水煤浆黏度计(图1),它是带有微电脑的同轴圆筒上旋式黏度计。开展试验的试样和测试工况

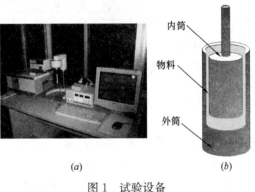

图1　试验设备
(a)试验仪器；(b)物料筒示意

见表2。

试验试样 表2

编号	成分组成	质量配比	试验环境
1	500#硅油		温度为10℃、25℃（常温）、30℃、50℃、80℃时，剪切速率分别为10、20、40、60、80、100 s^{-1}
2	500#硅油+石膏晶须粉末	5%、8%、10%、15%	温度为25℃时，其中10%和15%的试样还测试了温度在45℃、60℃、80℃时，剪切速率分别为10、20、40、60、80、100 s^{-1}
3	500#硅油+KS粉	5%、10%、15%、20%	温度为25℃时，剪切速率分别为10、20、40、60、80、100s^{-1}
4	500#硅油+硅石粉	5%、10%、15%	温度为25℃时，剪切速率分别为10、20、40、60、80、100s^{-1}

2.3 试验结果与分析

表3列出了常温下（25℃），表中各试样剪切速率为100s^{-1}时的表观黏度η（mPa·s）和剪切应力τ（Pa）。由表3可知，常温下，硅油在剪切速率为100s^{-1}时，表观黏度为568 mPa·s，而含有10%KS粉时，黏度提高到659 mPa·s，相同含量时，硅石粉提高到894 mPa·s，而石膏晶可提高到957 mPa·s。因此，细微粉末对硅油的黏度提高很明显，并且硅油从牛顿流体转变为剪切稀化流体。表3中k值随着所含颗粒粉末的比例的提高而增大，而n值随着所含颗粒粉末的比例的提高而减小，说明加入细微颗粒越多，混合物的黏度越大，且剪切稀化越明显，即随着剪切速率的提高，流体黏度下降很快，表现为n越小。

剪切速率为100s^{-1}时的表观黏度和剪切应力 表3

试样编号	η	τ	试样编号	η	τ
1（25℃）	568	56.8	1（50℃）	371	37.1
2（5%）	702	70.2	2（8%）	834	83.4
2（10%）	957	95.7	2（15%）	1329	132.9
3（5%）	622	62.2	3（10%）	659	65.9
3（15%）	767	76.7	3（20%）	862	86.2
4（5%）	682	68.2	4（10%）	894	89.4

研究表明，剪切稀化流体的本构关系可写成[19]：

$$\tau = k\gamma^n \tag{1}$$

其中，k为稠度系数，单位Ns^n/m^2；n为流动指数。图2给出不同颗粒粉末在质量比为10%温度为25℃时，测得数据的回归曲线，由曲线可知，所得规律基本符合式（1）表示的本构关系。表4给出了不同试样在不同工况下的本构关系参数拟合结果，可见通过颗粒添加物，能够使硅油在低剪切速率下具有非线性的力学特性。

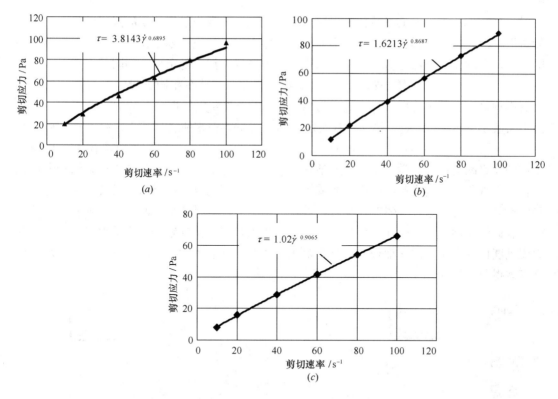

图 2 含 10%不同粉末混合物的本构关系回归曲线
(a) 10%石膏晶须；(b) 10%硅石粉；(c) 10%KS 粉

各试样本构关系曲线参数拟合　　　　　　　　　　　　　　　　　　　　　　表 4

试样编号	k	n	试样编号	k	n
1 (25℃)	0.5711	1	1 (50℃)	0.3717	1
2 (5%)	1.3938	0.8468	2 (8%)	2.4877	0.7542
2 (10%)	3.8143	0.6895	2 (15%)	8.4211	0.5869
3 (5%)	0.7448	0.9621	3 (10%)	1.02	0.9065
3 (15%)	1.3997	0.8673	3 (20%)	1.9485	0.8194
4 (5%)	0.9257	0.933	4 (10%)	1.6213	0.8687

剪切速率为 $10s^{-1}$ 时各温度下的黏度（单位：mPa·s）　　　　　　　　表 5

试样编号	25℃时	40℃时	60℃时	80℃时
1	575	495	375	222
2	1960	1675	1450	1216

由表 4 可知，500♯硅油在温度固定时，剪切速率较低时（$100s^{-1}$ 以下）表现为牛顿流体，即剪切速率与剪切应力呈线性关系（$n=1$）。温度较低时，硅油的性质比较稳定，随着温度的升高，黏度降低很快，80℃时有剪切稠化的现象，说明温度对该硅油的稳定性影响较明显（高标号的硅油受温度影响也很明显）。如表 5 所示，常温下（25℃）500♯硅油剪切速率为 $10s^{-1}$ 时的硅油黏度为 575mPa·s，而剪切速率为 $100s^{-1}$ 时的硅油黏度为 570mPa·s，而温度上升到 80℃时，剪切速率为 $10s^{-1}$ 时的硅油黏度为 218 mPa·s，而剪切速率为 $100s^{-1}$ 时硅油黏度为 222mPa·s，黏度下降大于 50%。由于阻尼器安装在结构中发挥作

用时，将结构的机械能最终转化为阻尼介质的内能，表现为阻尼介质的温度明显上升，将能量消耗掉，所以选择受温度影响不大的硅油作为阻尼介质，是制作黏滞流体阻尼器重要的因素。在制作阻尼器之前，一定要选择黏温关系好，温度稳定性高的阻尼介质。

图 3 给出了含 10％添加物的硅油常温时的特性，可见在相同含量时，石膏晶须对硅油的影响最大，其次为硅石粉，最后为 KS 粉，这与表 4 中的拟合参数的规律相同。这是因为松散密度 KS 粉＞硅石粉＞石膏晶须，相同质量的粉末所含体积 KS 粉＜硅石粉＜石膏晶须。另外硫酸钙晶须是纤维状颗粒，平均长径比可达 60，剪切稀化流体由于长链分子或颗粒之间的物理化学作用，形成某种松散的结构，随着剪切流动的进行，结构渐被破坏，使表观黏度随应变速度的增大而减小，非牛顿黏度的成因也是由于长链分子或颗粒本身的性质产生的。这种液体在静止时，长链分子细长纤维呈杂乱卷曲状态，随着流动的进行，它们沿流动方向排列起来，显然应变速度越大，定向排列整齐，流动阻力就愈小，其表观黏度也就愈小。因此细长的纤维在阻止流体间相互运动时，提高流体的黏度方面显然会表现得更优异些。

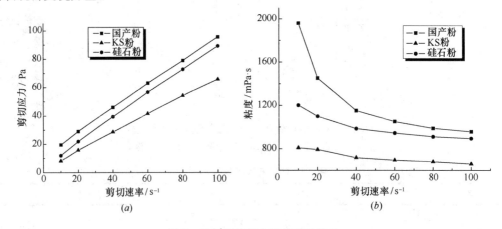

图 3　10％颗粒混合物的硅油特性
(a) 本构关系；(b) 剪切速率—表观黏度关系

由上述研究可知：(1) 通过附加颗粒物的方法能够有效地改善硅油的流动特性，但对温度影响作用较小，需要选用温度稳定性较高的黏滞阻尼材料或采取其他改进措施，才能够将材料应用到实际产品中。(2) 阻尼液体中含固体颗粒越多，混合物的剪切稀化越明显，更能满足阻尼器设计中非线性特性的要求；但试验时发现，加入固体粉末越多，介质越难混合均匀，致使混合物中粉末结成块状，考虑到阻尼器空隙细小，势必会影响介质的流动稳定性；对此可考虑采用分散剂、防沉剂等化学添加剂进行改进。(3) 虽然粉末含量越高，流动指数 n 越小，但介质的黏度显著增大，给设计阻尼器带来许多不确定因素；同时由阻尼器本构关系的推导过程可知，当流体的流动指数过小时，影响了流体阻尼器的阻尼系数，会降低阻尼器的耗能能力。所以在制作改性流体材料的阻尼器时，综合考虑多方面因素，选择合适的流体材料是设计的关键之一。

3　新型黏滞流体阻尼减振装置

笔者团队十余年来对黏滞流体阻尼材料进行了一系列研究，积累了一定的经验，并得到了合适的流体阻尼材料。在此基础上，开展了黏滞流体阻尼减振装置的设计、制作与试验研究。

3.1 线性黏滞流体阻尼器

3.1.1 双出杆黏滞流体阻尼器

黏滞流体阻尼器传统的单出杆形式在构造上存在缺陷，易造成油缸内压力的急剧变化，导致阻尼器产生的阻尼力是非常不稳定的[18]。对于该缺陷常用的方法是附加调节油缸和采用双出杆形式，但附加调节油缸的方法使得阻尼器构造和加工复杂，且不能提供较大的阻尼，不适合用于大型结构的控制，其应用前景受到限制。基于此，笔者团队设计了一种具有自主知识产权的双出杆黏滞流体器。其基本原理如图4（a），主缸内装满黏滞流体阻尼材料，副缸内无阻尼材料，当活塞向左运动时，原来在油缸外的部分活塞导杆进入阻尼器腔体内，而活塞背面同样体积的活塞导杆则被推出主缸而进入副缸，反之亦然。这样主缸内始终保持体积恒定，由于双出杆型流体阻尼器在活塞运动时，油缸内的总体积不会发生变化，这样油腔内的压强也不

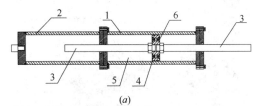

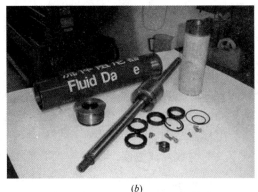

图4 双出杆黏滞流体阻尼器
1—主缸；2—副缸；3—导杆；
4—活塞；5—阻尼材料；6—阻尼孔
(a) 原理图；(b) 阻尼器样品

会产生过大变化，从而避免了前述单出杆流体阻尼器的弊病。

理论分析和研究表明[20]，黏滞流体阻尼器的阻尼力主要与活塞有效面积（指活塞横截面面积或主缸内截面面积减去导杆横截面积）、阻尼孔大小和长度、振动频率和位移幅值（二者实质上决定了活塞运动速度）、温度、阻尼材料性能（黏度、黏-温关系等）等因素有关，即活塞有效面积越大、阻尼孔越小、活塞运动速度越快、环境温度越低（在工作温度范围内）、阻尼材料黏性越大，阻尼力越大。为了结合理论推导获得阻尼器力学模型，同时为产品研发积累经验，笔者团队于2000年分别在东南大学混凝土和预应力混凝土结构教育部重点实验室、香港理工大学土木与结构工程学系结构试验室进行了三组阻尼器样品在不同频率、不同位移幅、不同黏度硅油和不同温度下的动力试验，其中一组缩尺模型试验（香港，图5（a）），两组足尺模型试验（香港和南京各一组，图5（b）），共24根不同参数规格的阻尼器[18,20]。

结合流体动力学原理，对试验数据处理后进行回归分析，则阻尼器的阻尼力与各影响因素之间的关系按照下式确定：

$$F = k'kk_u\rho \frac{A^3}{A_0^{1.5}} V \tag{2-1}$$

其中，F 为阻尼器输出阻尼力，单位为（N）；μ 为甲基硅油的动力黏度，单位为（Pa·

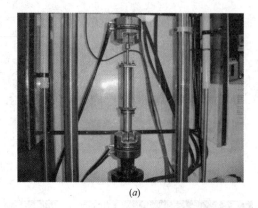

图 5　双出杆黏滞流体阻尼器试验
(a) 缩尺模型；(b) 足尺模型

s)；ρ 为甲基硅油的密度，单位为（kg/m³）；A 为活塞有效面积，是活塞总截面积扣除导杆截面积后的值，单位为（m²）；A_0 为所有阻尼孔截面积之和，单位为（m²）；V 为活塞运动速度，单位为（m/s）；k_μ 是与黏度有关的系数，与黏度呈幂函数关系；k' 为修正系数（与阻尼器结构、加工精度、阻尼材料等有关，根据试验测定，无量纲）；k_t 为一个与温度有关的常数，在温度 $-20\sim+40$℃范围内，可由下式插值确定：

$$k_t = -0.2112T + 57.1712 \tag{2-2}$$

式中，T 为阻尼介质的温度，单位为（℃）。将系数进行合并后，式（2-1）即可写成目前通用的模型形式：

$$F = CV \tag{2-3}$$

其中，C 是阻尼器的阻尼系数。事实上式（2-3）亦可由 Maxwell 模型结合试验结果得到[20]。

根据式（2-1）的计算与试验结果的对比（见图6）可知，计算结果与试验结果较为吻合，说明通常在采用黏滞流体阻尼器进行减振设计时，可根据设计需要，调整阻尼器的参数、油缸直径、活塞有效面积、阻尼孔大小和数量、阻尼材料黏度等，获得满足不同需要的阻尼器。

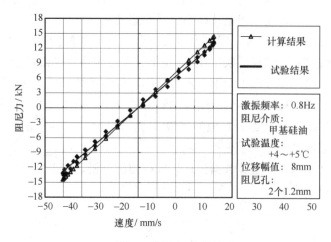

图 6　阻尼力-速度关系

3.1.2 调节阀式黏滞流体阻尼器

由式（2-3）可见，所设计的双出杆黏滞流体阻尼器是一种线性黏滞流体阻尼器，其构造简单、性能稳定、计算方便，在实际工程中得到较为广泛的应用，但是按常规小震或风振设计的线性阻尼器，在大震时的输出力过大，给结构节点和支撑系统的设计带来困难。为解决该问题，通常有两种方法：一是采用非线性黏滞流体阻尼器，该装置将在3.2节中进行介绍；另一方法是对线性阻尼器的最大输出力进行限制。

基于后一种方法，设计了三种调节阀式黏滞流体阻尼器[21, 22]，三种调节阀的构造原理如图7所示：当活塞运动速度较小时，缸筒内压力不能达到调节阀开启压力，阻尼器的工作性能与常规的黏滞阻尼器一致；活塞运动速度较大时，压力达到或超过调节阀的开启压力，一方面通过阻尼孔耗能，另一方面因调节阀的溢流作用，阻尼器输出力基本保持稳定。其力学模型如下：

$$F = \begin{cases} CV & (V < \dfrac{F_k}{C}) \\ C'V + F_k & (V \geqslant \dfrac{F_k}{C}) \end{cases} \tag{3}$$

其中，C，C' 分别为开启前后的阻尼系数；F_k 为调节阀开启时阻尼器最大输出阻尼力。

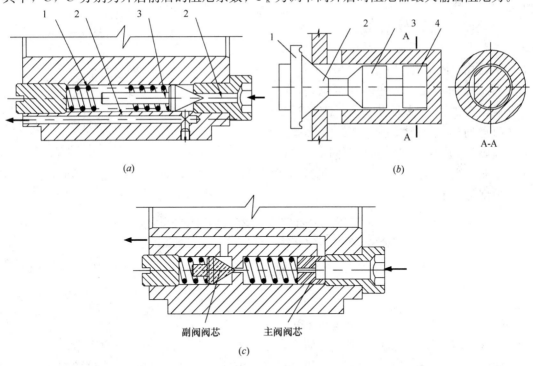

图7 调节阀原理图

(a) A1型调节阀
1—调压弹簧；2—阻尼孔；3—阀芯
(b) A2型调节阀
1—偏流盘；2—锥阀段；3—锥面段；4—阻尼活塞
(c) B型调节阀

为了研究调节阀式黏滞流体阻尼器的特性，并验证所推导力学模型，制作了分别对应三种调节阀的三个阻尼器样品（其余参数与构造相同），在东南大学结构试验室 MTS 1000kN 疲劳试验机开展了动力试验。限于篇幅，本文主要给出 A2 型调节阀式黏滞流体阻尼器加载频率为 0.25Hz 的试验与数值模拟结果对比（图 8）。由图可见：（1）所设计的调节阀式黏滞流体阻尼器能够满足大激励时限制输出力的要求；（2）所提出的力学模型能够较好地模拟该阻尼器的力学性能。

3.2 非线性黏滞流体阻尼器

相较于线性黏滞流体阻尼器，非线性黏滞流体阻尼器具有更好的耗能能力和高速时较低的输出力，在实际工程中亦有着较多的应用。要达到阻尼器非线性的目的，通常有两种思路：一是对阻尼材料进行改进（如第 2 节），二是对阻尼器的构造进行设计。本节主要介绍笔者团队近年来在后一方面的研究工作。

3.2.1 细长孔式黏滞流体阻尼器

本节研究的黏滞阻尼器在活塞上设有细长阻尼孔（阻尼孔构造见图 9 所示），当阻尼器工作时，随着活塞的往复运动，阻尼介质相应的由活塞两边的高压腔经过细长的阻尼通道流往低压腔。在黏滞流体反复流经阻尼通道的过程中，流体因克服摩

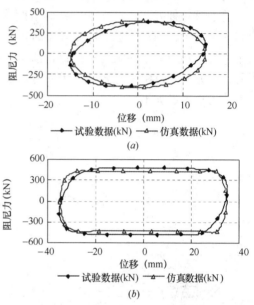

图 8 阻尼器力-位移曲线

(a) 15mm 位移幅值；(b) 35mm 位移幅值

擦等影响因素，而耗散外界输入的机械能或机械功。当以幂律流体为阻尼介质时，阻尼器力学模型可表示为[23]：

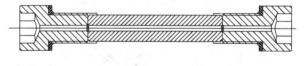

图 9 细长阻尼孔构造

$$F = CV^{\alpha} \tag{4}$$

其中，C 为阻尼系数；V 为活塞运动速度。

为进一步对所设计的阻尼器进行研究，制作了 16 根具有不同规格的样品，在东南大学结构试验室 MTS1000kN 疲劳试验机进行了动力试验（如图 10 所示），试验包括了不同频率（0.1Hz-1.5Hz）、不同幅值（15mm-50mm）的试验工况。图 11 给出了其中一个样品的力-位移曲线，可见细长孔式黏滞流体阻尼器具有明显的非线性特性，最大阻尼器超过 400kN。此外，总结试验结果可知，保持输入位移幅值、激励频率和阻尼介质不变，随着阻尼孔增大，最大阻尼力减小，滞回环逐渐趋于扁平，耗能能力降低。同时，借助本次试验结果，文献［23］对阻尼介质灌注质量和瞬时刚度进行了讨论，并得到了一些有益的经验。

图 10 细长孔式黏滞流体阻尼器试验

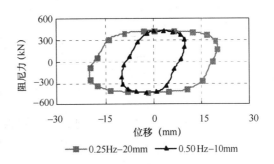

图 11 细长孔式黏滞流体阻尼器滞回曲线

3.2.2 螺旋孔式黏滞流体阻尼器

本节研究的黏滞阻尼器在活塞上设有螺旋式阻尼孔（阻尼孔构造见图12所示）。螺旋通道是一种曲线管道，在流体力学领域，通常把平面弯管、扭管、螺旋管道等通称为曲线管道。根据流体动力学原理，进行一定的简化，可以导出与式（4）相同的力学模型[23]，只是其中决定阻尼系数和阻尼指数具体数值的因素有所不同。

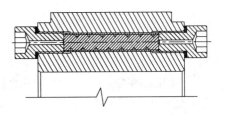

图 12 螺旋阻尼孔构造

试验用螺旋孔式黏滞流体阻尼器规格　　　　表 6

编号	主缸筒内直径	活塞导杆直径	阻尼孔直径
H1	180	90	3.0
H2	180	90	4.0

本文制作了2根（表6）具有不同规格的样品，在东南大学结构试验室 MTS1000kN 疲劳试验机进行了动力试验，试验方式同图10，试验工况与3.2.1节中相同。图13给出了两个试验样品的力-位移曲线，由图可见，本文提出的螺旋孔式黏滞流体阻尼器具有明显的非线性特性，最大阻尼器300kN左右。通过本次试验：（1）研究了活塞相对运动速度、环境温度、阻尼孔构造等因素对该型阻尼器力学性能的影响；（2）随着温度的变化，滞回环包络面积的大小有所变化，随着环境温度的升高，输出阻尼力有所减小，然而降幅不明显，最大输出阻尼力的波动范围在15%以内；（3）与细长孔式黏滞流体阻尼器对比可知，同一条件下，细长孔式阻尼器的最大输出阻尼力小于螺旋孔式阻尼器，后者的滞回曲线更饱满，耗能能力更强，随着加载频率的增大，两种阻尼器均逐步出现瞬时刚度，在相同激励下，后者的刚度要大于前者；（4）进行的疲劳试验表明，力-位移滞回曲线始终非常饱满，第500个循环与其后的第10000个循环以及第20000个循环相比，滞回环的形状和大小基本没有发生变化，阻尼器的拉、压最大输出阻尼力的波动在10%以内，具有较高的稳定性。

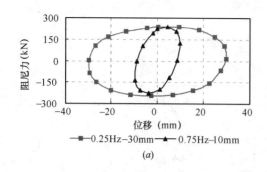 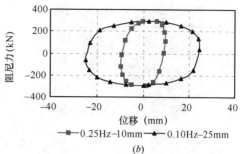

图 13 螺旋孔式黏滞流体阻尼器滞回曲线
(a) 阻尼器 H1；(b) 阻尼器 H2

3.3 其他黏滞阻尼减振装置

随着对黏滞阻尼减振技术认识的深入和实际工程的需要，除了上述线性黏滞流体阻尼器和非线性黏滞流体阻尼器以外，开展了其他黏滞阻尼减振装置的研发工作。

3.3.1 变阻尼黏滞流体阻尼器

提出了变阻尼黏滞流体阻尼器（如图14），对比图4可知，该阻尼器在双出杆黏滞流体阻尼器的基础上增加了可根据需要专门设计的阻尼棒[24]，工作时，通过活塞和阻尼棒共同控制阻尼的大小，活塞行程中的阻尼槽可以根据需要在阻尼棒上设定，达到控制阻尼变化，使得在位移较小时阻尼较小，位移较大时阻尼较大的目的。

同样，制作的阻尼器样品在东南大学结构试验室 MTS1000kN 疲劳试验机上进行了动力试验（如图15）。对试验数据的回归拟合可知，该变阻尼黏滞流体阻尼器的阻尼指数为 0.326，阻尼系数当位移小于 50mm 时为 $0.7818e6 N/(m/s)^{0.326}$，当位移大于 50mm 时为 $1.275e6 N/(m/s)^{0.326}$，成功地实现了自动变阻尼的功能。由图 15（b）可见，阻尼器滞回曲线饱满，结合其余的试验结果发现疲劳试验下阻尼器性能具有较高的稳定性[25]，有较高的实用价值。

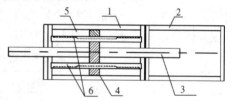

图 14 变阻尼黏滞流体阻尼器构造
1—主缸；2—副缸；3—导杆；4—活塞；
5—阻尼介质；6—带槽阻尼棒

3.3.2 黏滞阻尼墙

黏滞阻尼墙是一种主要应用于建筑结构的墙式速度相关型消能器，自 20 世纪 80 年代以来在国外已有一定的研究，但国内的研究起步较晚，相应的产品和试验研究开展的较少。

本文设计了一种新型黏滞阻尼墙（如图16），新型黏滞阻尼墙由外部钢箱、顶板、内钢板和黏滞阻尼介质组成。钢箱内装有黏滞阻尼介质，一块或多块横向内钢板置于钢箱和黏滞阻尼材料之间[26]。应用于实际工程时，为了满足结构空间和阻尼墙安装需求，可在阻尼墙下部设置支架，将支架的底部与下层楼面相连；一般还需在钢箱外部设置钢筋混凝土或防火材料保护墙体，以防黏滞阻尼墙受到撞击、腐蚀、火灾等因素影响。

设计制作的样品在南京丹普科技工程有限公司消能产品实验室进行了试验，加载装

 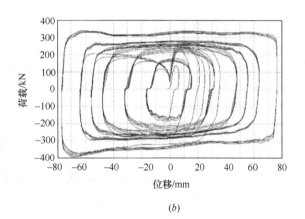

图 15 变阻尼黏滞流体阻尼器试验
(a)试验装置;(b)力-位移曲线

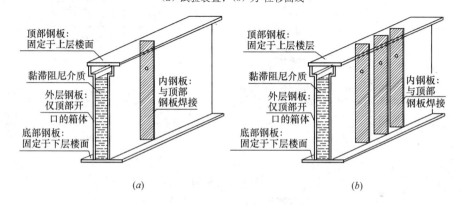

图 16 新型黏滞阻尼墙
(a)单片式;(b)多片式

置如图 17 所示。试验在常温条件下进行,试验时实测室温为 20℃。试验采用正弦波激励,以输入位移控制加载。通过施加不同频率和不同幅值的位移,分别测得新型黏滞阻尼墙的位移、阻尼力与对应的时间,从而得到黏滞阻尼墙阻尼力随加载频率、位移幅值变化的动力特性。

图 18 给出了加载频率为 0.1Hz 和 1.0Hz 时,黏滞阻尼墙的力-位移曲线。试验得到的阻尼力-位移滞回曲线较为饱满,滞回曲线形状类似于圆角矩形(介于椭圆和矩形之间)或椭圆,大致上关于原点对称,说明新型黏滞阻尼墙的耗能效果较好。同时,还进行了阻尼墙的耐久性试验,结果表明,地震作用控制时,阻尼墙的阻尼力变化较小,每 10 次循环记录的阻尼力变化幅度在 5% 以内,说明新型黏滞阻尼墙在地震作用下的动力性能稳定;风荷载作用控制时,1000 次循环阻尼墙的阻尼力衰减在 10% 以内。

4 黏滞阻尼减振设计理论与试验

上述研究分别从材料和构件的层次对黏滞阻尼减振技术进行了研究,并导出了材料性能、构件构造尺寸与阻尼减振装置宏观特性的关系,建立了各类减振装置的力学模型。本

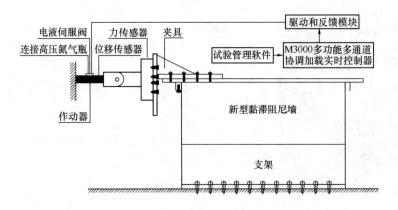

图 17 试验加载装置

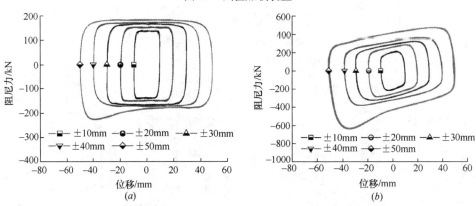

图 18 黏滞阻尼墙滞回曲线
(a) 加载频率 0.1Hz；(b) 加载频率 1.0Hz

节以这些成果为基础，对黏滞阻尼减振装置安装在工程结构中的计算和设计方法开展理论和试验研究，为进一步的推广应用奠定基础。

4.1 结构黏滞阻尼减振计算方法与试验

4.1.1 动力方程与分析技术

第 3 节的理论和试验研究表明，多数黏滞阻尼减振装置的力学模型均可以用公式（4）表示，其中当阻尼指数 $\alpha=1$ 时，为线性黏滞流体阻尼器，$\alpha<1$ 时，为非线性黏滞流体阻尼器。根据结构动力学原理，可知安装黏滞流体阻尼器的工程结构的动力方程为[20, 27]：

$$M\ddot{x}(t) + C\dot{x}(t) + Kx(t) + F_{ve} = F(t) \tag{5-1}$$

式中：M、C、K 分别为结构的质量、阻尼和刚度矩阵；I 为单位列向量；$\ddot{x}(t)$、$\dot{x}(t)$ 和 $x(t)$ 分别为质点加速度、速度和位移列阵，$F(t)$ 是外力作用，如地震作用、风荷载等；F_{ve} 为黏滞消能支撑的附加控制力列阵，当为多高层结构时，可按如下过程进行计算[28]：

$$F_{vej,i} = C_{dj,i}\xi_{j,1}^{1+\alpha_{j,i}} | \dot{X}_j - \dot{X}_{j-1} |^{\alpha_{j,i}} \mathrm{sgn}(\dot{X}_j - \dot{X}_{j-1}) \tag{5-2}$$

其中，$C_{dj,i}$、$\alpha_{j,i}$ 和 $\xi_{j,i}$ 分别为第 j 楼层的第 i 个阻尼器的附加阻尼、阻尼指数和变形系数；X_j、X_{j-1} 和 $\dot{X}_j$、$\dot{X}_{j-1}$ 分别表示 j 层和 $j-1$ 层相对地面的水平位移和速度。则，第 j 层的黏

滞消能支撑提供的总的水平控制力为：

$$F_{vej} = \sum_{i=1}^{m} F_{vej,i} \quad (5\text{-}3)$$

其中，m 为第 j 楼层阻尼器的总数。则，式（5-1）中阻尼器对结构的附加水平控制力列阵为：

$$F_{ve} = \{F_{ve1} \ F_{ve2} \cdots F_{vej} \cdots F_{ven}\}^T \quad (5\text{-}4)$$

其中，n 为楼层总数。

随着计算机的发展，结构动力响应的数值模拟技术已日渐成熟，面对越来越复杂的结构，人们习惯于应用有限元软件或者自行编制计算程序进行结构分析，黏滞消能减振结构也不例外。笔者团队对于一些工程结构和简化模型自行编制程序实现了动力响应的非线性分析过程[27,29,30]。同时，为了对复杂结构进行分析计算，已在 ANSYS、ETABS、SAP2000、Midas、Perform3D、Msc.Marc、Abaqus 等有限元软件中实现了对附加黏滞流体阻尼器的工程结构在风、地震、人行等荷载作用下的动力响应分析[20,31-34]。

4.1.2 结构黏滞阻尼减振试验

计算分析表明，黏滞流体阻尼器具有较强的耗能能力，能够有效地降低结构的动力响应，从而保护结构免受动力灾害的破坏。为了进一步对黏滞流体阻尼器的减振效果和计算分析方法进行验证，设计了一个 1∶3 缩尺的钢框架模型，在东南大学九龙湖校区土木工程实验室振动台开展了地震作用下钢框架结构黏滞阻尼减振的试验研究[35]（图19）。

试验选用了 3 条实际记录的地震波：II 类场地的 El-Centro 波，III 类场地的 Taft 波和 IV 类场地的 Washington 波。图 20（a）给出了 Washington 波（490gal）作用下模型顶层加速度放大系数曲线（此处仅截取了前15s 的试验结果），结合表 7 的层间位移峰值，可见安装黏滞流体阻尼器后，模型顶层加速度在整个加载时间范围内有较大程度的衰减（最高达到 42.0%）。图 20（b）给出了布置在模型第三层的阻尼器滞回曲线，这与之前对阻尼器单独进行的加载试验所得曲线基本吻合，曲线饱满，耗能能力较好。同时，对时程曲线的频谱分析可知安装阻尼器后结构特征周期变化较小，结合图 20（b）可知黏滞流体阻尼器附加刚度较小，对原结构动力特性的影响较小，仅仅是增加了结构阻尼，这为工程设计提供了便利。

图 19 钢框架结构黏滞阻尼减振试验

Washington 波（490gal）作用下结构层间位移峰值　单位：mm　表 7

楼层	1层	2层	3层
原结构	4.29	4.46	3.50
减振结构	2.85	2.79	2.03
衰减率	33.6%	37.4%	42.0%

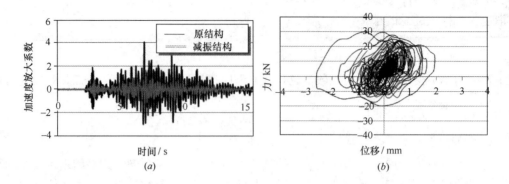

图 20 Washington 波作用下结构与阻尼器响应曲线
(a) 顶层加速度放大系数曲线 (490gal);
(b) 3 层阻尼器力-位移曲线 (660gal)

4.2 结构减振优化设计方法

在结构减振系统的设计中,减振装置的参数、布置数量、位置及方式等参数的选取决定了减振的效率,因此这些参数的优化是系统设计中的重要步骤。由于结构和减振装置的非线性,导致整个减振体系的动力方程必然为一非线性方程,难以通过数学方法得到理论上的最优参数解。而不适当地设置减振装置往往会造成经济上的浪费,有时甚至会对振动响应产生放大作用,因此如何采用一定数量适当参数的减振装置来实现最佳的控制效果成了投资者最为关心的问题之一,同时也成为振动控制领域学者研究的重要方向之一[27]。

一般的优化问题可以描述为满足一定的约束条件下,求解使得目标函数为极小(或极大)的设计变量,其标准数学模型可写为:

$$求 X \quad X = (x_1, x_2, \cdots, x_i, \cdots, x_n)^T \quad (6)$$
$$\min F(X) = \min [f_1(x), f_2(x), \cdots, f_m(x)]^T$$
$$\text{s.t. } g_j(X) \leqslant 0 \ (j = 1, 2, \cdots, m)$$
$$g_j(X) = 0 \ (j = 1, 2, \cdots, m)$$

其中,X 为设计变量;x_i 为第 i 个设计参数;$F(X)$ 为目标函数;$g_j(X) \leqslant 0$ 为不等式约束条件;$g_j(X) = 0$ 为等式约束条件。

对于黏滞流体阻尼器的优化设计,主要涉及阻尼器的布置位置和阻尼器的参数(阻尼系数、阻尼指数)两类设计变量,优化目标主要为结构位移、加速度及其他性能指标(如能量、损伤等)。对于一般工程结构的黏滞阻尼减振系统,其目标函数往往具有非线性特性,且无法表达为设计变量的具体函数形式,从而无法求得其关于设计变量的导数或梯度函数。这些使得工程结构黏滞阻尼减振优化必须解决两方面的问题:(1)在理论上,优化方法在搜索过程中可不依靠目标函数的梯度;(2)在技术上,优化算法程序与目标函数的有限元计算的无缝对接,实现对复杂目标函数的自动重复计算。

通常,对于阻尼器位置的优化主要依靠设计者的专业知识和经验,或者是反复计算寻找规律,这对于常规结构具有较高的可行性[36]。而对于复杂结构,直接判断阻尼器的最佳位置比较困难。因此,基于遗传算法(基本原理如图 21)研究了大跨空间结构消能支撑的位置优化研究[37],该方法在理论上采用二进制编码的改进遗传算法,在技术上编制

Matlab 程序和 ANSYS 模型进行对接，成功地实现了对大跨机库的消能支撑位置的优化研究。该方法充分利用了遗传算法二进制编码的特点，可以便利地推广到其他工程结构的减振优化设计中。

随后，针对阻尼器的参数，提出了基于响应面[38]、模式搜索算法[39]、遗传算法[27,40]等优化方法的减振优化设计方法，实现了技术上 Matlab 优化算法程序与 ANSYS[39]、SAP2000[40]以及自编动力分析子程序[27]的对接。然而，这些方法本质上单目标优化的特性，使得在对结构多个目标进行优化设计时需要对每个目标赋予权重，转化为单目标优化进行求解，随着目标数目的增加，权重具体数值的选取将越来越困难。而工程结构设计时往往涉及多个目标，为此，进一步以多目标遗传算法 NSGA-II 为基础，开展了黏滞阻尼减振的多目标优化研究。该方法的基本流程如图 22 所示，主要包括优化建模、结构建模和多目标遗传搜索三大步骤：优化建模

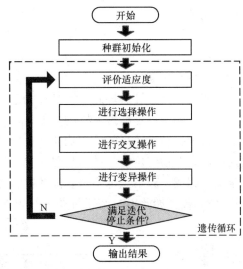

图 21 遗传算法基本计算流程

主要包括确定设计变量、优化目标、约束条件和变量初始值；结构建模主要包括建立分析模型和荷载模拟两部分；多目标遗传搜索在步骤上比一般遗传搜索多了种群合并、排序与修剪的过程，同时在具体步骤的细节中亦有所不同。

这些基于智能算法的优化设计理论与方法，具有普遍的适用性，能够在各类工程结构的黏滞阻尼减振设计中应用。但由于复杂结构一次非线性分析的耗时较长，且方法中需要大量的重计算，导致设计成本较高，其进一步的发展需要借助计算机技术的进步提高计算效率。

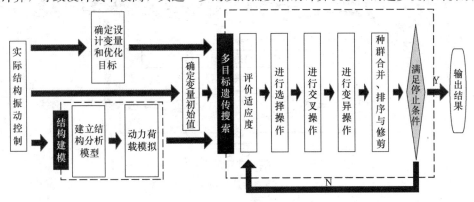

图 22 基于 NSGA-II 的减振优化设计方法

4.3 结构减振设计方法

4.3.1 简化设计方法

黏滞消能减振的简化设计方法在《建筑抗震设计规范》（GB 50011—2010）[41]有所提

及。笔者团队于2003年开始针对结构减振设计方法进行系统的研究工作[20]。其核心的思想是利用等效阻尼比,确定等效阻尼比的方法通常有能量法、强行解耦法和模态应变能法[1]。对于黏滞流体阻尼器,若考虑其指数的非线性,通常只能采用能量法进行求解。等效阻尼比的表达如下:

$$\zeta_e = W_d / (4\pi W_s) \tag{7}$$

其中,W_d为结构中减振装置在结构预期位移下往复一周所消耗的能量;W_s为设置减振装置的结构在预期位移下的总应变能。简化设计方法流程如图23所示,由图可见,该简化方法实际包括计算循环和设计循环两个循环流程:(1)计算原结构的动力特性与动力响应,并以此初步选定减振装置的参数;(2)选定初始的顶点最大位移,计算等效阻尼比,进而分析减振结构体系的动力响应,比较顶点位移,当不满足时不断按图中所示的方法往复迭代,直至预期位移与计算位移误差满足要求;(3)比较分析结果与设计目标,若满足要求设计完成,若不满足要求,按减振装置参数分析的规律或专业知识调整参数,重新进行第(2)步,如此不断循环往复,直至满足设计要求。

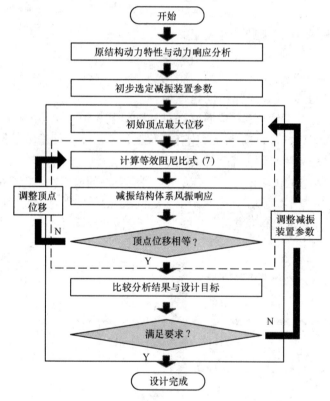

图23 结构消能减振体系设计流程

4.3.2 三阶段设计方法

上述的简化设计方法操作方便,但对设计人员的专业知识和经验要求较高,同时,对于部分复杂结构即使是经验丰富的设计人员也很难通过经验判断参数优化的方向,得到的设计参数往往离理论上的最优设计值较远。而4.2节中的智能优化设计方法具有较高的普适性,程序或软件完善后即使没有丰富的经验也能够得到较优设计参数,但是由于这类智能优化设计算法往往容易陷入局部最优造成收敛困难,设计成本无法控制,且最终设计参数在实际产品中是否存在也无法控制。

因此，提出了黏滞阻尼减振设计的三阶段设计法：

第一阶段，根据团队多年来在黏滞阻尼减振领域积累的实际工程数据库和团队成员的知识经验，初步确定装置布置位置和初始参数；

第二阶段，以结构知识对结构精细分析模型进行简化，采用4.2节中提出的智能优化设计方法进行设计，得到理论最优参数；

第三阶段，以结构精细化模型为基础，采用本节的简化设计方法，结合黏滞流体阻尼器产品目录，得到设计准最优参数。

该设计方法中：（1）第一阶段引入了团队积累的工程数据库，为经验提供了科学依据；（2）智能优化采用了第一阶段的参数作为初始值，避免了陷入局部最优，同时，采用简化模型，大大降低了重分析的耗时，提高了智能优化的速度；（3）简化设计过程中，采用了智能优化的结果，使得最终设计结果不会过于偏离理论最优，同时，考虑了阻尼器产品目录，使得设计与实际紧密结合。

5 黏滞阻尼减振技术的推广应用

随着理论的不断发展和技术的不断成熟，黏滞阻尼减振技术已日渐成为众多土木工程结构抵御动力灾害的首要选择之一。一项技术要得到推广应用，首先需要进行的就是技术标准化和实施示范工程。

5.1 黏滞阻尼减振技术标准化

在黏滞阻尼减振领域内，最早的规范可以追溯到1993年由北加州结构工程师协会（Structural Engineers Association of Northern California，简写SEAONC）所发表的针对消能减振结构的设计与安装指导[42]。这本规范的出现源于1989年北加州Loma Prieta地震之后人们对减振装置的关注以及一些减振生产厂商的重视。在国内，2001年发布的《建筑抗震设计规范》GB 50011—2001[43]增加了"隔震与消能减震设计"一章，首次将耗能减振技术的内容写入规范中。2013年12月，又推出了由广州大学主编的《建筑消能减震技术规程》JGJ 297—2013[44]，对包含黏滞流体阻尼器在内的消能器的技术性能、设计、部件的连接与构造、施工、验收和维护等方面进行了规定。

近年来，除了前述的研究以外，笔者团队在减振技术标准化过程中专门开展了一些有针对性的理论和试验研究[28,45]，从而为相关技术规程的编制提供依据。2007年，笔者团队作为主编单位编制了国家行业标准——《建筑消能阻尼器》JG/T 209，并于2012年对该规程进行了重新修订[46]，规程中对建筑消能阻尼器中的速度相关型阻尼器的术语与定义、分类与标记、技术要求、试验方法、检验规则、标志、包装、运输和储存进行了规定。2009年，笔者团队与中国电子工程设计院合作编制了《建筑结构消能减震（振）设计——国家建筑标准设计图集》09SG6610-2[47]，图集中介绍了黏滞消能器、黏弹性消能器、金属屈服型消能器和摩擦型消能器等四种常用消能减震器及其组成的消能部件，给出了各种消能器与混凝土结构、钢结构的连接详图以及黏滞消能部件计算示例。

5.2 实际工程应用

近年来，由于技术的进步，黏滞流体阻尼减振装置在国内外的应用越来越多[7,14,48]，已经成为抵御各种动力灾变的重要手段之一。多年来，笔者团队主持了一批黏滞流体阻尼减振项目的设计研究工作[14,48,49]，以下分别介绍黏滞流体阻尼器在大跨、高耸、高层结构中应用的三个代表性工程分别在人行荷载、风荷载和地震作用下的减振设计。

5.2.1 北京奥林匹克公园国家会议中心

北京奥林匹克公园国家会议中心（如图24）承担着2008年奥运会的国际会议大厅和新闻中心的主要职能，工程分为会议和展览两大区域，其中在会议区四层有一块区域为60m×81m跨度的钢结构楼盖，该区域在奥运会比赛时为比赛场地，赛后为5500人大会堂。由于该区域使用功能复杂，因此业主要求采用活荷载$7.5kN/m^2$（局部座椅仓储处为$10kN/m^2$），远远大于一般的楼盖活荷载（$2\sim3.5kN/m^2$），属于重荷载大跨度钢结构楼盖。通过模态分析得到：第一阶竖向振型自振频率2.7866Hz，第二阶竖向振型自振频率3.5039Hz。结构的第一自振频率和人正常行走、跳跃的频率（1.8Hz～2.7Hz）接近，容易产生共振。如果依靠增大截面和改变结构型式的方法，从技术、经济和空间利用的角度看是不现实且不合理的，因此决定采用多点悬吊调频质量阻尼器MTMD系统对结构响应进行控制。

图24 北京奥林匹克国家会议中心

会议中心在奥运会后作为高标准的国家级会议中心，需要满足各种各样的使用要求，在计算时采用了行走、跳跃、起立等分析工况。经过多次循环优化计算，楼盖共布置72套TMD减振装置（图25），其中阻尼器采用线性黏滞阻尼器。

经动力分析（结果如表8所示），可见：MTMD系统大大减小了楼盖的加速度响应，对快走和跳跃等频率接近结果自振频率的工况效果尤为明显；减振后的结构加速度响应基本满足规范对人体舒适度的要求。为考察所安装MTMD系统的有效性，验证分析结果的正确性，对该结构进行了现场动力测试，验证了减振效果和分析方法的正确性[49]。

5.2.2 南京奥体中心观光塔

南京奥体中心观光塔（图26（a））由主塔塔身、弧形桁架和观光平台三部分组成。主塔塔身则包括观光梯和救生梯。塔身顶点标高为110.2m，观光平台下底面标高为101.4m，主塔高宽比13.7，大大超过常规高耸结构高宽比限值。观光平台环向不对称悬挑，沿正立面左端悬挑长度 18.5m，右端悬挑 8m。

图 25　楼盖 TMD 布置图

结构竖向振动峰值加速度　　　　　　　　　　　　　　表 8

工况	未减振	减振后	减振率	加速度限值	备注
1	0.00969g	0.00494g	49.1%	0.005g（"办公"）	快走
2	0.0185g	0.0152g	17.4%	0.015g（"商业"）	慢走
3	0.0219g	0.0117g	46.7%	0.015g（"商业"）	跳动
4	0.0106g	0.0105g	1.3%	0.015g（"商业"）	慢速起立
5	0.0168g	0.0151g	10.3%	0.015g（"商业"）	快速起立
6	0.0339g	0.0325g	4.2%	0.04g（"仅节奏性活动"）	音乐会

在横风向脉动风荷载作用下，观光平台的加速度响应最大值超过专家审查会专家提出的关于人体舒适度的要求，因此，必须对南京奥体中心观光塔结构进行减振控制。为此，该工程在 105.7~88.4 米之间设置了 30 个黏滞阻尼器（图 26（b））。设置阻尼器后，结构的各项响应均有所下降，其中横风向脉动风荷载作用下观光平台处加速度峰值由 0.2235m/s² 下降到 0.1481m/s²，满足了人体舒适度（0.15m/s²）的要求；顺风向脉动风

(a)　　　　　　　　　　(b)

图 26　南京奥体光塔
(a) 结构；(b) 阻尼器

荷载作用下标高观光平台处位移峰值由 0.1920m 下降到 0.1592m；同时观光平台的扭转也得到了一定程度的改善。

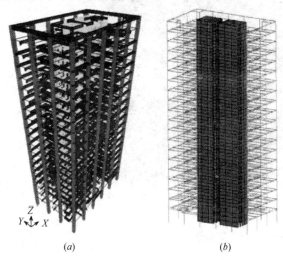

图 27 有限元模型
(a) Etabs 模型；(b) Mac 模型

5.2.3 宿迁民丰农村合作银行主楼

宿迁民丰农村合作银行位于江苏省宿迁市，主楼抗震设防类别为丙类，抗震设防烈度为 8 度，设计基本加速度为 0.30g。设计地震分组为第一组，场地类别为Ⅲ类场地。主楼高 82.1m，为 A 级高度钢筋混凝土高层建筑，地上 19 层，地下 1 层，采用框架－核心筒结构，其框架抗震等级为一级，剪力墙抗震等级为一级。

根据设计院提供的资料，分别建立了 Etabs 模型（图 27（a））和 Msc.Marc 模型（图 27（b）），前者主要用于多遇地震作用下的弹性分析，后者用于罕遇地震作用下的弹塑性分析。梁、柱采用纤维单元模拟，剪力墙用分层壳元进行模拟[33]。本工程沿结构的两个主轴方向分别设置黏滞流体阻尼器，阻尼器的参数取值见表 9，阻尼器各楼层布置见表 10，总共布置了 108 根黏滞流体阻尼器。

阻尼器参数　　　　表 9

阻尼器类型	阻尼指数 α	阻尼系数/(N·s/m)	数量	最大输出力/kN
A	0.25	1.0×10^6	88	850
B	0.25	0.9×10^6	20	850

黏滞阻尼器楼层布置　　　　表 10

楼层	X 向		楼层	Y 向	
	型号	数量/层		型号	数量/层
1～5, 10	A	2	1～5, 10	A	2
6～9, 11～14	A	4	6～9, 11～14	A	4
15～19	B	2	15～19	B	2

本工程选用 USER232 波、USER656 波和 USER845 波作为地震动输入进行时程分析，其中 USER232 波、USER656 波为天然波，USER845 波为根据Ⅲ类场地模拟的人工波。图 28（a）和（b）给出了 USER232 地震波罕遇地震作用下的结构层间位移，X 向最大减震率为 28%，Y 向最大减震率为 36%。图 28（c）给出了结构在 Y 向地震作用下塑性发展情况：(1) 灰色表示结构构件完好，保持弹性；(2) 构件（框架梁、框架柱、墙端暗柱）端部出现深灰色及黑色表示该构件端部出现塑性铰（构件中的钢筋纤维屈服），其中黑色表示构件塑性发展程度更大，构件中的钢筋纤维最大应变已达到屈服应变的 5 倍；(3) 单出现黑色表示该单元混凝土应变已经达到压碎应变。对比可见，罕遇地震作用下，

相对于原结构，减震结构中框架梁的塑性铰数量减小，墙端暗柱发生屈服的数量也有所减小，墙根部混凝土没有出现压碎的情况。因此安装黏滞流体阻尼器后，结构的塑性发展程度减小，结构整体抗震性能提高。

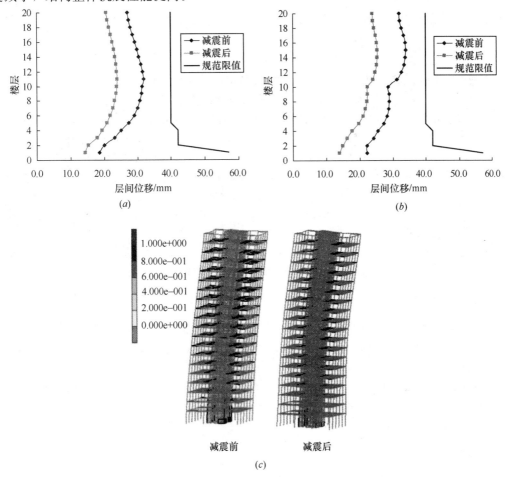

图 28　USER232 波罕遇地震作用下结构和阻尼器响应
（a）X 向层间位移（b）Y 向层间位移；（c）Y 向地震塑性发展

6　结语

本文主要介绍了笔者团队近 15 年来在黏滞阻尼减振技术领域的研究与应用工作，内容覆盖了材料、构件、结构三个层次，主要包括以下几个方面：

（1）系统研究了黏滞流体阻尼装置所采用的阻尼材料，在对比分析各阻尼材料的基础上，提出采用颗粒粉末对硅油进行改进，得到了流动指数小于 1 的具有非线性本构的流体阻尼材料。

（2）设计了 6 种黏滞流体阻尼减振装置，进行了理论和试验研究，并与有关企业合作进行了产品化研究，最终形成了黏滞流体阻尼技术的系列产品。

（3）建立了工程结构黏滞流体阻尼减振的分析方法和简化设计方法，提出了结构减振的智能优化设计方法和三阶段设计方法,并已在实际工程中进行了应用。

(4)主编了国家行业标准《建筑消能阻尼器》JG/T 209 和《建筑结构消能减震(振)设计——国家建筑标准设计图集》09SG6610-2,主持了一批黏滞阻尼减振工程(包括大跨、高耸、高层结构等)的设计研究工作。

参考文献

[1] 李爱群. 工程结构减振控制[M]. 北京:机械工业出版社,2007.

[2] 滕军. 结构振动控制的理论、技术和方法[M]. 北京:科学出版社,2009.

[3] 李宏男,肖诗云,霍林生. 汶川地震震害调查与启示[J]. 建筑结构学报. 2008,29(04):10-19. (Li Hongnan, Xiao Shiyun, Huo Linsheng. Damage investigation and analysis of engineering structures in the Wenchuan earthquake[J]. Journal of Building Structures. 2008,29(04):10-19(in Chinese).)

[4] Spencer B F, Nagarajaiah S. State of the art of structural control[J]. JOURNAL OF STRUCTURAL ENGINEERING-ASCE. 2003,129(7):845-856.

[5] Housner G W, Caughey T K, Chassiakos A G, et al. Structural Control: Past, Present, and Future [J]. Journal of Engineering Mechanics. 1997,123(9):897-971.

[6] Soong T T, Spencer B F. Supplemental energy dissipation: state-of-the-art and state-of-the practice [J]. ENGINEERING STRUCTURES. 2002,24(3):243-259.

[7] Symans M D, Charney F A, Whittaker A S, et al. Energy dissipation systems for seismic applications: Current practice and recent developments[J]. JOURNAL OF STRUCTURAL ENGINEERING-ASCE. 2008,134(1):3-21.

[8] Soong T T, Dargush F G. Passive Energy Dissipation Systems in Structural Engineering[M]. New York: John Wiley & Sons, Inc., 1996.

[9] Constantinou M C, Symans M D. Experimental and analytical investigation of seismic response of structures with supplemental fluid viscous dampers[R]. National Center for Earthquake Engineering Research, State Universisty of New York at Buffalo, 1992.

[10] 欧进萍,吴波. 被动耗能减振系统的研究与应用进展[J]. 地震工程与工程振动. 1996(03):72-961996(03):72-96(in Chinese).

[11] 周福霖. 结构减震控制[M]. 北京:地震出版社,1997.

[12] 王肇民. 高耸结构振动控制[M]. 上海:同济大学出版社,1997.

[13] 瞿伟廉. 高层建筑和高耸结构的风振控制设计[M]. 武汉:武汉测绘科技大学出版社,1991.

[14] 张志强,李爱群. 建筑结构黏滞阻尼减震设计[M]. 北京:中国建筑工业出版社,2012.

[15] 王亚勇,薛彦涛,欧进萍等. 北京饭店等重要建筑的消能减震抗震加固设计方法[J]. 建筑结构学报. 2001,22(02):35-39. (Wang Yayong, Xue Yantao, Ou Jinping 等. Structural Analyses and Design of Seismic Retrofitting with Energy Dissipation Dampers for Beijing Hotel and Other Key Buildings in Beijing[J]. Journal of Building Structures. 2001,22(02):35-39(in Chinese).)

[16] 北京火车站抗震加固与改造设计项目组. 北京火车站抗震鉴定与加固技术[J]. 工程抗震. 2003(01):25-292003(01):25-29(in Chinese).

[17] 欧进萍,丁建华. 油缸间隙式黏滞阻尼器理论与性能试验[J]. 地震工程与工程振动. 1999,4(04):82-89Theory and performance experiment of viscous damper of clearance hydrocylinder[J]. 1999,4(04):82-89(in Chinese).

[18] 叶正强,李爱群,程文瀼等. 采用黏滞流体阻尼器的工程结构减振设计研究[J]. 建筑结构学报. 2001,22(04):61-66. (Ye Zhengqiang, Li Aiqun, Cheng Wenrang Yang Guohua 等. Study on Vi-

bration Energy Dissipation Design of Structures with Fluid Viscous Dampers[J]. Journal of Building Structures. 2001, 22(04): 61-66(in Chinese).)

[19] 刘斌. 改性黏滞流体及其阻尼器的研究[D]. 东南大学, 2007(in Chinese).

[20] 叶正强. 黏滞流体阻尼器消能减振技术的理论、试验与应用研究[D]. 东南大学, 2003(in Chinese).

[21] 黄镇, 李爱群, 陆飞. 调节阀式黏滞阻尼器原理与性能试验[J]. 东南大学学报（自然科学版）. 2007, 37(03): 517-520. (Huang Zhen Li Aiqun Lu. Theoretical research and performance experiment of viscous damper with pressure adjustment valve[J]. Journal of Southeast University(Natural Science Edition). 2007, 37(03): 517-520(in Chinese).)

[22] 黄镇, 李爱群. 新型黏滞阻尼器原理与试验研究[J]. 土木工程学报. 2009, 42(06): 61-65. (Huang Zhen, Li Aiqun. Experimental study on a new type of viscous damper[J]. China Civil Engineering Journal. 2009, 42(06): 61-65(in Chinese).)

[23] 黄镇. 非线性黏滞阻尼器理论与试验研究[D]. 东南大学, 2007(in Chinese).

[24] 梁沙河, 李爱群, 彭枫北. 变阻尼黏滞阻尼器的减震原理和力学模型分析[J]. 特种结构. 2009, 26(03): 48-53. (Liang Shahe Li Aiqun Peng. Mechanical Model and Earthquake Resistance Analysis of Controllable Damping-viscous Damper[J]. Special Structures. 2009, 26(03): 48-53(in Chinese).

[25] 梁沙河, 李爱群, 彭枫北. 幂率流体变阻尼黏滞阻尼器的动力性能试验研究[J]. 工业建筑. 2010(05): 39-422010(05): 39-42(in Chinese).

[26] 夏冬平, 张志强, 李爱群等. 新型黏滞阻尼墙动力性能试验研究[J]. 建筑结构. 2013(13): 46-502013(13): 46-50(in Chinese).

[27] 陈鑫. 高耸钢烟囱风振控制理论与试验研究[D]. 南京: 东南大学, 2012. (Chen Xin. Theoretical and experimental study on vibration control of high-rise steel chimneys under wind load[D]. Nanjing: Southeast University, 2012(in Chinese).

[28] 陈鑫, 李爱群, 刘涛等. 黏滞消能支撑的设计方法与参数分析[J]. 工程抗震与加固改造. 2008, 30(02): 10-15. (Chen Xin, Li Ai-Qun, Liu Tao 等. Design Methods and Parameter Analysis of Viscous Energy Dissipation Braces[J]. Earthquake Resistant Engineering and Retrofitting. 2008, 30(02): 10-15(in Chinese).

[29] 黄瑞新, 李爱群, 张志强等. 北京奥林匹克中心演播塔TMD风振控制[J]. 东南大学学报（自然科学版）. 2009, 39(03): 519-524. (Huang Ruixin, Li Aiqun, Zhang Zhiqiang 等. TMD vibration control of Beijing Olympic Center Broadcast Tower under fluctuating wind load[J]. Journal of Southeast University(Natural Science Edition). 2009, 39(03): 519-524(in Chinese).

[30] 孙广俊, 李爱群, 张志强. 基于等效层模型的静动力抗震分析及其在剪力墙减震设计中的应用[J]. 振动工程学报. 2013(01): 75-82(in Chinese).

[31] 焦常科. 大跨双层斜拉桥地震响应分析与减震控制研究[D]. 东南大学, 2012(in Chinese).

[32] 徐庆阳. 大跨机库结构减震分析与优化研究[D]. 东南大学, 2008(in Chinese).

[33] 裴赵云, 缪志伟, 李爱群. 高烈度区某框架-核心筒结构耗能减震控制研究[J]. 建筑结构. 2013(01): 5-9(in Chinese).

[34] 张志强, 赵翔, 马斐. 消能减震框架结构罕遇地震弹塑性时程分析[J]. 江苏建筑. 2013(1): 13-16. (Zhang Zhi-Qiang, Zhao Xiang, Ma Fei.《Jiangsu Construction》. 2013(1): 13-16(in Chinese).

[35] 王健. 速度型与位移型阻尼器匹配减震设计研究[D]. 南京: 东南大学, 2013(in Chinese).

[36] 赵伯友, 陈鑫, 李爱群, 沈顺高. 基于纤维模型的空间网架支承方案优化设计与分析[J]. 防灾减灾工程学报. 2009(in Chinese).

[37] 徐庆阳, 李爱群, 丁幼亮等. 基于改进遗传算法的大跨机库柱间消能支撑的位置优化研究[J]. 土

木工程学报．2013(06)：35-43(in Chinese)．

[38] 孙传智，李爱群，缪长青等．减震结构黏滞阻尼器参数优化分析[J]．土木建筑与环境工程．2013(01)：80-85(in Chinese)．

[39] 徐庆阳，李爱群，丁幼亮等．基于模式搜索算法的结构被动控制系统参数优化研究[J]．振动与冲击．2013(10)：175-180(in Chinese)．

[40] 乌兰，李爱群，沈顺高．软钢阻尼器在中国妇女活动中心酒店结构中优化研究[J]．工业建筑．2013(S1)：255-260(in Chinese)．

[41] 中华人民共和国住房和城乡建设部．《建筑抗震设计规范》GB 50011—2010[S]．北京，2010．

[42] Whittaker A S, Aiken I D, Bergman D, et al. Code requirements for the design and implementation of passive energy dissipation systems[C]// Proc., of ATC 17-l Seminar on Seismic Isolation, Passive Energy Dissipation, and Active Control, California：1993，2：497-508.

[43] 中华人民共和国建设部．《建筑抗震设计规范》GB 50011—2001[S]．北京，2001．

[44] 中华人民共和国住房和城乡建设部．《建筑消能减震技术规程》JGJ 297—2013[S]．北京，2013．

[45] 张志强，刘文文，李爱群等．大承载力黏滞流体阻尼器的动力性能试验研究[J]．建筑结构．2011(S1)：178-180(in Chinese)．

[46] 中华人民共和国建设部．《建筑消能阻尼器》JG/T 209—2012[S]．北京，2012．

[47] 中国建筑标准设计研究院．建筑结构消能减震(振)设计——《国家建筑标准设计图集》09SG6610-2[S]．北京，2009．

[48] 李爱群．黏滞流体阻尼器在高层建筑减振设计中的应用研究[J]．徐州工程学院学报．2005，20(01)：7-14.(Li Ai-Qun. Research on Energy Dissipation Design of High-rise Building using Fluid Viscous Damper[J]．2005，20(01)：7-14(in Chinese)．

[49] 李爱群，陈鑫，张志强．大跨楼盖结构减振设计与分析[J]．建筑结构学报．2010，31(6)：160-170.(Li Aiqun, Chen Xin, Zhang Zhiqiang. Design and analysis on vibration control of long-span floor structures[J]. Journal of Building Structures. 2010，31(6)：160-170(in Chinese).)

城市高架桥抗震性能评估方法研究

任伟新，陈　亮，王佐才

摘　要：结合我国地震工程与桥梁抗震方面的相关规范和标准，建立一套符合安徽省特点且操作性较强的城市高架桥抗震性能评估方法，该方法主要分为两个阶段：第一个阶段是抗震性能的初步评估，第二个阶段是抗震性能的详细评估。其中，第一阶段是对某一区域内的城市高架桥进行初步评估，找出抗震性能不足的高架桥，并对抗震性能的缺陷程度进行合理量化，从而确定需要进一步详细评估的桥梁以及加固优先级。第二阶段是对初步评估阶段选出的抗震性能缺陷程度较严重的高架桥进行详细评估，并将基于性能的地震工程学理论应用于其中，从而确定高架桥重要易损构件的抗震性能是否满足要求，为制定合理的抗震维修加固方案提供重要依据。

关键词：城市高架桥；抗震性能评估；初步评估；详细评估；基于性能的地震工程

Abstract: The method for seismic performance evaluation of the city viaduct, which is practical and applicable to Anhui province, is developed based on national codes and standards on earthquake engineering and seismic design of bridges. In general, the evalution process has two steps. The first step is a systematic procedure for the preliminary seismic evaluation and ranking of city viaducts in a region according to their seismic vulnerability that can be rationally quantified so as to indentify and prioritize viaducts for secondary evaluation and seismic retrofit. Based on this preliminary investigation, the viaducts with relatively high ranking are given the first priority for the detailed seismic evaluation in the second step. Furthermore, performance-based earthquake engineering can be used for detiled evaluations to assess the seismic performance of important and vulnerable components for seismic retrofit measure or other course of action.

Key words: city viaduct; seismic performance evaluation; preliminary seismic evaluation; detailed seismic evaluation; performance-based earthquake engineering

1. 概述

随着我国交通基础建设的高速发展，桥梁建设迎来了一个新的高峰期，特别是城市高架桥的建设，更如雨后春笋般迅速涌现出来。然而，近几年我国破坏性强震频发，例如汶川地震、玉树地震等，城市高架桥作为城市生命线系统中的枢纽工程，一旦在地震中遭受破坏，会造成生命财产的巨大损失，给抗震救灾工作带来巨大困难，加重次生灾害，图1为城市高架桥在历次大地震中遭受到的典型震害。

历次大地震对城市高架桥的抗震性能提出了更加严格的要求，特别是已建成的城市高架桥中，有很多建成年代较早，在设计时采用的是旧的抗震设计规范，甚至没有进行有效的抗震设计，也未能采取较好的抗震构造措施，加之运营时间较长，桥梁整体状况较差，在地震作用下极易遭受严重破坏。因此，既有城市高架桥抗震性能的评估也就显得极

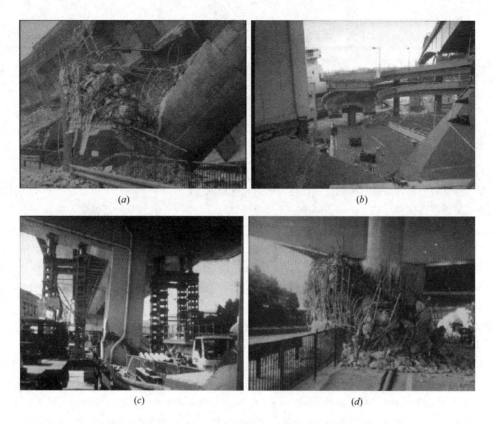

图1　地震作用下城市高架桥的典型破坏
(a) 高架桥总体坍塌；(b) 立交桥垮塌　(c) 墩柱剪坏；(d) 桥墩钢筋压屈

为重要。

然而，目前国内外对于已建城市高架桥抗震性能评估的方法和相关规范还较少。在国内，《公路桥梁加固设计规范》JTG/T J22—2008 中提到了部分有关桥梁抗震评估方面的内容，但只是要求综合考虑桥梁结构重要性、结构本身特点及结构的地震易损性、基础及场地的特征、桥梁的在地区的抗震设防烈度和结构建造的年代及加固经济评价等方面的内容进行评估，并给出了一些步骤和一个评估流程图，其采用的主要评估方法是构件的能力需求比法，其评估流程具体见图2。该部分内容主要涉及的桥梁抗震性能评估方法较为笼统，人为经验性较强，没有一个较为明确的评判标准和较易实现的详细评定流程，可操作性较差。而且，其结果只是确定出某一桥梁是否需要加固，对于抗震性能到底有多大程度上的不足，国内规范在该方面并没有给出详细的量化指标。因此，我国规范给出的评估方法虽然在思路上具有重要参考意义，但评估流程仍然较为粗糙，其中许多地方需要改进和细化。

对于国外各国的评估方法，美国这方面做得要好一些。在《Seismic Retrofitting Manual for Highway Structures》指南以及美国各州制定的相关评估指南中，提到了一些抗震评估的方法和思路，将桥梁抗震性能评估分为两个阶段，即初步评估阶段和详细评估阶段。其中，进行初步评估时用到的地震分级法，有较为清晰的步骤和流程，操作性相对要强一些，而且有较明确的量化评判标准，这样就能对桥梁抗震性能的缺陷程度进行量化

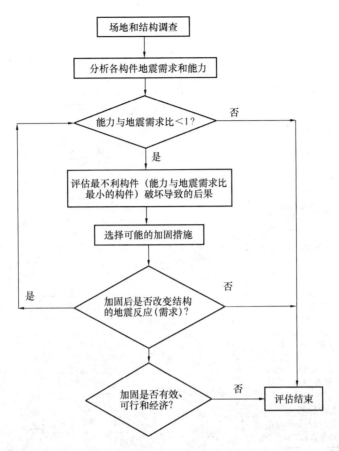

图 2 《公路桥梁加固设计规范》
JTG/T J22—2008 抗震性能评估流程

和评定。同时，该指南对详细评估也提供了较多的方法。但是，该指南是针对美国公路桥梁，而国内的桥梁设计规范、设防分类、设计使用年限、场地、地震动区划、地震烈度和设计加速度反应谱等都与美国有较大差异，这样就无法直接使用这些方法来评估国内的桥梁，更不能直接用于城市高架桥的抗震性能评估。

综上所述，国内外的桥梁抗震性能方法各有利弊。因此，我们在总结和凝练国内外相关评估方法的基础上，结合我国地震方面的相关规范、标准以及公路桥梁和城市桥梁抗震设计规范，通过大量分析、计算，得出一套符合安徽省特点且操作性强的城市高架桥抗震性能评估方法，并能经过适当调整推广应用于全国范围内。

在该方法中，我们将城市高架桥抗震性能评估工作分为两个主要阶段：第一个阶段是抗震性能的初步评估，第二个阶段是抗震性能的详细评估。其中，第一阶段是对某一区域内的城市高架桥进行初步评估，找出抗震性能不足的高架桥，并对抗震性能的缺陷程度进行合理量化，从而确定是否需要进行加固以及加固优先级。第二阶段是对初步评估阶段选出的抗震性能缺陷程度较严重的高架桥进行细部构件的详细评估，从而确定高架桥重要构件和易损构件的抗震性能是否满足要求，从而为其后的抗震维修加固方案的合理制定提供重要依据。

2. 抗震性能初步评估方法

由于地震分级的方法标准明确、流程清晰、便于操作,所以我们采用该方法进行安徽省城市高架桥的抗震性能初步评估。经过不断改进,最终确定了抗震性能的初步评估方法,其流程如图3所示。

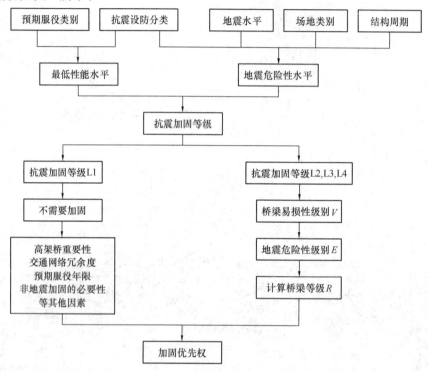

图3 抗震性能初步评估流程图

从流程图中可以看出,初步评估主要分为两大部分,前一部分是抗震加固等级的确定,后一部分是加固优先权的确定。前一部分的工作是为了判定某一高架桥从整体上来讲是否需要加固,分别给予了4种加固等级,用以表征需要加固的必要性,同时也为后面确定加固优先权服务。后一部分工作是为了确定加固优先权,若抗震加固等级为L1,则无须加固,也不再进行桥梁等级的计算,;若抗震加固等级超过L1,那么就要计算桥梁易损性级别,再计算出桥梁等级,由此确定加固优先权。

其中,桥梁等级R是对桥梁抗震性能的定量评定,R的取值是由桥梁易损性级别V和地震危险性级别E两者相乘得到的,即$R=VE$。由于V和E的范围都是0到10,所以桥梁等级R的取值范围是0到100,数值越大,说明抗震性能越不足,越达不到所要求的抗震性能目标,加固的必要性就越大。通过计算分析我们发现,按照合肥地区的抗震设防烈度,大部分城市高架桥的评定结果都在50以内,接近50的基本可以判定为抗震性能远远达不到预期的抗震性能目标,亟待加固。

按照桥梁等级R的数值大小,不仅可以看出每座桥梁抗震性能不足的程度,而且可以对某地区内所有城市高架桥的抗震性能进行一个排序,为以后加固的优先顺序提供可靠依据。在考虑桥梁等级的同时，还需要考虑其他技术、社会和经济等方面的因素，根据这些因素考虑是否需要变动桥梁加固的优先顺序，也就是对前面定量评定得出的结论进行合理的人为修正，从而得到最终的高架桥抗震加固优先级。

该初步评估方法适用于以梁桥为主要桥型的规则和非规则城市高架桥，但暂不适用于结构复杂的悬索桥、斜拉桥、单跨跨度超过150m的特大跨径梁桥和拱桥等特殊桥梁。但是该方法的思路和原则仍然可以应用于这些桥梁，但需针对特定的桥型进行相应的改进和调整。为便于该方法的使用，对应每一步的操作，我们都给出了用来判定的表格、详细流程和一些需要特别注意的地方，从而大大提高了可操作性和使用效率，方便了该方法的推广应用。

3. 抗震性能初步评估实例

根据以上初步评估方法，我们给出了三座运营时间和服役状况水平各异的桥梁的抗震性能初步评估实例。

第一个评估对象取自包河大道高架工程分项之一的繁华大道A匝道第五联连续梁桥，机动车单向双车道。其上部结构为预应力钢筋混凝土箱梁，共4跨，每跨30m，总长120m，桥面总宽9m，车行道净宽8m；下部结构为1.5m×1.8m的矩形单柱墩，中墩固结。各桥墩的桩基础均采用2根直径1.6m的钻孔灌注桩，均按摩擦桩设计，具体图4和图5。

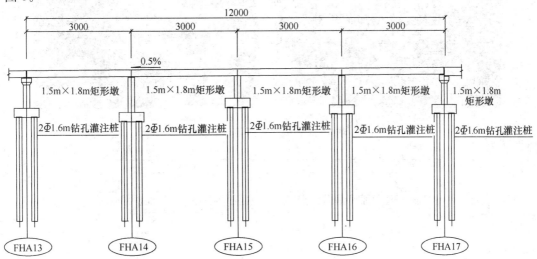

图4 繁华大道A匝道第五联连续梁桥总体布置图（单位：cm）

第二个评估对象取自巢湖市东塘路陆家河桥，该桥为城市主干路上的一座跨河桥，结构形式为钢筋混凝土简支梁桥，跨径为16m，双向六车道，四条机动车道，两条非机动车道，中间未设置隔离带，两侧为人行道，具体见图6。

该桥主梁形式为混凝土空心板梁，总共26片，单片板梁尺寸为1.18m×1m×16m，桥面总宽31.6m，车行道净宽19.6m，人行道净宽6m×2，伸缩缝设于桥头两端，两伸缩缝距离20m，具体见图7和图8。

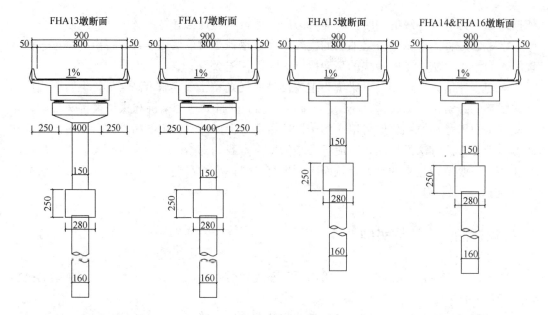

图 5 桥梁横断面图（单位：cm）

图 6 东塘路陆家河桥外观视图

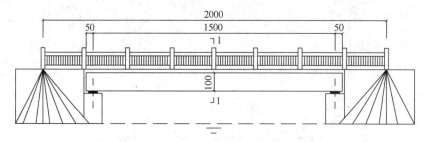

图 7 东塘路陆家河桥侧视图（单位：cm）

第三个评估对象取自裕溪路公跨铁立交桥中的三跨。裕溪路公跨铁立交桥工程于2005年开工，2006年竣工通车，结构形式为简支梁桥，共9跨，全长218m，双幅路，中间设防撞护栏，每幅三车道单向行驶，具体见图9。

该桥主梁为预应力钢筋混凝土空心板梁，选其中三跨作为评估对象，每跨22m，共66m，单幅桥面总宽18.5m，人行道2m，非机动车道3.5m，机动车道12m，机非隔离带

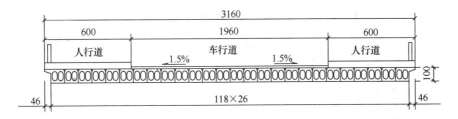

图 8 东塘路陆家河桥横断面图（单位：cm）

和防撞墙均为 0.5m。其上部结构是简支梁，下部结构为轻型柱式桥墩，盖梁采用预应力钢筋混凝土结构。桩基础采用扩大基础，具体见图 10 和图 11。

图 9 裕溪路公跨铁立交桥外观视图

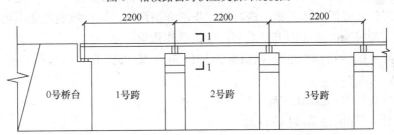

图 10 裕溪路公跨铁立交桥三跨侧视图（单位：cm）

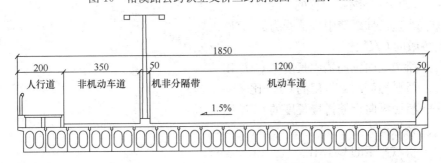

图 11 裕溪路公跨铁立交桥上部结构横断面图（单位：cm）

按照抗震初步评估的方法对上面三座桥梁进行初步评估，得到的桥梁等级 R 值见表 1 所示。

初步评估桥梁等级 R 表 1

桥 名 \ 地震作用	地震作用 E_1	地震作用 E_2
繁华立交工程	4.8	18
东塘路陆家河桥	1.8	8.65
裕溪路公路铁立交桥	9.6	36

从最终的打分可以得到三座桥梁的桥梁等级由大到小依次为：裕溪路公路铁立交桥、繁华立交工程、东塘路陆家河桥。

裕溪路公跨铁立交桥的桥龄较长，但并非很旧，重要性较高，抗震设防类别较高，通过评估后，分数最高，考虑优先进行加固；繁华立交工程属于新桥，但其重要性较高，抗震设防类别较高，通过评估后，分数排在第二位，考虑第二个进行加固；东塘路陆家河桥桥龄较长，但是属于小桥，重要性较低，抗震设防类别较低，通过评估后，分数最低，考虑最后进行加固。

通过三个工程实例，验证了该方法具有较强的可操作性，对合肥地区的城市桥梁具有较好的适应性，并且可以将该方法推广到其他地区的桥梁抗震性能评估中去，对该领域的工程技术人员具有一定的参考价值。

4. 基于能力/需求比（C/D）的抗震性能详细评估方法

在抗震性能详细评估阶段，主要是针对初步评估筛选出的抗震性能不足，即桥梁等级 R 较高的高架桥，根据具体要求，进行进一步的详细评估，确定抗震性能不足的具体部位和易损构件，为将来维修加固方案的制定提供具体依据。

对于合肥地区城市高架桥，我们选用构件能力/需求比的方法进行抗震性能的详细评估。

计算的构件能力/需求比主要有以下 11 个：

(1) 支承长度和限位器的 C/D 比

r_{ad}——桥台位移的 C/D 比

r_{bd}——支座或伸缩缝位移的 C/D 比

r_{bf}——支座或伸缩缝限位器受力的 C/D 比

(2) 桥墩的 C/D 比

r_{ca}——桥墩纵向钢筋锚固长度的 C/D 比

r_{cc}——桥墩箍筋约束作用的 C/D 比

r_{cs}——桥墩纵向钢筋搭接长度的 C/D 比

r_{cv}——桥墩剪力的 C/D 比

r_{ec}——桥墩弯矩的 C/D 比

(3) 基础的 C/D 比

r_{ef}——基础弯矩的 C/D 比

r_{fr}——基础转动的 C/D 比

(4) 场地土的 C/D 比

r_{sl}——场地土液化作用的 C/D 比

5. 基于性能的抗震能力详细评估方法

将基于性能的地震工程理论（Performance-Based Earthquake Engineering，PBEE）运用于城市桥梁抗震性能评估中，需要量化工程场地桥梁结构在地震作用下的响应，获得桥梁结构地震需求（Engineering Demand Parameter，EDP）在不同的地面运动强度（Intensity Measure，IM）水平下超越某一特定值的概率，再根据能力/需求比获得桥梁结构在不同的 IM 水平下超越某一特定破坏状态的概率，从而获得桥梁构件及其整体的地震易损性曲线，并最终通过地震易损性曲线确定城市高架桥重要构件和易损构件的抗震性能是否满足要求，从而为其后的抗震维修加固方案、城市防灾减灾计划等的合理制定提供重要依据[1-5]。

基于性能的桥梁抗震性能详细评估方法可以分为四个阶段：

（1）地震危险性分析（Hazard analysis）——用来描述工程场地的地震活动性；

（2）结构分析（Structural analysis）——用来获取在不同水准的地震作用下，结构的地震力和地震位移等响应参数；

（3）破坏分析（Damage analysis）——将结构分析中所获取的地震反应参数转化为结构的物理破坏状态；

（4）损失分析（Loss analysis）——将破坏状态与社会、经济损失结合起来，从而确定重要或易损构件的加固优先级、维修加固方案等。

对于结构工程师而言，如果仅从结构角度为主要出发点，则可将以上基于性能的桥梁抗震能力详细评估方法更为简化地分为三个阶段，即分析阶段（1）～（3）。

首先，通过工程场地的概率地震危险性分析（Probabilistic Seismic Hazard Analysis，PSHA），获得工程场地的地震危险性曲线。然后，通过概率地震需求分析（Probabilistic Seismic Demand Analysis，PSDA）预计桥梁结构工程需求参数（EDP）超越某一个特定值的年平均频率（Mean Annual Frequency，MAF），并估计结构在未来地震灾害下的抗震能力。对于桥梁结构，这里的 EDP 可以是墩顶最大位移和漂移比、梁端最大位移、支座最大位移、标准化滞回能量和 Park and Ang 累积破坏参数等。PSDA 将场地特定的地面运动危险性曲线与通过结构动力分析获得的地震反应联系在一起。PSDA 的最终结果就是概率地震需求模型（Probabilistic Seismic Demand Model，PSDM），并以对应于多重地震需求水平（从弹性到倒塌）的结构地震需求危险性曲线来表示。EDP 危险性曲线对于结构多重性能目标的预计至关重要。

在概率地震破坏分析阶段，可以通过震害调查、检测或结构动力试验，确定桥梁结构不同破坏状态所对应的 EDP 水平，例如可以根据混凝土剥落的数量、纵向钢筋的鼓胀变形程度、横向钢筋的断裂、伸缩缝的偏移量以及支座位移等 EDP 水平，将桥梁的破坏状态分为完好、轻微破坏、中等破坏、重度破坏和倒塌等不同水平，通过能力/需求比来获得桥梁结构在不同的 IM 水平下超越某一特定破坏状态的概率，即地震易损性曲线。

通过对城市高架桥地震易损性曲线的研究，可以获得桥梁不同重要构件的易损性以及不同设计参数对城市高架桥地震反应和抗震性能的影响，不仅能够对城市高架桥抗震设计起到指导作用；而且可以用来确定不同构件的震后修复和加固优先级，可以根据不同构件的易损性采取合理的结构抗震和加固措施，降低重要构件的地震易损性；预计不同地震强度水平对于桥梁结构的破坏，以及评估不同的破坏状态可能造成的直接社会、经济损失和城市交通系统功能的失效情况，从而制定出确实、有效的地震防灾减灾计划。

我们将通过安徽省合肥市包河大道城市高架桥的抗震性能评估工作说明以上方法的具体应用。

5.1 包河大道高架桥工程场地地震危险性分析

根据工程场地的地震动参数值和场地条件，结合《城市桥梁抗震设计规范》CJJ 166—2011[6]，本项目场地水平设计地震动加速度反应谱（阻尼比为0.05）取为：

$$S = \begin{cases} S_{max}(5.5T+0.45) & T_1 < 0.1s \\ \eta_2 S_{max} & 0.1s \leqslant T \leqslant T_g \\ \eta_2 S_{max} \left(\dfrac{T_g}{T}\right)^\gamma & T_g < T < 5T_g \\ [\eta_2 0.2^\gamma - \eta_1(T-5T_g)]S_{max} & 5T_g \leqslant T \leqslant 6s \end{cases} \quad (1)$$

$$S_{max} = 2.25 A_{max} \quad (2)$$

式中，S_{max}——水平设计加速度反应谱最大值；

T_g——特征周期（s），具体见表1；

η_2——结构的阻尼调整系数，取1.0；

A_{max}——水平向设计基本地震动加速度峰值，具体见表1；

γ——衰减指数，取0.9；

η_1——下降斜率，取0.02；

T——结构自振周期（s）。

水平设计地震动加速度反应谱（阻尼比为0.05）可见图12。

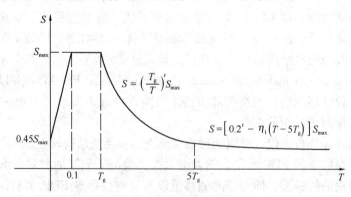

图12 水平设计地震动加速度反应谱

工程场地水平向设计地震动峰值加速度及反应谱（5%阻尼比）参数值见表2；工程场地不同超越概率下的水平设计地震动加速度反应谱（阻尼比为5%）见图13，以下的分析将以这四条设计谱作为目标谱。

场地地表水平向设计地震动峰值加速度及反应谱（5%阻尼比）参数值　　表2

概率值	T_1 (sec)	T_g (sec)	β_m	A_{max} (gal)
50年63%	0.1	0.44	2.5	43
50年10%	0.1	0.48	2.5	130
50年5%	0.1	0.54	2.5	178
50年2%	0.1	0.58	2.5	260

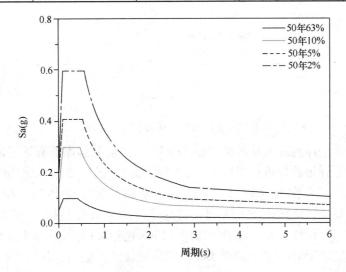

图13　工程场地水平设计地震动加速度反应谱（阻尼比为5%）

5.1.1　建立地震危险性曲线

在地震工程学中，50年63%、50年10%、50年5%和50年2%超越概率下的地震危险性水平对应的平均回归期分别是50年、474年、975年和2475年。本文选择峰值地面运动加速度PGA作为地面运动强度参数（IM），由于IM假定服从对数正态分布，根据回归期与IM在对数空间的直线拟合公式可以获得工程场地的地震危险性曲线，具体见公式（3）和图14，即地震灾害模型[1,3]。

$$\lambda(IM) = k_0(IM)^{-k} \tag{3}$$

式中：k_0和k为回归系数。

综上所述，包河大道高架桥工程场地的地震危险性模型为：

$$\lambda(IM) = 0.0000254(IM)^{-2.14563} \tag{4}$$

式中：回归系数$k_0=0.0000254$，$k=2.14563$。

5.1.2　实际地震波的选择

本文根据震级M、距离R和局部场地土壤条件S等重要地震参数来选择实际地震波，尽量将所选地震波的M、R和S等重要条件的差异限制在一个较小的范围之内，以减小所选地震波之间的变异性，并能更好地符合桥址处的实际地震灾害环境[7-9]。

本文所选地震波的重要地震参数均符合以下条件：震级>M6.0；工程场地距离破裂带的最近距离和J-B距离在20～100km；30m表层土的平均剪切波速在200～500m/s。

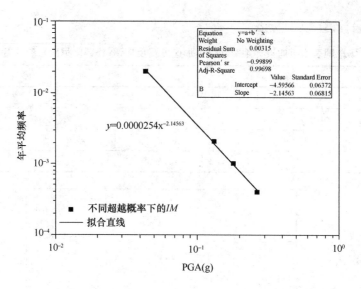

图 14　工程场地的地震危险性曲线

本文对美国太平洋地震工程研究中心（PEER）强震数据库所有站点记录的地震波两个水平分量进行了详细的分析和识别[10]，最终根据图 3 的四条水平设计地震动加速度反应谱（阻尼比为 5%）作为目标谱，共选择四组原始的实际地震波即四个地震波库（Bin），分别模拟不同超越概率地震水准下的地震动，即 Bin1 模拟 50 年 63% 超越概率的地震、Bin2 模拟 50 年 10% 超越概率的地震、Bin3 模拟 50 年 5% 超越概率的地震、Bin4 模拟 50 年 2% 超越概率的地震，四个 Bin 的实际地震波反应谱、几何平均值谱见图 15～图 18。

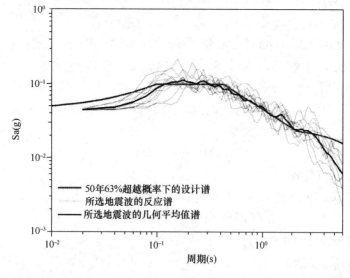

图 15　50 年 63% 超越概率（Bin1）

分析图可知，通过将 PGA 一次调幅到目标水平，即 50 年 63%（0.044g）、10%（0.133g）、5%（0.181g）和 2%（0.265g）的超越概率水平后，四个 Bin 的几何平均值谱在 0.1～4.0s 周期范围内与目标谱匹配较好，可以代表工程场地在不同水准地震作用下

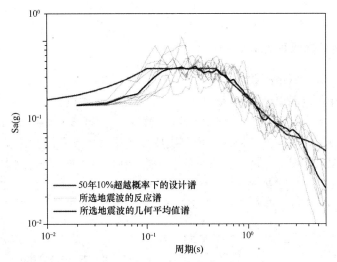

图 16 50年10%超越概率（Bin2）

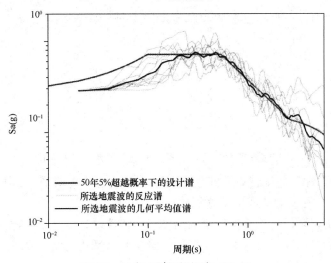

图 17 50年5%超越概率（Bin3）

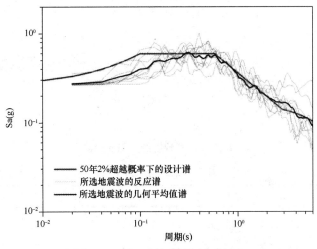

图 18 50年2%超越概率（Bin4）

的地震动。

5.2 包河大道高架桥概率地震需求分析

本文选择包河大道城市高架桥中的繁华大道A匝道第五联连续梁桥作为算例桥梁，该桥为4×30m的预应力混凝土连续箱梁桥，1号～5号桥墩的高度分别为6.218、9.378、6.008、7.778和5.688m，桥墩尺寸均为1.5m（横桥向）×1.8（纵桥向）m，其中3号桥墩为固定墩，其他桥墩纵向设置活动盆式橡胶支座，桥梁总体布置图具体见图19。本文主要针对纵桥向进行地震易损性分析，以说明基于性能的桥梁抗震性能详细评估方法的应用。

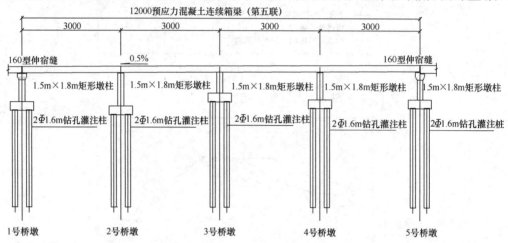

图19 繁华大道A匝道第五联连续梁桥总体布置图

算例桥梁的弹性和非线性有限元模型的建立和动力分析均采用美国太平洋地震工程研究中心（PEER）开发的地震反应分析软件OpenSees来实现[11]，有限元模型具体见图20。通过模态分析，可以得到算例桥梁的结构动力特性，计算结果具体见表3，其中只列出了对桥梁纵向和横向地震反应影响显著的三阶主要振型的周期和频率。

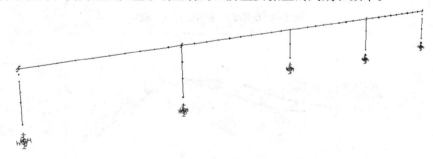

图20 算例桥梁的有限元模型

算例桥梁结构动力特性　　　　　　表3

振型阶数	周期（s）	频率（Hz）	振型描述
1	0.55	1.81	横向对称振动
2	0.48	2.08	纵向振动
3	0.40	2.50	横向反对称振动

5.2.1 工程需求参数（EDP）的选择

本文主要针对连续梁桥最易损的构件：支座和桥墩，并考虑到落梁震害的问题，以纵桥向地震反应为主，选择三个工程需求参数（EDP）作为研究对象，即支座纵桥向最大位移、梁端纵桥向最大位移和纵桥向固定墩墩顶漂移比，分别代表桥梁不同构件在地震作用下的位移需求，以下简称为固定墩墩顶漂移比、梁端最大位移以及各支座最大位移。其中，在1号、2号、4号和5号四个桥墩处的支座以下均简称为支座1、支座2、支座4和支座5。所选择EDP的具体定义如下。

（1）支座纵桥向最大位移：纵桥向支座的最大位移，可以考虑到支座的破坏以及可能产生的落梁震害。

（2）梁端纵桥向最大位移：连续梁梁端在纵桥向的最大位移，因为梁端位移过大时可能会导致落梁震害。

（3）墩顶漂移比：定义为墩顶的最大绝对位移与桥墩高度之比，反映了桥墩结构在地震作用下的最大变形能力，见公式（5）[7]。

$$d = \frac{|u|_{\max}}{H} \quad (5)$$

其中，$|u|_{\max}$ 为墩顶的最大绝对位移，H 为墩高。

5.2.2 算例桥梁的概率地震需求分析

运用繁华大道A匝道第五联连续梁桥的非线性有限元模型，以及分别对应工程场地50年63%、10%、5%和2%超越概率地震水准的四组实际地震波，通过调幅，对算例桥梁进行非线性动力时程分析，并且假定在特定IM水平下的结构地震反应（EDP）水平服从对数正态分布[12-14]。

根据结构动力分析的计算结果，将不同IM水平下，结构达到的地震反应（EDP）水平在对数空间内进行直线拟合，可以推导出IM与EDP之间的数学关系（具体见公式（6）），即概率地震需求模型（Probabilistic Seismic Demand Model，PSDM）。本文的PSDM具体见图21。

$$EDP = a(IM)^b, \text{或 } \ln EDP = \ln a + b\ln(IM) \quad (6)$$

由于在不同的IM水平下，结构EDP假定服从对数正态分布，因此，IM在特定强度水平im下即特定的地震风险水平下，EDP超越某一特定值edp的条件概率可用公式（7）来计算，并绘于图22～图24。

$$P(EDP > edp \mid IM = im) = 1 - \Phi\left(\frac{\ln(EDP) - \mu_{\ln EDP \mid IM}}{\sigma_{\ln EDP \mid IM}}\right) \quad (7)$$

式中，$\Phi(\ )$ 为标准正态分布函数；

$\mu_{\ln EDP \mid IM}$ 为EDP的自然对数均值；

$\sigma_{\ln EDP \mid IM}$ 为EDP的自然对数标准差。

分析图22～图24，我们可以非常清晰地获得：在特定的地震危险性水平（$IM=im$）下，①结构EDP超越某一特定值的概率；②对于某一特定的超越概率，结构EDP可能达到的水平；③对于结构EDP水平的极限值可以有一个初步的判定。

对于本文的工程场地，50年2%超越概率下的结构地震需求水平要明显大于其他三个地震水准，且地震风险水平越高，EDP的超越概率也越大，结构越有可能发生破坏。

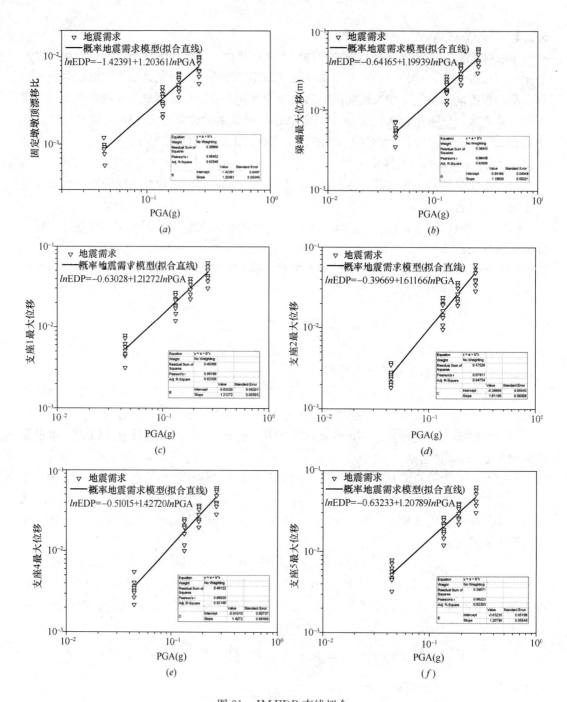

图 21　IM-EDP 直线拟合
(a) 固定墩墩顶漂移比；(b) 梁端最大位移；(c) 支座 1 最大位移
(d) 支座 2 最大位移；(e) 支座 4 最大位移；(f) 支座 5 最大位移

分析图 24 可知，在地震危险性或 IM 水平较小的情况下，例如 50 年 63％超越概率地震危险性水平，矮墩处的支座相对位移比高墩处的要大；但随着地震危险性水平或地面运动强度水平的提高，高墩处的支座相对位移开始逐渐变大，甚至超过矮墩。通过图 24 的对比分析，可以对不同位置处的同一构件地震需求水平大小及其变化规律进行初步判别。

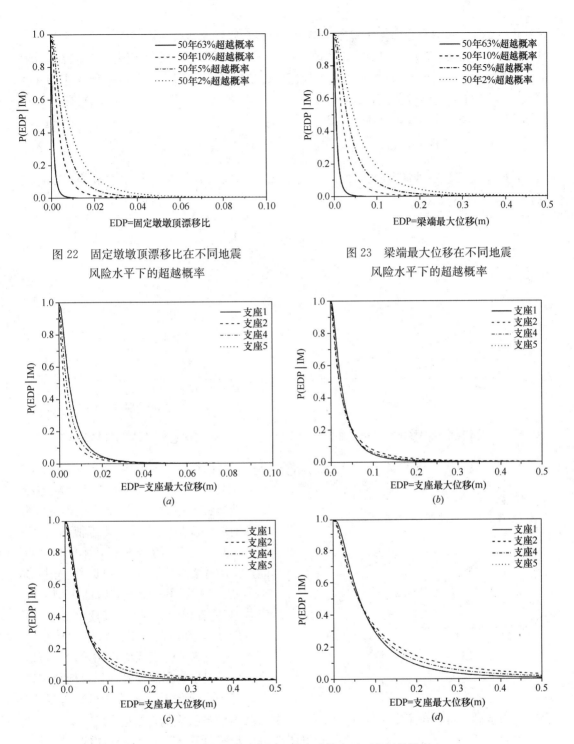

图22 固定墩墩顶漂移比在不同地震风险水平下的超越概率

图23 梁端最大位移在不同地震风险水平下的超越概率

图24 各支座位移在不同地震危险性水平下的超越概率

(a) 50年63%超越概率地震危险性水平；(b) 50年10%超越概率地震危险性水平；
(c) 50年5%超越概率地震危险性水平；(d) 50年2%超越概率地震危险性水平

5.2.3 建立算例桥梁的 EDP 危险性曲线

根据以上对于不同 EDP 参数推导出的 IM-EDP 数学关系即 PSDM：$EDP = a$

$(IM)^b$，并基于包河大道高架工程所在场地的地震危险性曲线公式：$\lambda(IM)=k_0(IM)^{-k}$，其中 $k_0=0.0000254$，$k=2.14563$，能够在所感兴趣的整个地震风险范围内（即所感兴趣的 IM 范围内），以闭合解的形式直接推导出桥址工程场地处 EDP 超越某一特定限值 edp 的概率，即 EDP 危险性曲线，具体见公式（8）。将各个参数值代入公式（8）即可获得不同 EDP 的危险性曲线，具体见图 25～图 27。

$$\lambda(EDP>edp\mid IM)=k_0\left[\left(\frac{EDP}{a}\right)^{\frac{1}{b}}\right]^{-k}\exp\left[\frac{1}{2}\frac{k^2}{b^2}\sigma^2_{\ln EDP\mid IM}\right] \tag{8}$$

式中：k 和 k_0 可以由地震危险性曲线获得；a 和 b 可以由 IM-EDP 样本的最佳拟合直线获得。

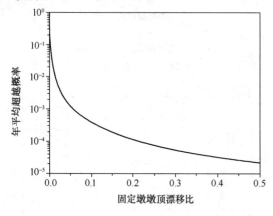

图 25 固定墩墩顶漂移比的危险性曲线

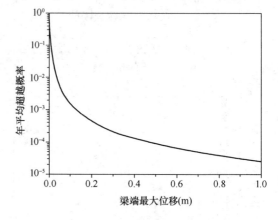

图 26 梁端最大位移的危险性曲线

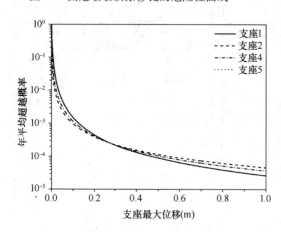

图 27 各支座最大位移的危险性曲线

分析图 25～图 27 的 EDP 危险性曲线，我们可以很容易获知，在工程场地特定的地震灾害环境下，待分析桥梁中不同构件的地震反应超越某一特定值的概率。或者，对于特定的超越概率，不同构件可能产生的地震反应。并且能够对比分析处于不同位置的同一种桥梁构件（例如支座），其地震反应的大小及其变化规律。

EDP 危险性曲线对于桥梁结构抗震性能评估有着非常重要的作用，因为可以让工程师根据实际需要（例如目标超越概率），预测桥梁结构可能发生的地震反应，并以此为基础，进行桥梁结构抗震加固设计，并设置合理的抗震构造措施。

需要注意的一点是，采用公式（8）以闭合解的形式计算 EDP 危险性曲线，在其计算过程中有一个隐含的假定：同方差假定，即假定在 IM-EDP 的直线拟合过程中，对于所有的 IM 水平，EDP 计算数据的方差是相同的。这一假定在较小的 IM 范围内是可行的，但在整个 IM 范围内，EDP 的方差一般不会是常数。因此，可以在不同的 IM 范围内进行分段直线拟合或采用数值积分方法求解 $PSDM$，以获得更为精确的结果[2,12]。

5.2.4 定义破坏状态

随着基于性能的抗震设计思想的提出，各国学者相继开展了如何合理确定结构抗震性能目标的研究。一般而言，性能目标与结构的损伤状态密切相关，而结构的损伤状态通常可以采用位移或损伤指标来确定。本文主要针对桥梁结构中最易损的构件：桥墩和支座，并考虑落梁震害，对桥梁构件的破坏状态进行定义和量化。

需要注意的是，由于各国桥梁的结构形式、设计规范等还存在较大差异，且国内外对于桥梁结构整体及其不同构件的抗震性能目标的研究成果还较少，故对于破坏状态的定义及其量化指标还没有统一的标准，但不管采用何种破坏状态定义及其量化指标，都不影响本文提出的抗震性能详细评估方法的实际应用。

1. 桥墩结构的破坏状态定义

本文在对比分析国内外大量桥墩试验数据的基础上，并利用 Pushover 分析，以墩顶漂移比作为 EDP 参数对算例桥梁固定墩的破坏状态进行划分和量化[2,4,5,15-18]，具体见表4。

固定墩破坏状态定义 表4

破坏状态	破坏描述	破坏机制	固定墩的墩顶漂移比（%）
DM1	完好	弹性	——
DM2	轻微破坏	纵向钢筋首次屈服	0.50
DM3	中度破坏	保护层混凝土出现裂缝	0.62
DM4	重度破坏	保护层混凝土出现大面积剥落	1.88
DM5	完全破坏/倒塌	纵向钢筋屈曲或拉断，核心混凝土压碎	6.52

2. 支座的破坏状态定义

本文根据中国规范对地震作用下盆式橡胶支座允许相对位移的具体规定，并参考国内外相关文献和桥墩变形破坏准则[6,19-22]，用支座相对位移定义了盆式橡胶支座的5种损伤状态，具体见表5。

盆式橡胶支座破坏状态定义 表5

支座位置	破坏状态（支座最大位移 m）				
	DM1	DM2	DM3	DM4	DM5
	完好	轻微破坏	中度破坏	重度破坏	完全破坏
1号桥墩	——	0.003	0.1	0.15	0.2
2号桥墩	——	0.003	0.1	0.15	0.2
4号桥墩	——	0.003	0.1	0.15	0.2
5号桥墩	——	0.003	0.1	0.15	0.2

3. 落梁的破坏状态定义

一般情况下，对于连续梁桥，由于地震作用，主梁在桥台处由于纵桥向位移过大，超过了梁端至台帽边缘的最小距离 a，即会发生落梁[6,21]，具体见图28。但落梁震害一般很难像以上桥墩和支座那样将破坏状态再细分为几个等级。因此，本文只将落梁破坏状态分为两个等级，即完好和落梁（即倒塌），具体见表6。

落梁破坏状态定义 表6

位置	落梁或倒塌（梁端最大位移 m）	位置	落梁或倒塌（梁端最大位移 m）
13号桥台	1.32	17号桥台	1.32

5.2.5 算例桥梁的地震易损性分析

通过概率地震需求分析，已经获得了在不同的 IM 水平下，结构地震需求的均值和标准差，以及 IM 与 EDP 之间的数学关系，即 PSDM：$EDP=a(IM)^b$，具体见表7～表9。

桥梁的地震易损性曲线表示在不同地震动强度即 IM 水平下，结构地震需求（Demand，D）超越不同破坏状态所对应的特定反应值（Capacity，C，即结构抗震能力）的条件概率。由于假定地震需求和抗震能力均服从对数正态分布，因此，地震易损性曲线可用公式（9）进行定义。

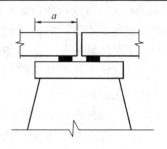

图28 梁端至墩、台帽或盖梁边缘的最小距离 a

基于墩顶漂移比的 PSDM 表7

墩顶漂移比	$IM-EDP$ 关系	离散度
固定墩	$\ln(EDP)=-1.42391+1.20361\ln(IM)$	0.85

基于支座最大位移的 PSDM 表8

支座位置	$IM-EDP$ 关系	离散度
13号桥台支座	$\ln(EDP)=-0.63028+1.21272\ln(IM)$	0.86
14号桥墩支座	$\ln(EDP)=-0.39669+1.61166\ln(IM)$	1.12
16号桥墩支座	$\ln(EDP)=-0.51015+1.42720\ln(IM)$	1.00
17号桥台支座	$\ln(EDP)=-0.63233+1.20789\ln(IM)$	0.85

基于梁端最大位移的 PSDM 表9

梁端最大位移	$IM-EDP$ 关系	离散度
位移最大的桥台	$\ln(EDP)=-0.64165+1.19939\ln(IM)$	0.84

$$P(D>C\mid IM)=\Phi\left[\frac{\ln\left(\frac{\mu_d}{\mu_c}\right)}{\sqrt{\beta_c^2+\beta_d^2}}\right]=\Phi\left[\frac{\ln\left(\frac{a(IM)^b}{\mu_c}\right)}{\sqrt{\beta_c^2+\beta_d^2}}\right] \quad (9)$$

式中，μ_d——结构地震需求的均值预计；

μ_c——结构抗震能力的均值预计；

β_d——结构地震需求的自然对数标准差；

β_c——结构抗震能力的自然对数标准差；

其他符号同前。

根据美国 HAZUS99 规定，当易损性曲线以谱加速度 Sa 作为 IM 时，$\sqrt{\beta_c^2+\beta_d^2}$ 取为 0.4；以 PGA 作为 IM 时，$\sqrt{\beta_c^2+\beta_d^2}$ 取为 0.5。将对应于不同的 IM 值时，结构破坏状态的失效概率计算出来，即可绘制成地震易损性曲线。图中，横坐标表示 IM 的大小，纵坐标表示地震作用下结构反应超越不同破坏状态的条件概率。

5.2.6 桥梁构件地震易损性曲线

根据以上介绍的桥梁结构地震易损性曲线计算方法，我们获得了合肥市包河大道高架桥不同构件对于各种破坏状态的超越概率，具体见图 29 和图 30。通过地震易损性曲线，我们可以清晰分析在不同的地面运动强度水平下，桥梁不同构件超越各种破坏状态的概率，即发生各种破坏的概率。

因此，对于已建城市高架桥，首先通过地震部门对于工程场地未来可能发生的地面运动强度水平进行合理预计；然后，工程师即可根据以上步骤进行城市高架桥的地震易损性分析，通过易损性曲线可以得到在不同水准的地震作用下，桥梁各重要构件最有可能发生哪种破坏状态及其相应概率，以及不同构件发生破坏的可能顺序等重要信息，从而获得已建桥梁在不同水准地震作用下的抗震性能水平；并能以易损性为依据，对各桥梁构件的维修加固优先级进行确定，进而实施有针对性的桥梁抗震加固设计，并设置合理的抗震构造措施，以避免桥梁发生严重震害。

例如，通过分析图 29（b）可知，连续梁桥的主梁在纵桥向水平地震作用下，由于两个边墩的墩高、刚度、支座类型等都比较接近，故在边墩处引起的梁端最大位移也非常接近，因此，引起落梁震害的概率也基本一致。通过分析图 29 的（c）～（f）可知，不同桥墩处的支座，达到不同破坏状态的超越概率各不相同，这主要与桥墩之间的高度比、刚度比、配筋率等设计要素密切相关，可以通过优化这些设计要素、对最先破坏或最易损的支座进行优先更换、采用抗震性能更好的抗震支座等方法提高各支座的抗震能力，降低支座的易损性，甚至可能通过这些方法有效控制各支座的破坏顺序，避免发生严重震害。

进一步，我们可以对比分析不同位置处的同一种桥梁构件达到相同破坏状态的超越概率，以寻找抗震性能较低的薄弱点。分析图 30 可知，对于同一种破坏状态，边墩（矮墩/刚性墩）处单向活动支座的地震易损性基本一致，且发生轻微破坏和中度破坏的概率要大于中墩（高墩/柔性墩）处的单向活动支座；但随着地面运动强度水平的提高，中墩支座达到重度破坏和完全破坏的概率开始提高，甚至大于边墩支座。因此，在小震和中震情况下，我们应重点关注刚度较大的边墩处的支座，优先对其进行维护或更换；但在大震情况下，则关注重点应转移到刚度较小的中墩处的支座。

5.2.7 桥梁结构地震易损性曲线

为了比较桥墩和支座这两个构件的相对易损程度，并考虑落梁震害，我们将所有构件的地震易损性曲线放到一起进行分析，可以获得更多的有用信息，具体见图 31。

通过对桥梁各构件地震易损性曲线的综合对比分析，可以获得算例桥梁整体结构地震易损性的诸多重要信息，从而对算例桥梁的抗震性能进行合理评估，从中我们可以得出以下一些有益的结论：

（1）在纵桥向地震作用下，对于整个连续梁桥结构而言，如果桥墩配筋（纵向钢筋及其配箍率）符合相关设计规范，那么支座是最易损的构件，其次是固定墩，而梁端只要留有足够的搭接长度，发生落梁的可能性最小；

（2）对于同一种破坏状态，在刚度较大的边墩处，其单向活动支座地震易损性基本一致，且发生轻微破坏和中度破坏的概率要大于刚度较小的中墩处的单向活动支座；

（3）随着地面运动强度水平的提高，中墩处支座达到重度破坏和完全破坏的概率会逐渐提高，甚至大于边墩处的支座；

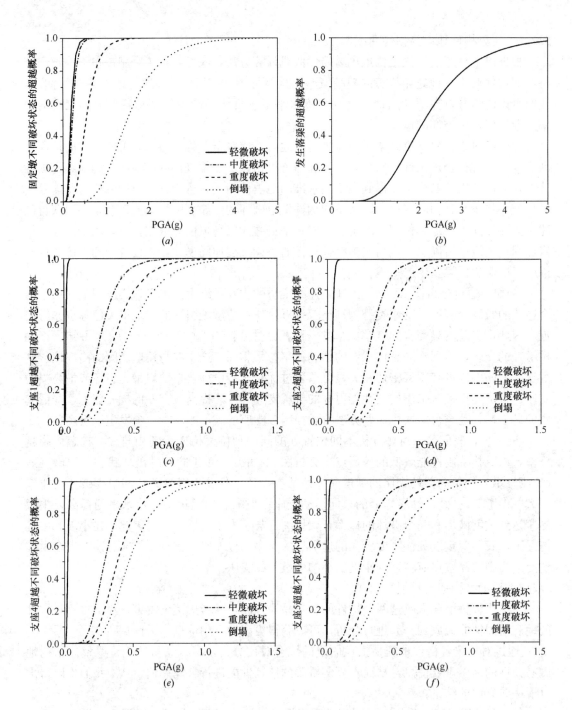

图 29　桥梁重要构件达到不同破坏状态的超越概率
(a) 固定墩；(b) 落梁震害；(c) 1#桥墩支座；(d) 2#桥墩支座；
(e) 4#桥墩支座；(f) 5#桥墩支座

因此，根据地震易损性曲线对于算例桥梁抗震性能的合理评估，我们可以采取以下有针对性的处理措施，以提高算例桥梁的抗震性能：

（1）在小震情况下，首先需要维修边墩处的支座，其次才是中墩上的支座；对于桥墩，只需对固定墩进行常规检测，以防止有其他原因导致的破坏，例如汽车的撞击等；

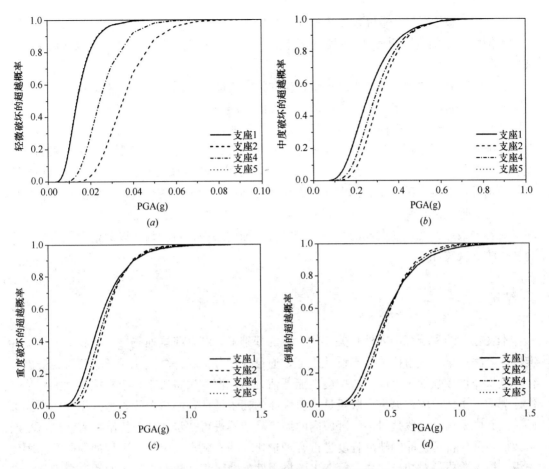

图 30 各支座达到相同破坏状态的超越概率
(a) 轻微破坏；(b) 中度破坏；(c) 重度破坏；(d) 完全破坏

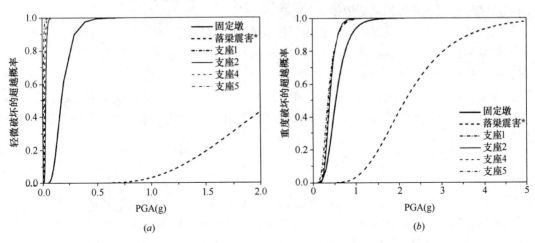

图 31 不同构件达到相同破坏状态的超越概率

（2）在中震情况下，首先需要检测维修所有桥墩上的支座以及固定墩，对于其他设有活动支座的桥墩主要进行常规检测，对于刚性墩特别是矮墩，则需要进行仔细检测，防止其发生以剪切为主的破坏；

(3) 在大震情况下，首先需要更换所有桥墩上的支座并对固定墩进行维修加固，防止其发生倒塌，特别是算例桥梁的固定墩比较矮，且直径较大，要防止其发生以剪切为主的破坏；

(4) 可以通过选择合理的支座类型，例如抗震支座，降低支座的破坏概率；

(5) 可以通过合理布置支座，降低各桥墩的地震力；

(6) 采用合理的抗震构造措施，降低重要构件的地震易损性。

综上所述，桥梁地震易损性分析的结果对于抗震性能评估的意义非常显著。根据基于性能的地震工程学理论，利用地震易损性分析，可以寻找到城市高架桥中的抗震薄弱点，以及各构件达到相应破坏状态的顺序，即桥梁在地震作用下的破坏规律。并能根据各构件结构形式、地震易损性及其破坏状态的特点，确定相应的震后维修加固优先级及其方案，同时对各种破坏状态可能引起的社会、经济等方面的损失进行预计，以制定出相应的城市高架桥防灾减灾计划。

6. 结论

本文在总结和凝练国内外相关评估方法的基础上，结合我国地震方面的相关规范、标准以及公路桥梁和城市桥梁抗震设计规范，通过大量分析、计算，建立了一套符合安徽省特点且操作性较强的城市高架桥抗震性能评估方法，该方法将城市高架桥抗震性能评估工作分为两个主要阶段：第一个阶段是抗震性能的初步评估，第二个阶段是抗震性能的详细评估。其中，第一阶段是对某一区域内的城市高架桥进行初步评估，找出抗震性能不足的高架桥，并对抗震性能的缺陷程度进行合理量化，从而确定是否需要进行加固以及加固优先级。第二阶段是对初步评估阶段选出的抗震性能缺陷程度较严重的高架桥进行细部构件的详细评估，从而确定高架桥重要构件和易损构件的抗震性能是否满足要求，为制定合理的抗震维修加固方案提供重要依据。

该评估方法适用于以梁桥为主要桥型的规则和非规则城市高架桥，但暂不适用于结构复杂的悬索桥、斜拉桥、单跨跨度超过 150m 的特大跨径梁桥和拱桥等特殊桥梁。但是该方法的思路和原则仍然可以应用于这些桥梁，但需针对特定的桥型进行相应的改进和调整。为便于该方法的使用，对应每一步的操作，我们都给出了用来判定的表格、详细流程和一些需要特别注意的地方，从而大大提高了可操作性和使用效率，方便了该方法的推广应用。

参考文献

[1] S. K. Kunnath. Application of the PEER PBEE Methodology to the I-880 Viaduct [R]. Pacific Earthquake Engineering Research Center, University of California, Berkeley, CA. 2007.

[2] Kevin M and Bozidar S. Seismic Demands of Performance-Based Design of Bridges [R]. Pacific Earthquake Engineering Research Center, University of California, Berkeley, CA. 2003.

[3] T. H. Lee and K. M. Mosalam. Probabilistic Seismic Evaluation of Reinforced Concrete Structural Components and Systems [R]. Pacific Earthquake Engineering Research Center, University of California, Berkeley, CA. 2006.

[4] B. G. Nielson and R. DesRoches. Seismic fragility methodology of highway bridges [C]. Structure Congress 2006：Structural Engineering and Public Safety.

[5] B. G. Nielson and R. DesRoches. Seismic fragility methodology for highway bridges using a component level approach [J]. *Earthquake Engng Struct. Dyn.* 2007. 36：823 – 839.

[6] 中华人民共和国住房和城乡建设部.《城市桥梁抗震设计规范》CJJ 166—2011 [S]. 2011.

[7] N. Shome，C. A. Cornell，P. Bazzurro，and J. E. Carballo. Earthquakes，records and nonlinear responses [J]. Earthquake Spectral. 1998，14(3)：469－500.

[8] I. Iervolinoa and C. A Cornell. Record Selection for Nonlinear Seismic Analysis of Structures [J]. Earthquake Spectra. 2005，21(3)：685 – 713.

[9] B. A. Bolt. Seismic Input Motions for Nonlinear Structural Analysis [J]. ISET Journal of Earthquake Technology. 2004. 41：223-232.

[10] PEER Strong Ground Motion Database [OL]. http：//peer. berkeley. edu/NGA. 2014.

[11] OpenSees-The Open System for Earthquake Engineering Simulation [OL]. http：//opensees. berkeley. edu/. 2014.

[12] Baker J. W.，Vector-Valued Ground Motion Intensity Measures for probabilistic Seismic Demand Analysis [D]. Ph. D. Thesis，Stanford University，2005.

[13] P. Giovenale, C. A. Cornell and L. Esteva. Comparing the adequacy of alternative ground motion-intensity measures for the estimation of structural responses [J]. *Earthquake Engng Struct. Dyn.* 2004；33：951-979.

[14] 胡聿贤. 地震工程学(第二版) [M]. 北京：地震出版社. 2006.

[15] Column Database [OL]. http：//nisee. berkeley. edu/spd/. 2014.

[16] Y. J. Park and A. H. S. Ang. Mechanistic seismic damage model for reinforced concrete [J]. Journar of structural engineering ASCE. 1985. 111(4)：722-739.

[17] W. C. Stone and A. W. Taylor. Seismic performance of circular brige columns designed in accordance with AASHTO/CALTRANS standards [R]. Rep. BSS-170. Ntional Institute of Standards and Technology，Gaithersburg，Md. 1993.

[18] H. H. M Hwang，J. B. Jernigan and Y. W. Lin. Evaluation of seismic demage to mamphis bridges and highway systems [J]. Journal of Bridge Engineering ASCE. 2000. 5(4)：322-330.

[19] 李立峰，吴文鹏，黄佳梅等人. 板式橡胶支座地震易损性分析 [J]. 湖南大学学报(自然科学版). 2011. 38(11)：1-6.

[20] 叶爱君，胡世德，范立础. 桥梁支座抗震性能分析的模拟 [J]. 同济大学学报. 2001. 29(1)：6-9.

[21] 中华人民共和国交通部.《公路桥梁抗震设计细则》JTGT B02-01—2008 [S]. 北京：人民交通出版社. 2008.

[22] 中华人民共和国交通部.《公路桥梁盆式橡胶支座技术标准》(JT/T 391—1999) [S]. 北京：人民交通出版社. 1999.

Low Cycle Fatigue Behavior of Confined Concrete Filled Tubular Columns

Y. Xiao and P. Qin

(National Center for International Research of Building Safety and Environment, Hunan University, Changsha, Hunan, 410082, **China.**)

Abstract: In order to improve shear strength and ductility of reinforced concrete short columns, Tomii, Sakino and Xiao et al. proposed and studied transversely confined reinforced concrete column or tubed column system, which was first published at an international conference. Through a large number of cyclic loading tests, they validated the superior seismic behavior of tubed columns with stable load carrying capacity, large energy dissipation capacity and ductility. In early 90s, Priestley, Seible and the author et al. adopted the tubed column concept in seismic retrofit of bridge columns, and the research formed the design guidelines for seismic retrofit practices of vast number of bridges in California and elsewhere. Xiao et al further investigated prefabricated fiber reinforced plastic (FRP) jacketing for seismic retrofit of existing reinforced concrete columns. The system was successfully applied in actual bridge retrofit, including retrofitting more than three thousands columns in the eight km long Yolo Causeway Bridge near Sacramento, California. Based many years of research accumulation on transversely confined reinforced concrete and concrete filled tubular columns, the author proposed a novel column system, confined concrete filled tubular (CCFT) column, and published at a conference in 2003. Tests and analyses validated that by providing external transverse confinement, the local buckling of steel tube can be effectively prevented or delayed. As consequences, the load carrying capacity, low cycle fatigue rupture resistance and ductility of concrete filled tubular columns can be improved.

Key words: transverse confinement; concrete filled tubes; low cycle fatigue; tubed columns; FRP; CFT; CCFT

约束钢管混凝土柱的低周疲劳性能 *

肖 岩,秦 鹏

(湖南大学 建筑安全与环境国际级国际联合研究中心,湖南 长沙 410082)

摘 要:为提高钢筋混凝土柱的抗剪性能和延性,富井、崎野和作者等人研究并于1985年在国际上首

---

* 项目基金:国家自然科学基金资助项目(51178174)。

　作者简介:肖岩,男,博士,教授,博导,长江学者,千人计划学者,湖南大学土木工程学院,E-mail:yanxiao @hnu.edu.cn。

次发表了钢管横向约束混凝土柱（tubed column）的成果。通过大量低周往复荷载试验，验证了钢管横向约束混凝土柱的优异的抗震承载力、耗能性能和延性。Priestley，Seible 和作者等人进一步采用钢管横向约束对既有钢筋混凝土桥柱进行了抗震加固的实验研究，其成果指导了美国加利福尼亚州等地的大量既有桥梁的抗震加固实践。作者等人九十年代后期研发了预制型玻璃纤维增强塑料（FRP）管壳加固混凝土柱的方法并得到了工程应用。在多年有关钢管、FRP 约束混凝土柱和钢管混凝土柱研究基础上，作者于 2003 年在国际上首次发表了约束钢管混凝土柱的研究成果。试验结果标明，通过施加横向约束可以推迟或防止钢管混凝土的局部屈曲，增大钢管混凝土柱子的延性。本文进一步针对这种新型体系在地震荷载作用下的低周疲劳性能进行实验研究。

关键词：钢管横向约束；FRP 横向约束；钢管混凝土；低周疲劳；约束钢管混凝土

中图分类号：TP

1. Introduction

Concrete filled tubular (CFT of CFST) column are recently widely used in building structures. In a conventional CFT column system, concrete is filled in steel tubes, which typically continue throughout several stories or the full-height of a building. The steel tube is expected to carry stresses primarily in longitudinal direction caused by axial loading and moments, as well as in transverse direction caused by shear and the internal passive pressure due to concrete dialation, i.e., the confining stress.

The concept of using steel tube as primarily transverse reinforcement for reinforced concrete (RC) columns was first studied by a research group lead by Tomii (Tomii, Sakino and Xiao et al. 1985; Xiao et al. 1986). The terminology of "tubed column" first adopted by Tomii et al. (1985), refers to the function of the tube as that of the hoops in a hooped RC column. Thus, the composite action between the steel tube and concrete is primarily expected in transverse direction only for a tubed column.

Jacketing retrofit of existing deficient RC columns can also be considered as the application of tubed column concept. For most cases of jacketing retrofit, the jacket is used to provide additional transverse reinforcement to increase the capacity and to improve the ductility of an existing column. This is achieved by welding steel plate shells or wrapping fiber reinforced plastics (FRP) to enclose an existing column to form a tubed system.

Based on the research on CFT columns and tubed columns, Xiao proposed the so-called confined concrete filled tubes or CCFT (Xiao et al. 2003; Xiao et al. 2005), as shown in Fig. 1. The CCFT column is expected to overcome disadvantages of the conventional CFT column and to provide the ideal choice for structural design of tall buildings or bridges, particularly for seismic regions. Comparing with with the conventional CFT columns, the CCFT columns can have the following merits：

i. In a CCFT column, the functions of the through-tube (similar to the tube in a conventional CFT column) and the additional transverse reinforcement are separated, with the former mainly resists longitudinal stresses caused by axial load and moment as well as shear in the middle portion of the column, whereas the additional reinforcement mainly enhances the potential plastic hinge regions.

ii. The additional transverse reinforcement can effectively prevent or delay the local buckling of the through-tube in the plastic hinge regions of a CFT column, thus improving its seismic performance with stable load carrying capacity and ductility.

iii. The concrete in the column plastic hinge regions can be more efficiently confined by the additional transverse reinforcement, and as a consequence, the ductility of the column can be assured.

iv. Due to the additional transverse confinement, the through-tube in the compression zone of the plastic hinge region is subjected to biaxial compressive stresses (strictly speaking, should be triaxial compression). This is a more efficient working state for steel tube as compared with the combination of axial compression and transverse tension, which is the working stress state of the tube in compression zones of a conventional CFT column.

v. In a conventional CFT column, in order to prevent the local buckling of the steel tube in the plastic hinge regions, relatively thicker steel tube is required, and typically such thickness is provided throughout the length of the column, particularly for columns with a rectangular section. On the other hand, in a CCFT column, the through-tube is designed mainly as longitudinal reinforcement to resist axial load and moment, and is enhanced transversely by the additional transverse reinforcement in the potential plastic hinge regions. The secondary function of the through-tube is to resist shear in the middle portion of the column, and this can typically be achieved by using the same thickness of the through-tube. Efficient details and the confining effects from the joining beams also relaxed the need of thick steel tube through the connection region. Thus, it is expected that even with the addition of the transverse reinforcement for the potential hinge regions, the total amount of steel usage in a CCFT column may be less than the identical CFT column.

vi. Since the additional transverse reinforcement is only provided for the potential plastic hinge regions of the columns, a structural system using CCFT columns remains essentially the same as a CFT structure. Thus, design details such as connections developed for conventional CFT structures are still applicable to the proposed CCFT system.

For the design of the additional transverse reinforcement, steel plates, angles, tubes or pipes which have larger transverse stiffness and resistance can be used, similar as those proved to be effective to enhance the retrofitting efficiency of rectangular jacketing by the first author (Xiao and Wu 2003). In addition, fiber-reinforced plastics (FRP) or reinforced concrete can be used (Xiao et al. 2005).

This paper reports the progress of a large experimental program designed to investigate the low-cycle fatigue and the improvement.

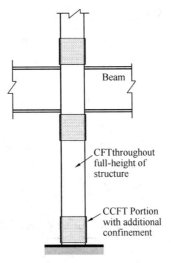

Fig. 1 Concept of confined concrete
filled tubular (CCFT) column
(Xiao 2003, Xiao 2005)

2. Experimental Programs

2.1 Model CFT and CCFT Columns

Large-scale model CFT and CCFT columns were designed to simulate typical columns in multi-story buildings in seismic regions. The basic details of the CFT model columns are illustrated in Fig. 2. The model column specimens had a circular section with a diameter of 325 mm, and a height of 1,500 mm from the point of lateral loading to the top of the footing. Thickness of steel tube was 3 mm for two columns and 6 mm for the others.

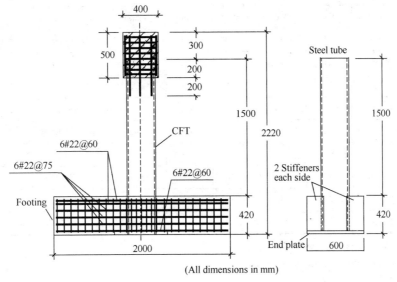

(All dimensions in mm)
Fig. 2 Model CFT column details

Fig. 3 Transverse confinement in model CCFT circular columns
(3mm or 6mm Steel tube; Concrete Infill; CFRP Wraps with 1mm foam tape underlay)

A total of 20 CFT and CCFT model columns have been tested to date with testing matrix shown in Table 1. The CFT columns were tested to provide basic information for conventional circular CFT columns subjected to cyclic loading, at different drift amplitudes or with increased drift amplitude. The confined concrete filled tubular (CCFT) model columns were essentially similar to the counterpart CFT columns shown in Fig. 1, except with the additional transverse confinement at the column end. The confined section details are exhibited in Fig. 3.

Testing matrix of circular CFT model columns Table 1

Specimen	D/t ratio	D (mm)	f'_c (MPa)	f_y (MPa)	Loading pattern	Comments
CFT57-1300-V	57	330	37	293	Variable drift amplitude	CFT57-V *
CFT57-1300-C2	57	330	37	293	Constant drift amplitude at 2%	CFT57-C2 *
CFT57-1300-C4	57	330	37	293	Constant drift amplitude at 4%	CFT57-C4 *
CFT57-1300-C6	57	330	37	293	Constant drift amplitude at 6%	CFT57-C6 *
CFT110-1300-C2-1	110	330	34	356	Constant drift amplitude at 2%	CFT110-C2-1 *
CFT110-1300-C2-2	110	330	34	356	Constant drift amplitude at 2%	CFT110-C2-2 *
CFT110-1300-C4	110	330	34	356	Constant drift amplitude at 4%	CFT110-C4 *
CFT110-1300-C6-1	110	330	34	356	Constant drift amplitude at 6%	CFT110-C6-1 *
CFT110-1300-C6-2	110	330	34	356	Constant drift amplitude at 6%	CFT110-C6-2 *
CFT112-1300-V1	112	336	39	249	Variable drift amplitude	CFT112-V *
CFT112-1300-C2	112	336	39	249	Constant drift amplitude at 2%	CFT112-C2 *
CFT112-1300-C4	112	336	39	249	Constant drift amplitude at 4%	CFT112-C4 *
CFT54-2000-V	54	325	28.0	312	Variable drift amplitude	C3-CFT6 * *
CCFT54-2000-V	54	325	28.0	312	Variable drift amplitude	C4-CCFT6 * *
CFT112-2000-V	112	336	39.1	303	Variable drift amplitude	C1-CFT3 * *
CCFT112-2000-V	112	336	39.1	303	Variable drift amplitude	C2-CCFT3 * *
CFT112-1300-V2	112	336	39.1	303	Variable drift amplitude	—
CFT112-1300-C6	112	336	39	303	Constant drift amplitude at 6%	—
CCFT112-1300-C6	112	336	39	303	Constant drift amplitude at 6%	—
CCFT112-1300-V	112	336	39	303	Variable drift amplitude	—
CCFT83-P-C or V	83	330	35	300	(various drift ratios and axial load)	In progress

Note: * : model columns reported in Zhang et al. ;
* * : model columns reported in Xiao et al.

2.2 Test setup

All the model columns were tested using the test setup shown in Fig. 4. Lateral force was applied to the top of the model column through a 630 kN (142 kip) capacity pseudo controlled hydraulic actuator. Constant axial load was applied to the column through post-tensioning two 50 mm (1.97 in.) diameter high-strength steel rods using two 1,500 kN (337 kip) capacity hydraulic hollow jacks. The forces of the rods were transferred to the model column by a cross beam mounted on top of the load stub. A load cell was connected under each jack to monitor the axial load during the test. A pump and a pressure relieve valve were used in conjunction to minimize the variation of the axial load due to the shifting of the column neutral axis during testing. In order to eliminate bending of the high-strength rods, a specially designed pin was connected to the lower end of each rod. The imposed lateral displacement was measured with both the displacement transducer of the actuator and a linearly variable differential transducer (LVDT). The built-in load cell of the actuator recorded the corresponding lateral force. Electrical resistance strain gauges were attached on the surface of the steel tube to infer stress states through measured strains at various locations of the steel tube.

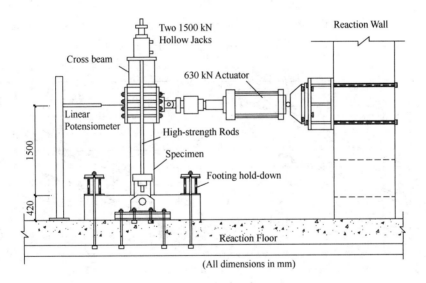

Fig. 4 Test setup

3. Experimental Results

CFT Column Tests

It was found that the induced damage in critical regions of a concrete-filled-tube (CFT) column could be attributed to cumulative damage of the steel tube caused by repeated cyclic loading in the post-yield strain range. Based on the results of initial tests on

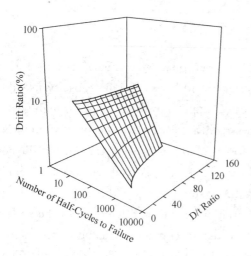

Fig. 5 Fatigue life surface for CFT columns

twelve large-scale model column specimens, representing two types of circular CFT columns with different steel tube diameter-to-thickness ratios, Zhang et al. (2009) indicated that the low cycle fatigue behavior of CFT columns depends on the thickness of the steel tube. Fatigue life expressions useful for application in damage-based seismic design were developed and utilized to predict the damage index for additional CFT columns based on experimental data reported in the literature. Fig. 5 shows the fatigue life in terms of number of half cycles before failure, in relationships with the drift ratio, diameter to thickness ratio.

CCFT Column Tests

The improved cyclic behavior of CCFT column can be demonstrated by comparing with the CFT column. As shown in Fig. 6 for the hysteresis loops of two counterpart model columns CFT112-1300-C6 and CCFT112-1300-C6, subjected to cyclic lateral loading with a constant drift amplitude of 6%, the number of cycles experienced before failure in the CCFT model column was more than that of the CFT model column.

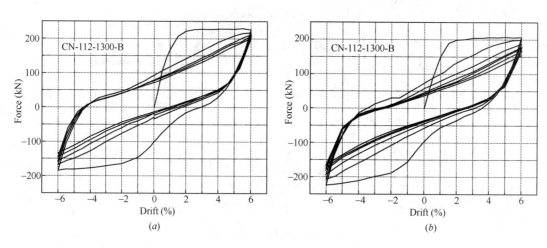

Fig. 6 Comparison of hysteresis loops of CFT and CCFT model columns
(a) CFT112-1300-C6; (b) CCFT112-1300-C6

By observing the failure processes and patterns of the two model columns, it was found that the CFRP confinement had positive effect in delaying low cycle fatigue in the CCFT model column. For the conventional CFT model column, buckling occurred on the compression face of the lower portion at 60mm above the footing upon loading to achieve the first peak drift at 6%. The buckled portion had a length of about 130mm. When the load was reversed, the tube buckled corresponding to a drift ratio of about 2%. In the subsequent loading cycles, buckling was intensified and formed an "elephant foot" type of ring at the end of the column, 20mm above the footing. In the fifth cycles, when the specimen reached the maximum negative displacement, the steel tube was fractured along the weld seem line and the length of the rip was 50mm from the bottom. After this stage, the loading was essentially not stable as the column could not hold the axial load. The buckling of the steel tube was effectively delayed in the model CCFT column, which endured more loading cycles corresponding to the peak drift ratio of 6%. In the second cycle, the carbon fibers in compression region were entirely split along hoop direction, about 30 mm from the bottom. In the cycles from the third to the sixth, the buckling of the steel tube was squeezed out from the circumferential crack of the CFRP. In the seventh cycle, when the specimen reached the maximum positive direction displacement, the steel tube in the tension region was ruptured.

4. Concluding Remarks

Experimental program is underway to investigate the hysteretic behavior of concrete filled tubular columns and confined concrete filled tubular columns subjecting to cyclic lateral loading. The tests exhibited the combined effects of the repeat of post yield local buckling and straightening of the steel tube in the critical column section. Such effects eventually result in the low-cycle fatigue of the steel tube and then the CFT column failure. Based on the testing results to date, it is found that providing additional transverse confinement to the potential plastic hinge region of the CFT columns, in other words, using CCFT column design, the local buckling of the steel tube can be delayed and the cyclic behavior can be improved.

Acknowledgements

The research described in this paper was supported by the National Natural Science Foundation of China (NSFC) (project No. 51178174).

References

M. Tomii, K. Sakino, Y. Xiao et al (1985), Earthquake Resisting Hysteretic Behavior of Reinforced Concrete Short Columns Confined by Steel Tube, Proceedings of the International Speciality Conference on Concrete Filled Steel Tubular Structures, pp. 119~125, China.

Y. Xiao, M. Tomii. and K. Sakino (1986), Experimental Study on Design Method to Prevent Shear Failure of Reinforced Concrete Short Circular Columns by Confining in Steel Tube, Transactions of Japan Concrete Institute, Vol. 8, pp. 535~542.

Y. Xiao, W. H. He, X. Y. Mao, et al (2003), Confinement Design of CFT Columns for Improved Seismic Performance, Proceedings of the International Workshop on Steel and Concrete Composite Construction (IWSCCC-2003), pp. 217-226, Taipei, R. China.

Y. Xiao, W. H. He and K. G. Choi (2005), Confined Concrete Filled Tubular (CCFT) Columns, ASCE Journal of Structural Engineering, Vol, 131, pp. 488-497.

Xiao, Y. ; H. Wu, (2000), "Compressive Behavior of Concrete Stub Columns Confined by Fiber Composite Jackets", Journal of Materials, ASCE Vol. No. May 2000.

G. W. Zhang, Y. Xiao, and S. Kunnath (2009), Low-cycle fatigue damage of circular concrete-filled-tube columns, ACI Structural Journal, Vol, 106(2), pp. 151-159.

面向抗震设计的钢板件和钢截面分类

童根树[1]，付 波[2]

([1] 浙江大学高性能材料和结构研究所，教授，杭州，310058，tonggs@zju.edu.cn；
[2] 浙江大学高性能材料和结构研究所，博士，杭州，310058，tonggs@zju.edu.cn)

摘 要：目前各国规范均根据宽厚比对钢截面进行分类的规定，但是这些分类的标准并非专门为抗震设计而制定。钢构件的抗震性能更为重要的方面是延性。本文提出了基于结构影响系数的大小对钢构件的截面进行分类的建议：从结构影响系数，反推对钢构件侧移的延性要求，再获得对截面曲率延性的要求，通过研究截面宽厚比与截面曲率延性或者压缩延性的关系，由延性要求计算出对截面宽厚比的要求。本文研究了单块板件、工字形截面和箱形截面在结构影响系数为 0.25，0.3，0.35，0.45，0.55 情况下的宽厚比限值，分考虑结构的超强系数和不考虑结构的超强系数，给出了宽厚比限值的计算公式。

关键词：宽厚比限值；结构影响系数；抗震设计；钢结构

Classification of Steel Cross-Sections for Seismic design of Steel Structures

TONG Genshu, FU Bo

(Professor, Department of Civil Engineering, Zhejiang University, 310058)

Abstract: Currently classification of cross-sections are presented in various codes for design of steel structures, but they are not proposed specifically for seismic design of steel structures. For seismic design, the more important aspect is the ductility of the cross-sections. The paper proposed a new philosophy in that the classification of the cross-section is based on the seismic force reduction factors (SFRF). Firstly the ductility demand on the steel members is derived from the SFRF, and then the curvature ductility or compressive ductility demand on the cross-section is calculated. Based on the relations of cross-section ductility and the width-to-thickness ratio of steel plate and cross-section, the width-to-thickness limitations for a given SFDF is thus obtained. The present paper made a study of width-to-thickness limitations for three types of plates and H- and Box- cross sections , and for five SFRFs：0.25, 0.3, 0.35, 0.45, 0.55, and formulas are presented which related the width-to-thickness ratio to the SFRFs in a continuous way for both cases: taking the overstrength factors od the structures into consideration or not.

Ke ywords: width-to-thickness ratio; seismic force reduction factor; seismic design; steel structure

1. 引言

板件是钢结构中受压和压弯构件的组成部分。为充分发挥钢材的受力性能，板件厚度通常用得较小,但是板越薄,就越容易产生局部失稳。板件过早失稳会降低截面的塑性转动能力,从而影响到结构的延性。根据截面承载力,塑性转动变形能力的不同,欧洲规范EC3[1]和EC8[2]将钢构件的截面分为以下四类:

Ⅰ类截面:截面不仅能够形成塑性铰,而且还要有一定的塑性转动能力,这类截面称为塑性设计截面,也称特厚实截面。

Ⅱ类截面:截面能够形成塑性铰,但不要求很大的塑性转动能力,这类截面称为弹塑性设计截面,也称厚实截面。

Ⅲ类截面:钢截面的边缘最大应力不超过钢材的屈服强度,这类截面称为弹性设计截面,也称非厚实截面。

Ⅳ类截面:允许截面发生局部屈曲,这类截面称为超屈曲设计截面,也称纤细截面或薄柔截面。

美国规范[3,4]的截面分类与欧洲规范相同。我国现行的《建筑抗震设计规范》GB 50011—2010[5]和《钢结构设计规范》GB 50017—2003[6]还没有相类似的规定,但正在修订中的 GB 50017—2003(征求意见稿)[7]增加了截面分类的相关条文:对于受弯及压弯构件根据局部屈曲制约截面承载力和转动能力的程度,设计截面分为 S1、S2、S3、S4、S5 共 5 级。

S1 级,塑性设计截面。可达全截面塑性,保证塑性铰具有塑性设计要求的转动能力,且在转动过程中承载力不降低。

S2 级,塑性屈服强度截面。可达全截面塑性,但由于局部屈曲,塑性铰的转动能力有限。

S3 级,部分塑性开展的截面。翼缘全部屈服,腹板可发展不超过 1/4 截面高度的塑性。

S4 级,弹性屈服强度截面。即边缘纤维屈服截面,边缘纤维可达屈服强度,但由于局部屈曲而不能发展塑性。

S5 级,超屈曲设计截面。在边缘纤维达屈服应力前,腹板可能发生局部屈曲。

从两种分类的表述可知,我国即将采用的截面分类同欧美相比增加了一类部分塑性开展的截面。这是因为我国钢结构设计规范在对构件进行强度设计时,允许截面部分开展塑性,故在截面分类中引入了 S3 类截面,这是我国特有的分类。另外的 S1,S2,S4,S5 截面实际上是同欧美的Ⅰ～Ⅳ类截面相对应,所以我国对截面分类的方法可认为是同欧美相一致的。

日本建筑学会出版的钢结构极限状态设计指针 AIJ2010[8]也将钢构件截面分为 P-Ⅰ-1,P-Ⅰ-2,P-Ⅱ,P-Ⅲ四类,分别与 EC3 的Ⅰ～Ⅳ类截面相对应。与欧美和中国规范不同的是,AIJ2010 考虑了翼缘和腹板的相互作用,对工字形截面的板件宽厚比限值采用了翼缘和腹板的相关关系式来表示,但是对箱形截面没有给出相类似的宽厚比限值相关关系式。

上述分类方法与截面的抗震性能有关,但也不完全如此,因为抗震设计的关键因素是延性,而上面的分类除了第Ⅰ类(A类),其他都跟承载力相联系。板件的宽厚比限值是根据承载力和对达到极限承载力以后的变形能力的定性描述进行的,并没有给出板件延性与宽厚比的关系,缺乏截面分类的具体定量指标。

Mazzolani[9,10]等研究了压力作用下箱形截面铝合金构件的板件屈曲性能，通过试验和理论分析，拟合了截面延性系数 μ_ϕ 和腹板及翼缘宽厚比的关系式。在此基础上给出了截面分类的 μ_ϕ 限值如下：Ⅰ类截面：$\mu_\phi \geqslant 12.3$；Ⅱ类截面：$4.7 \leqslant \mu_\phi \leqslant 12.3$；Ⅲ类截面：$2.3 \leqslant \mu_\phi \leqslant 4.7$；Ⅳ类截面：$\mu_\phi \leqslant 2.3$。这种分类改进了以往的做法，将截面延性需求与分类直接联系了起来。由于材料性质的不同，铝合金构件的限值不能直接用于钢截面的分类。

本文将根据地震力计算的结构影响系数的大小（五个值）对截面进行分类。从结构影响系数，区分考虑和不考虑结构的超强系数，得到结构的延性系数，反推对截面的延性系数要求，然后根据第 2～5 章拟合的板件和截面的延性系数公式，得到各类截面宽厚比分界值，并拟合了宽厚比限值和结构影响系数的相关公式。其中对工字形截面和箱形截面的宽厚比限值，给出的是翼缘宽厚比和腹板宽厚比的相关关系式。

2. 面向抗震设计的截面分类方法

根据延性地震力理论[11]，同一种材料，同一种结构体系，决定地震力取值的结构影响系数相同，则要求结构所应具有的延性也相同。目前结构影响系数即地震力调整系数的研究大都采用基于杆件层面的位移延性系数。从本质上来说，构件的延性是建立在截面的延性基础之上的，而截面延性是确定截面板件宽厚比的重要因素。这样推论可知，设计钢构件时采用哪一类截面，对结构的延性有重要的影响，并将决定钢结构的地震力取值。因此本文尝试提出面向抗震设计的截面分类方法。其出发点是：以结构影响系数 $C=0.25$，0.30，0.35，0.45 和 0.55 来确定截面的分类，并记为Ⅰ-E，Ⅱ-E，Ⅲ-E，Ⅳ-E，Ⅴ-E，加字母 E 表示面向抗震。

从世界范围看，构成结构影响系数的因素是结构的延性系数和超强系数[12]。超强系数在一定程度上代表了抗震结构的超静定次数和多道抗震防线，其包含的因素变化很大[11]，不好把握。下面分两种情况来考虑截面的分类。一种是不考虑超强系数，由等位移准则可推出对结构的延性需求是：

$$\mu_{d,req} = C^{-1} \tag{1}$$

第 2 种认为，按照薄壁构件设计的结构，结构影响系数取 0.55，此时对截面的宽厚比无特殊要求，相当于用超强系数来抵抗地震作用，对结构的延性要求是 1.0。据此，反推超强系数取 $1/0.55=1.82$，并把这个系数应用到前面的三类。这样对结构的延性需求是：

$$\mu_{d,req} = 0.55 C^{-1} \tag{2}$$

有了对结构的延性需求，可反推对构件截面的曲率延性需求。文献［13］采用悬臂柱模型和具有曲线段的弯矩-曲率滞回模型，分析了基于曲率延性、位移延性的地震力调整系数，考察了曲线段、后期刚度系数对地震力调整系数的影响，文中拟合了曲率延性和位移延性的关系式如下：

$$\mu_\phi = A(\mu_d - 1)^B + 1 \tag{3}$$

$$A = (0.91 - 0.16r)\alpha^{-0.5} + 0.7 \tag{4}$$

$$B = 0.56 + 0.14(r-1)^2 \tag{5}$$

式中：A、B 为计算系数，为弯矩-曲率曲线的强化段刚度和线弹性段刚度的比值，也叫后期刚度系数，r 为全截面屈服弯矩和材料非线性开始出现时的弯矩比值。

当截面无残余应力时，r 应等于截面的塑性开展系数。工字形截面和箱形截面的塑性开展系数同翼缘与腹板在整个截面积中所占的相对比重有关。腹板所占比重越大，发展塑性的能力越大。对焊接工字梁，r 通常大于 1.1；对轧制的工字钢和 H 形钢，r 可高达 1.2[14]。当截面存在残余应力时，r 值会大于截面的塑性开展系数。取 $r=1.1, 1.5, 2$，当后期刚度系数 $\alpha=0.01$ 时：

$$\begin{cases} r=1.1: \mu_\phi = 8.04(\mu_d-1)^{0.56}+1 \\ r=1.5: \mu_\phi = 7.4(\mu_d-1)^{0.6}+1 \\ r=2: \mu_\phi = 6.6(\mu_d-1)^{0.7}+1 \end{cases} \quad (6)$$

后期刚度系数 $\alpha=0.03$ 时：

$$\begin{cases} r=1.1: \mu_\phi = 4.94(\mu_d-1)^{0.56}+1 \\ r=1.5: \mu_\phi = 4.57(\mu_d-1)^{0.6}+1 \\ r=2: \mu_\phi = 4.11(\mu_d-1)^{0.7}+1 \end{cases} \quad (7)$$

根据式（6）、（7）可得到每一类截面所对应的截面延性需求如表1和表2所示。由文献［16］提出的板件和截面的延性系数计算公式即可给出各类截面所对应的板件宽厚比限值。

不考虑超强系数时抗震五类截面的延性需求 表1

α	r	类别	I-E	II-E	III-E	IV-E	V-E
		C	0.25	0.30	0.35	0.45	0.55
		$\mu_{d,Req}$	4	3.33	2.86	2.22	1.82
0.01	1.1	$\mu_{\phi,req}$	15.90	13.93	12.39	9.99	8.19
	1.5	$\mu_{\phi,req}$	15.23	13.24	11.71	9.33	7.58
	2.0	$\mu_{\phi,req}$	15.24	12.93	11.19	8.59	6.74
0.03	1.1	$\mu_{\phi,req}$	10.15	8.94	8	6.52	5.42
	1.5	$\mu_{\phi,req}$	9.78	8.56	7.61	6.14	5.06
	2.0	$\mu_{\phi,req}$	9.86	8.42	7.34	5.72	4.57

考虑超强系数时抗震五类截面的延性需求 表2

α	r	类别	I-E	II-E	III-E	IV-E	V-E
		C	0.25	0.30	0.35	0.45	0.55
		$\mu_{d,Req}$	2.2	1.83	1.57	1.22	1
0.01	1.1	$\mu_{\phi,req}$	9.91	8.24	6.86	4.44	1
	1.5	$\mu_{\phi,req}$	9.25	7.62	6.30	4.01	1
	2.0	$\mu_{\phi,req}$	8.50	6.79	5.45	3.29	1
0.03	1.1	$\mu_{\phi,req}$	6.47	5.45	4.60	3.11	1
	1.5	$\mu_{\phi,req}$	6.09	5.09	4.27	2.86	1
	2.0	$\mu_{\phi,req}$	5.67	4.60	3.77	2.42	1

3. 不考虑板件间相互作用的截面宽厚比分界

对受压的四边简支板、三边简支一边自由板和受压弯荷载的四边简支板，宽厚比与正则化宽厚比的关系分别为（Q235 钢材）：

$$\begin{cases} b_0/t = 56.29\lambda \\ b_1/t = 18.36\lambda \\ h_w/t_w = 28.14\sqrt{4+2\beta+2\beta^3}\lambda \end{cases} \quad (8)$$

这三种板件的延性系数公式分别为

$$\mu_{css} = 50 - 49\tanh\frac{4.2(\lambda-0.215)}{\lambda^{0.2}} \leqslant 50$$

$$\mu_{csf} = 50 - 49\tanh\frac{4(\lambda-0.2)}{\lambda^{0.65}} \leqslant 50$$

$$\mu_{bc2} = 50 - (48+p)\tanh(\chi + 2.4p^{p^2+0.62}) \leqslant 50$$

根据各类截面的延性需求即可反推相应的板件宽厚比限值。表 3～表 6 列出了不考虑超强和考虑超强时各类截面相对应的板件宽厚比限值。

不考虑超强系数时受压板件的宽厚比限值 表 3

α	r	类别	I-E	II-E	III-E	IV-E	V-E
0.01	1.1	$\mu_{\phi,req}$	15.90	13.93	12.39	9.99	8.19
		b_0/t	21.6	22.6	23.5	25.2	26.7
		b_1/t_f	5.5	5.7	5.9	6.3	6.8
	1.5	$\mu_{\phi,req}$	15.23	13.24	11.71	9.33	7.58
		b_0/t	21.9	23	24	25.7	27.3
		b_1/t_f	5.6	5.8	6	6.5	6.9
	2	$\mu_{\phi,req}$	15.24	12.93	11.19	8.59	6.74
		b_0/t	21.9	23.2	24.3	26.4	28.5
		b_1/t_f	5.6	5.9	6.1	6.7	7.2
0.03	1.1	$\mu_{\phi,req}$	10.15	8.94	8	6.52	5.42
		b_0/t	25	26.1	26.9	28.6	30.2
		b_1/t_f	6.3	6.6	6.8	7.2	7.7
	1.5	$\mu_{\phi,req}$	9.78	8.56	7.61	6.14	5.06
		b_0/t	25.3	26.4	27.3	29.2	30.7
		b_1/t_f	6.4	6.7	6.9	7.4	7.9
	2	$\mu_{\phi,req}$	9.86	8.42	7.34	5.72	4.57
		b_0/t	25.3	26.6	27.6	29.7	31.8
		b_1/t_f	6.4	6.7	7	7.6	8.2

考虑超强系数时受压板件的宽厚比限值 表 4

α	r	类别	I-E	II-E	III-E	IV-E	V-E
0.01	1.1	$\mu_{\phi,req}$	9.91	8.24	6.86	4.44	1
		b_0/t	25.2	26.6	28.1	32	40
		b_1/t_f	6.3	6.8	7.1	8.2	15
	1.5	$\mu_{\phi,req}$	9.25	7.62	6.30	4.01	1
		b_0/t	25.8	27.2	28.9	33	40
		b_1/t_f	6.5	6.9	7.3	8.5	15
	2	$\mu_{\phi,req}$	8.50	6.79	5.45	3.29	1
		b_0/t	26.5	28.2	30.1	34.8	40
		b_1/t_f	6.7	7.2	7.7	9.2	15
0.03	1.1	$\mu_{\phi,req}$	6.47	5.45	4.60	3.11	1
		b_0/t	28.6	30	31.6	35.5	40
		b_1/t_f	7.3	7.7	8.1	9.4	15
	1.5	$\mu_{\phi,req}$	6.09	5.09	4.27	2.86	1
		b_0/t	29.2	30.7	32.2	36.3	40
		b_1/t_f	7.4	7.8	8.3	9.7	15
	2	$\mu_{\phi,req}$	5.67	4.60	3.77	2.42	1
		b_0/t	29.7	31.6	33.4	38.2	40
		b_1/t_f	7.6	8.1	8.7	10.5	15

不考虑超强系数时压弯板件的宽厚比限值 表 5

α	r	类别	I-E	II-E	III-E	IV-E	V-E
0.01	1.1	$\mu_{\phi,req}$	15.90	13.93	12.39	9.99	8.19
		$P/P_y=0$	59.6	61.5	63.3	66.8	70.3
		$P/P_y=0.25$	43.8	45.3	46.8	49.5	52.2
		$P/P_y=0.5$	37	38.4	39.7	42.1	44.5
		$P/P_y=0.75$	30.4	31.7	32.7	34.8	36.9
		$P/P_y=1$	21.6	22.6	23.5	25.2	26.7
	1.5	$\mu_{\phi,req}$	15.23	13.24	11.71	9.33	7.58
		$P/P_y=0$	60.2	62.3	64.1	68	71.8
		$P/P_y=0.25$	44.3	45.9	47.4	50.3	53.3
		$P/P_y=0.5$	37.5	39	40.3	42.9	45.5
		$P/P_y=0.75$	30.8	32.1	33.3	35.5	37.8
		$P/P_y=1$	21.9	23	24	25.7	27.3
	2	$\mu_{\phi,req}$	15.24	12.93	11.19	8.59	6.74
		$P/P_y=0$	60.2	62.6	64.9	69.4	74.2
		$P/P_y=0.25$	44.3	46.2	48	51.5	55
		$P/P_y=0.5$	37.5	39.2	40.8	43.9	47
		$P/P_y=0.75$	30.8	32.3	33.7	36.5	39
		$P/P_y=1$	21.9	23.2	24.3	26.4	28.5

续表

α	r	类别	Ⅰ-E	Ⅱ-E	Ⅲ-E	Ⅳ-E	Ⅴ-E
0.03	1.1	$\mu_{\phi,req}$	10.15	8.94	8	6.52	5.42
		$P/P_y=0$	66.5	68.6	70.8	75	79.6
		$P/P_y=0.25$	49.3	51	52.5	55.6	58.7
		$P/P_y=0.5$	41.9	43.4	44.8	47.5	50.2
		$P/P_y=0.75$	34.7	36	37.1	39.5	41.7
		$P/P_y=1$	25	26.1	26.9	28.6	30.2
	1.5	$\mu_{\phi,req}$	9.78	8.56	7.61	6.14	5.06
		$P/P_y=0$	67	69.5	71.8	76.3	81.5
		$P/P_y=0.25$	49.7	51.5	53.2	56.7	60
		$P/P_y=0.5$	42.3	43.9	45.4	48.4	51.3
		$P/P_y=0.75$	35	36.4	37.7	40.2	42.6
		$P/P_y=1$	25.3	26.4	27.3	29.2	30.7
	2	$\mu_{\phi,req}$	9.86	8.42	7.34	5.72	4.57
		$P/P_y=0$	67	69.8	72.5	78.2	84.5
		$P/P_y=0.25$	49.7	51.8	53.8	57.8	62.1
		$P/P_y=0.5$	42.3	44.2	45.9	49.2	52.9
		$P/P_y=0.75$	35	36.6	38.1	41	44
		$P/P_y=1$	25.3	26.6	27.6	29.7	31.8

考虑超强系数时压弯板件的宽厚比限值   表6

α	r	类别	Ⅰ-E	Ⅱ-E	Ⅲ-E	Ⅳ-E	Ⅴ-E
0.01	1.1	$\mu_{\phi,req}$	9.91	8.24	6.86	4.44	1
		$P/P_y=0$	66.9	70.3	73.9	85.5	150
		$P/P_y=0.25$	49.5	52	54.8	62.7	115
		$P/P_y=0.5$	42.2	44.4	46.9	53.4	90
		$P/P_y=0.75$	34.9	36.9	38.9	44.5	65
		$P/P_y=1$	25.2	26.6	28.1	32	40
	1.5	$\mu_{\phi,req}$	9.25	7.62	6.30	4.01	1
		$P/P_y=0$	68.1	71.8	75.9	89.3	150
		$P/P_y=0.25$	50.5	53.2	56.2	65	115
		$P/P_y=0.5$	43	45.4	47.9	55.2	90
		$P/P_y=0.75$	35.6	37.7	39.9	45.9	65
		$P/P_y=1$	25.8	27.2	28.9	33	40
	2	$\mu_{\phi,req}$	8.50	6.79	5.45	3.29	1
		$P/P_y=0$	69.6	74.1	79.5	99	150
		$P/P_y=0.25$	51.6	55	58.7	70.5	115
		$P/P_y=0.5$	44	46.9	50	59.5	90
		$P/P_y=0.75$	36.5	39	41.7	49.2	65
		$P/P_y=1$	26.5	28.2	30.1	34.8	40

续表

α	r	类别	I-E	II-E	III-E	IV-E	V-E
0.03	1.1	$\mu_{\phi,req}$	6.47	5.45	4.60	3.11	1
		$P/P_y=0$	75.2	79.5	84.3	101	150
		$P/P_y=0.25$	55.7	58.8	62	72	115
		$P/P_y=0.5$	47.6	50	52.8	60.8	90
		$P/P_y=0.75$	39.5	41.7	44	50.2	65
		$P/P_y=1$	28.6	30	31.6	35.5	40
	1.5	$\mu_{\phi,req}$	6.09	5.09	4.27	2.86	1
		$P/P_y=0$	76.7	81.4	87	110	150
		$P/P_y=0.25$	56.7	60	63.5	75	115
		$P/P_y=0.5$	48.5	51.1	54.1	63	90
		$P/P_y=0.75$	40.3	42.5	45	52	65
		$P/P_y=1$	29.2	30.7	32.2	36.3	40
	2	$\mu_{\phi,req}$	5.67	4.60	3.77	2.42	1
		$P/P_y=0$	78.4	84.4	91.8	130	150
		$P/P_y=0.25$	57.9	61.9	66.5	83.5	115
		$P/P_y=0.5$	49.5	52.8	56.5	68.4	90
		$P/P_y=0.75$	41.1	43.9	46.9	55.7	65
		$P/P_y=1$	29.7	31.6	33.4	38.2	40

从表 3～表 6 可看出，考虑超强系数的板件宽厚比限值要大于不考虑超强系数的情况。在相同情况下，增大 r 或后期刚度系数 α 均可减小对截面的延性需求，相应的板件宽厚比限值也得到放宽。

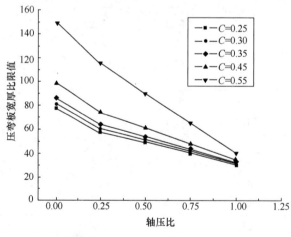

图 1 压弯板宽厚比限值（考虑超强，$r=1.1$，$\alpha=0.03$）

对照表 3、表 5 和 EC3 对四类截面的划分规定，即使按照后期刚度为 0.03 的数据，表中的四类划分界限也比 EC3 的划分界限严格很多，实际工程将无法接受。考虑超强系数后的四类截面划分界限如表 4、表 6 所示。此时后期刚度是 0.03 的数值，已经处在目前工程技术人员可以接受的范围，但是纯压荷载作用下的三边简支一边自由的板件分类界限，本文的方法还是更加严格。考虑超强系数，$r=1.1$，$\alpha=0.03$ 的压弯板件宽厚比限值可用式（9）来近似计算，公式计算值如图 1 所示，同表 6 中数据相比最大误差不超过 5%，式中 $p=P/P_y$。

$$\left[\frac{b}{t}\right] = 140 - 100p + 6.5\tanh(1-p)^{10} - (130 - 115\tanh p^{1.2} - 28\tanh p^{1.4})\tanh|(C-0.55)|^{0.38+0.22\tanh p^{20}} \tag{9}$$

造成截面分界过严的主要原因是确定宽厚比限值时没有考虑板件间的相互作用。从文献［16］可知，将板件间相互约束简化为理想的简支边界条件所得到的板件延性最小，特别是三边支承板件的非加载边为简支和固支时，延性相差很大，故根据简支边界条件的延性系数公式得到的分界过于保守。

4 工字形截面的宽厚比分界

根据 2 节提出的面向抗震设计的截面分类思路对工字形截面进行分类。按照表 1、表 2 给出的各类截面延性需求可反推相应的正则化宽厚比限值 λ（Q235）。工字形截面的延性系数计算公式是[16]。

$$\mu_{bc-I} = 50 - (49 + p^2)\tanh\left[\frac{3.7(\lambda - 0.22 + 0.066\tanh p^{0.01})}{\lambda^{0.15}}\right] \leqslant 50$$

式中 $p = P/P_y$，λ 是工字形截面的正则化长细比。

$$\lambda = \sqrt{\frac{f_y}{\sigma_{cr}}} = \frac{1.052}{\sqrt{k}}\left(\frac{h}{t_w}\right)\sqrt{\frac{f_y}{E}}$$

式中 h, t_w 是工字形截面的腹板高度和厚度，k 是按照腹板屈曲计算的工字形截面屈曲系数。

表 7 和表 8 列出了不考虑超强和考虑超强时的各类截面延性需求以及相对应的正则化宽厚比限值 λ（Q235）。对比表 7 和表 8 的数据可发现，考虑超强系数后相对应的 λ 限值均增大，说明超强系数的存在可以减小对截面的延性需求，增大 r 或 α 也能达到同样的效果。

不考虑超强系数时抗震五类截面的 λ（Q235）限值　　　　表 7

α	r	类别	Ⅰ-E	Ⅱ-E	Ⅲ-E	Ⅳ-E	Ⅴ-E
0.01	1.1	$\mu_{\phi,req}$	15.90	13.93	12.39	9.99	8.19
		$P/P_y=0$	0.424	0.445	0.464	0.499	0.531
		$P/P_y=0.2$	0.370	0.391	0.409	0.444	0.476
		$P/P_y=0.4$	0.369	0.390	0.408	0.442	0.474
		$P/P_y=0.6$	0.367	0.388	0.406	0.439	0.470
		$P/P_y=0.8$	0.365	0.385	0.403	0.436	0.466
	1.5	$\mu_{\phi,req}$	15.23	13.24	11.71	9.33	7.58
		$P/P_y=0$	0.431	0.453	0.473	0.510	0.544
		$P/P_y=0.2$	0.377	0.399	0.418	0.455	0.489
		$P/P_y=0.4$	0.376	0.398	0.417	0.452	0.486
		$P/P_y=0.6$	0.374	0.396	0.415	0.450	0.484
		$P/P_y=0.8$	0.372	0.393	0.412	0.446	0.478
	2	$\mu_{\phi,req}$	15.24	12.93	11.19	8.59	6.74
		$P/P_y=0$	0.431	0.457	0.481	0.524	0.564
		$P/P_y=0.2$	0.377	0.402	0.426	0.468	0.508
		$P/P_y=0.4$	0.376	0.401	0.425	0.467	0.506
		$P/P_y=0.6$	0.374	0.399	0.423	0.464	0.502
		$P/P_y=0.8$	0.372	0.397	0.419	0.460	0.496

续表

α	r	类别	Ⅰ-E	Ⅱ-E	Ⅲ-E	Ⅳ-E	Ⅴ-E
0.03	1.1	$\mu_{\phi,req}$	10.15	8.94	8	6.52	5.42
		$P/P_y=0$	0.496	0.517	0.535	0.570	0.602
		$P/P_y=0.2$	0.441	0.462	0.480	0.514	0.545
		$P/P_y=0.4$	0.439	0.459	0.477	0.511	0.542
		$P/P_y=0.6$	0.437	0.456	0.474	0.506	0.536
		$P/P_y=0.8$	0.434	0.452	0.469	0.500	0.529
	1.5	$\mu_{\phi,req}$	9.78	8.56	7.61	6.14	5.06
		$P/P_y=0$	0.502	0.524	0.544	0.580	0.614
		$P/P_y=0.2$	0.447	0.469	0.487	0.523	0.557
		$P/P_y=0.4$	0.445	0.467	0.484	0.520	0.553
		$P/P_y=0.6$	0.442	0.464	0.481	0.516	0.548
		$P/P_y=0.8$	0.438	0.460	0.477	0.510	0.539
	2	$\mu_{\phi,req}$	9.86	8.42	7.34	5.72	4.57
		$P/P_y=0$	0.502	0.527	0.549	0.591	0.633
		$P/P_y=0.2$	0.447	0.472	0.494	0.535	0.576
		$P/P_y=0.4$	0.445	0.470	0.491	0.531	0.571
		$P/P_y=0.6$	0.442	0.466	0.487	0.526	0.564
		$P/P_y=0.8$	0.438	0.461	0.482	0.520	0.556

考虑超强系数时抗震五类截面的 λ（Q235）限值　　　　表8

α	r	类别	Ⅰ-E	Ⅱ-E	Ⅲ-E	Ⅳ-E	Ⅴ-E
0.01	1.1	$\mu_{\phi,req}$	9.91	8.24	6.86	4.44	1
		$P/P_y=0$	0.500	0.530	0.561	0.638	0.800
		$P/P_y=0.2$	0.445	0.475	0.505	0.581	0.790
		$P/P_y=0.4$	0.443	0.472	0.502	0.577	0.780
		$P/P_y=0.6$	0.440	0.469	0.499	0.569	0.770
		$P/P_y=0.8$	0.437	0.465	0.493	0.560	0.760
	1.5	$\mu_{\phi,req}$	9.25	7.62	6.30	4.01	1
		$P/P_y=0$	0.512	0.544	0.575	0.658	0.800
		$P/P_y=0.2$	0.456	0.487	0.520	0.600	0.790
		$P/P_y=0.4$	0.454	0.485	0.517	0.595	0.780
		$P/P_y=0.6$	0.452	0.482	0.512	0.587	0.770
		$P/P_y=0.8$	0.447	0.477	0.505	0.575	0.760
	2	$\mu_{\phi,req}$	8.50	6.79	5.45	3.29	1
		$P/P_y=0$	0.525	0.563	0.601	0.698	0.800
		$P/P_y=0.2$	0.470	0.507	0.545	0.640	0.790
		$P/P_y=0.4$	0.468	0.505	0.542	0.632	0.780
		$P/P_y=0.6$	0.464	0.500	0.536	0.622	0.770
		$P/P_y=0.8$	0.460	0.494	0.528	0.608	0.760

续表

α	r	类别	Ⅰ-E	Ⅱ-E	Ⅲ-E	Ⅳ-E	Ⅴ-E
0.03	1.1	$\mu_{\phi,req}$	6.47	5.45	4.60	3.11	1
		$P/P_y=0$	0.571	0.601	0.631	0.709	0.800
		$P/P_y=0.2$	0.515	0.545	0.575	0.651	0.790
		$P/P_y=0.4$	0.512	0.542	0.570	0.644	0.780
		$P/P_y=0.6$	0.508	0.536	0.563	0.632	0.770
		$P/P_y=0.8$	0.502	0.528	0.554	0.617	0.760
	1.5	$\mu_{\phi,req}$	6.09	5.09	4.27	2.86	1
		$P/P_y=0$	0.582	0.613	0.645	0.728	0.800
		$P/P_y=0.2$	0.525	0.557	0.588	0.669	0.790
		$P/P_y=0.4$	0.522	0.553	0.583	0.661	0.780
		$P/P_y=0.6$	0.517	0.546	0.575	0.648	0.770
		$P/P_y=0.8$	0.511	0.538	0.566	0.631	0.760
	2	$\mu_{\phi,req}$	5.67	4.60	3.77	2.42	1
		$P/P_y=0$	0.594	0.631	0.670	0.768	0.800
		$P/P_y=0.2$	0.538	0.575	0.612	0.707	0.790
		$P/P_y=0.4$	0.535	0.570	0.606	0.696	0.780
		$P/P_y=0.6$	0.529	0.563	0.598	0.680	0.770
		$P/P_y=0.8$	0.522	0.554	0.588	0.659	0.760

表7和表8中的正则化宽厚比限值λ不便于实际工程应用，需要将λ与板件宽厚比联系起来。定义参数η，并将宽厚比与λ相互联系，得到式（11），式中$\varepsilon_k=\sqrt{f_y/235}$。

$$\eta = \frac{h}{b} \cdot \frac{t_f}{t_w} \tag{10}$$

$$\left(\frac{h}{t_w}\right)\varepsilon_k = \sqrt{\frac{kE}{235}} \frac{\lambda}{1.052} \tag{11}$$

文献[15]拟合的工形截面屈曲系数k的表达式是h/b，t_f/t_w，应力分布参数β的函数。给定应力状态参数β（$\beta=0$表示轴压，$=2$表示纯弯），从表7和表8可查到对应的宽厚比限值λ，取任意的h/b，t_f/t_w得到k代入式（11）可算出$\left(\frac{h}{t_w}\right)\varepsilon_k$，再由式（10）即可得到：

$$\frac{b}{t_f}\varepsilon_k = \frac{1}{\eta} \frac{h}{t_w}\varepsilon_k \tag{12}$$

取h/b的变化范围为1.25～6，t_f/t_w的变化范围为1～4，通过上述步骤即可得到对应于给定λ的$\frac{h}{t_w}\varepsilon_k - \frac{b}{t_f}\varepsilon_k$相关关系。图2～图6分别给出了不考虑超强和考虑超强的五组$\frac{h}{t_w}\varepsilon_k - \frac{b}{t_f}\varepsilon_k$相关关系。

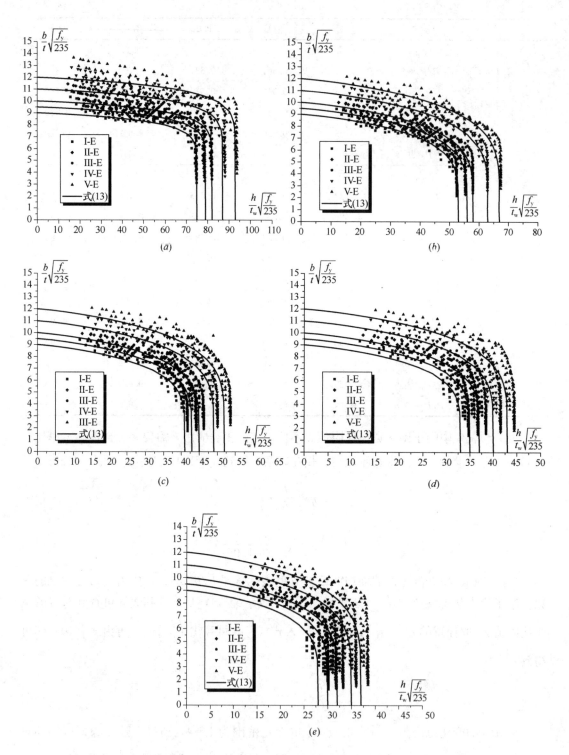

图 2 工字形截面 $b/t_f - h/t_w$ 相关关系（不考虑超强，$r=1.1$，$\alpha=0.01$）
(a) $P/P_y=0$; (b) $P/P_y=0.2$; (c) $P/P_y=0.4$; (d) $P/P_y=0.6$; (e) $P/P_y=0.8$

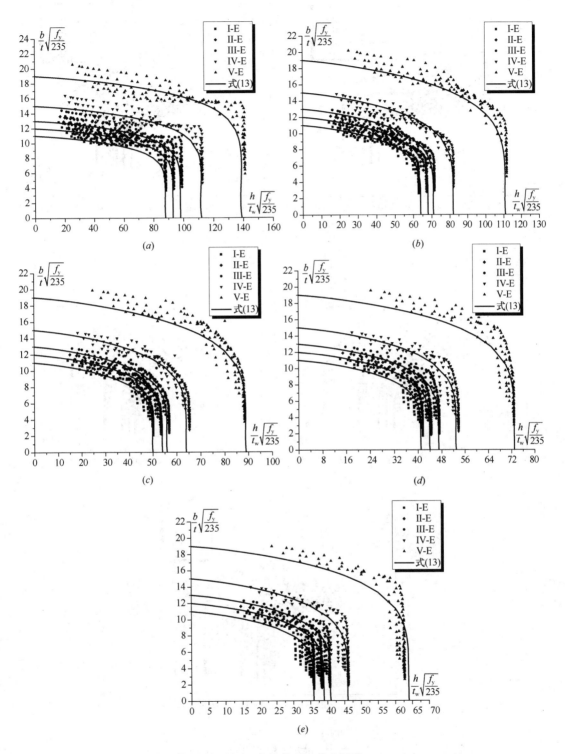

图 3 工字形截面 $b/t_f - h/t_w$ 相关关系（考虑超强，$r=1.1$，$\alpha=0.01$）
(a) $P/P_y=0$；(b) $P/P_y=0.2$；(c) $P/P_y=0.4$；(d) $P/P_y=0.6$；(e) $P/P_y=0.8$

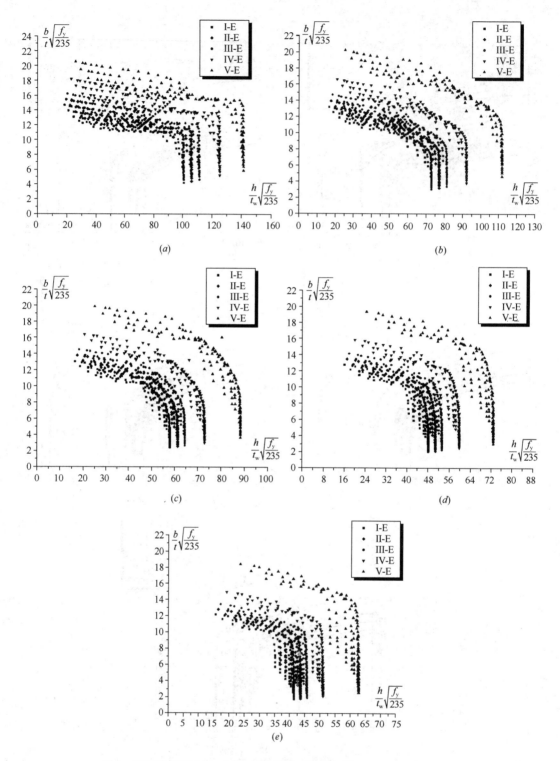

图 4　工字形截面 $b/t_f - h/t_w$ 相关关系（考虑超强，$r=1.1$，$\alpha=0.03$）
(a) $P/P_y=0$；(b) $P/P_y=0.2$；(c) $P/P_y=0.4$；(d) $P/P_y=0.6$；(e) $P/P_y=0.8$

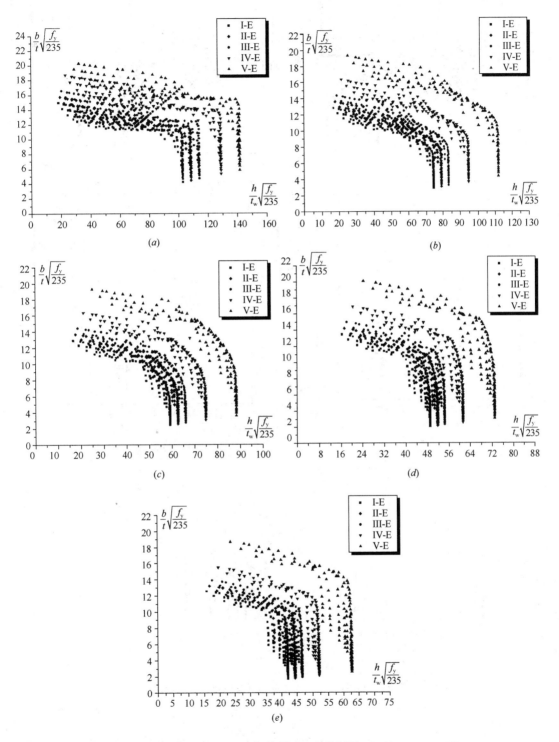

图 5 工字形截面 $b/t_f - h/t_w$ 相关关系（考虑超强，$r=1.5$，$\alpha=0.03$）
(a) $P/P_y=0$; (b) $P/P_y=0.2$; (c) $P/P_y=0.4$; (d) $P/P_y=0.6$; (e) $P/P_y=0.8$

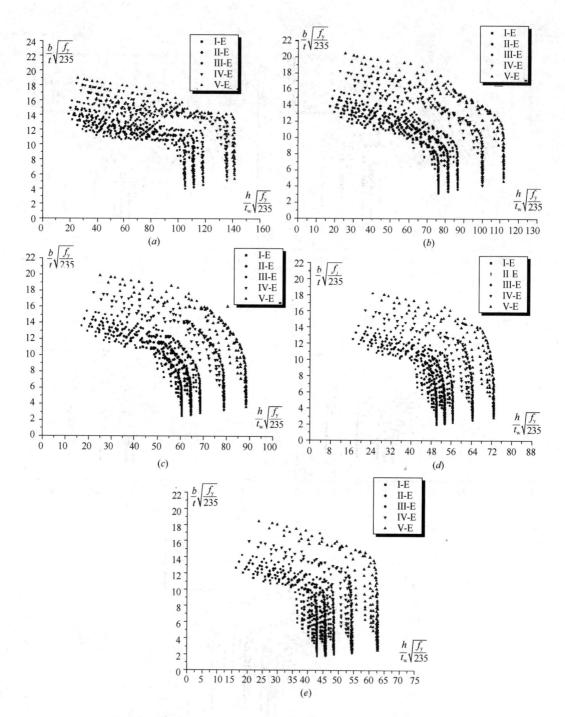

图 6 工字形截面 $b/t_f - h/t_w$ 相关关系（考虑超强，$r=2$，$\alpha=0.03$）
(a) $P/P_y=0$；(b) $P/P_y=0.2$；(c) $P/P_y=0.4$；(d) $P/P_y=0.6$；(e) $P/P_y=0.8$

从图上可知，$\frac{h}{t_w}\varepsilon_k - \frac{b}{t_f}\varepsilon_k$ 的相关关系外凸很明显。腹板宽厚比为 0 时，翼缘宽厚比限值不受轴压比影响，只和结构影响系数 C 有关，表现在随结构影响系数的增大，翼缘宽厚比限值逐渐增大。这是因为在受弯和压弯状态下，受压翼缘板为均匀受压的应力状态，所以轴压比的变化不影响其宽厚比限值。而当翼缘宽厚比为 0 时，结构影响系数和轴压比的变化均会影响到腹板宽厚比的取值。轴压比一定时，结构影响系数越大，腹板宽厚比限值越大。结构影响系数一定时，轴压比越大，腹板宽厚比限值越小。当 r 和 α 相同时，考虑超强系数之后，宽厚比限值均比不考虑超强系数时放宽了很多。其他条件相同，增大 r 或 α 也可以使板件宽厚比限值得到放宽。经过数据拟合，可以用式（13）来近似计算工形截面的板件宽厚比相关关系，对图 2 和图 3，公式中系数分别为式（14）和式（15）。

$$\frac{1}{A^m}\left(\frac{b}{t_f}\varepsilon_k\right)^m + \frac{1}{B^{1.2}}\left(\frac{h}{t_w}\varepsilon_k\right)^{1.2} = 1 \tag{13}$$

不考虑超强系数时（$r=1.1, \alpha=0.01$）：

$$\begin{cases} m = 12 - 9.21\tanh p^{0.01} \\ A = 6.5 + 10C \\ B = 75 + 54\tanh(C-0.25)^{0.9} - \\ \quad (51 + 29\tanh(C-0.025)^{0.9})\tanh(p^{1.6} + p^{0.6}) \end{cases} \tag{14}$$

考虑超强系数时（$r=1.1, \alpha=0.01$）：

$$\begin{cases} m = 7 - 2.64\tanh p^{0.01} \\ A = 6 + 20C + 830\tanh[(C-0.25)^5] \\ B = 88 + 100\tanh(C-0.25) + 2700\tanh(C-0.25)^4 - \\ \quad (57 + 140\tanh(C-0.25)^{1.4})\tanh(p^{1.6} + p^{0.6+930\tanh(C-0.25)^7}) \end{cases} \tag{15}$$

图 7 分别给出了 EC3 和 AIJ2010 对工字梁板件的四类截面分类曲线，其中 AIJ2010 也是采用相关关系来表示腹板和翼缘的宽厚比限值，图上同时绘出了式（13）（$r=1.1, \alpha=0.01$）的曲线。不考虑超强系数时，本文的 I-E～V-E 类截面宽厚比限值要比 EC3 和 AIJ2010 严格，大致只相当于 EC3 和 AIJ2010 的前两类截面。这是因为 EC3 和 AIJ2010 的后两类截面不考虑发展塑性，所以其限值放得比较宽。考虑了超强系数之后，本文的截面分界得到放宽，总体来看宽厚比限值比 EC3 的宽松，腹板宽厚比限值接近于 AIJ2010

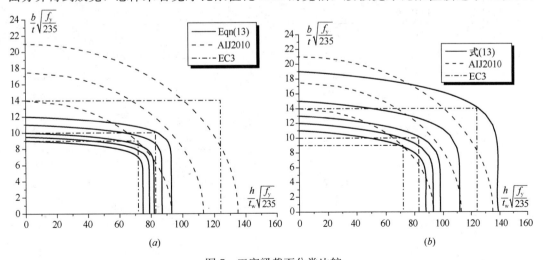

图 7　工字梁截面分类比较

(a) 不考虑超强（$r=1.1, \alpha=0.01$）；(b) 考虑超强（$r=1.1, \alpha=0.01$）

的限值水平，翼缘宽厚比限值比 AIJ2010 偏保守。由图 4～图 6 可知，增大 r 或 α 值，翼缘宽厚比限值可进一步得到放宽。

5. 箱形截面的宽厚比分界

根据面向抗震设计的截面分类思路，按照表 1、表 2 给出的各类截面延性需求反推箱形截面相应的正则化宽厚比限值 λ（Q235）。箱形截面的延性系数计算公式是[16]

$$\mu_{bc-b} = 50 - (48 + p^2)\tanh\left[\frac{4.5(\lambda - 0.24)}{\lambda^{0.4 - 0.6\tanh p^{2.8}}}\right] \leqslant 50$$

表 9 和表 10 列出了不考虑超强和考虑超强时的各类截面延性需求以及相对应的正则化宽厚比限值 λ（Q235）。与前面的情况相同，考虑超强系数或增大 r 和 α 均可使限值 λ 得到放宽。

不考虑超强系数时抗震五类截面的 λ（Q235）限值　　　　表 9

α	r	类别	I-E	II-E	III-E	IV-E	V-E
0.01	1.1	$\mu_{\phi,req}$	15.90	13.93	12.39	9.99	8.19
		$P/P_y=0$	0.373	0.388	0.403	0.43	0.457
		$P/P_y=0.2$	0.374	0.389	0.404	0.431	0.458
		$P/P_y=0.4$	0.379	0.395	0.410	0.438	0.464
		$P/P_y=0.6$	0.393	0.410	0.425	0.454	0.481
		$P/P_y=0.8$	0.416	0.434	0.45	0.479	0.507
	1.5	$\mu_{\phi,req}$	15.23	13.24	11.71	9.33	7.58
		$P/P_y=0$	0.378	0.395	0.410	0.439	0.468
		$P/P_y=0.2$	0.379	0.396	0.411	0.44	0.469
		$P/P_y=0.4$	0.384	0.401	0.417	0.447	0.476
		$P/P_y=0.6$	0.398	0.416	0.432	0.462	0.493
		$P/P_y=0.8$	0.422	0.441	0.457	0.489	0.518
	2	$\mu_{\phi,req}$	15.24	12.93	11.19	8.59	6.74
		$P/P_y=0$	0.378	0.397	0.416	0.451	0.487
		$P/P_y=0.2$	0.379	0.398	0.417	0.452	0.487
		$P/P_y=0.4$	0.384	0.404	0.423	0.459	0.492
		$P/P_y=0.6$	0.398	0.419	0.439	0.475	0.509
		$P/P_y=0.8$	0.422	0.444	0.464	0.501	0.535
0.03	1.1	$\mu_{\phi,req}$	10.15	8.94	8	6.52	5.42
		$P/P_y=0$	0.428	0.445	0.460	0.491	0.523
		$P/P_y=0.2$	0.429	0.446	0.461	0.492	0.523
		$P/P_y=0.4$	0.435	0.452	0.467	0.498	0.529
		$P/P_y=0.6$	0.451	0.469	0.484	0.514	0.544
		$P/P_y=0.8$	0.477	0.495	0.510	0.540	0.568

续表

α	r	类别	Ⅰ-E	Ⅱ-E	Ⅲ-E	Ⅳ-E	Ⅴ-E
0.03	1.5	$\mu_{\phi,req}$	9.78	8.56	7.61	6.14	5.06
		$P/P_y=0$	0.433	0.451	0.468	0.501	0.535
		$P/P_y=0.2$	0.434	0.452	0.468	0.502	0.535
		$P/P_y=0.4$	0.440	0.459	0.475	0.508	0.541
		$P/P_y=0.6$	0.457	0.475	0.491	0.524	0.555
		$P/P_y=0.8$	0.483	0.501	0.517	0.549	0.579
	2	$\mu_{\phi,req}$	9.86	8.42	7.34	5.72	4.57
		$P/P_y=0$	0.432	0.453	0.473	0.513	0.555
		$P/P_y=0.2$	0.433	0.454	0.474	0.514	0.555
		$P/P_y=0.4$	0.439	0.461	0.481	0.519	0.560
		$P/P_y=0.6$	0.455	0.477	0.497	0.535	0.574
		$P/P_y=0.8$	0.481	0.503	0.523	0.560	0.595

考虑超强系数时抗震五类截面的 λ（Q235）限值　　表10

α	r	类别	Ⅰ-E	Ⅱ-E	Ⅲ-E	Ⅳ-E	Ⅴ-E
0.01	1.1	$\mu_{\phi,req}$	9.91	8.24	6.86	4.44	1
		$P/P_y=0$	0.431	0.456	0.483	0.561	0.800
		$P/P_y=0.2$	0.432	0.457	0.484	0.562	0.790
		$P/P_y=0.4$	0.438	0.463	0.491	0.566	0.780
		$P/P_y=0.6$	0.454	0.480	0.507	0.580	0.770
		$P/P_y=0.8$	0.480	0.506	0.533	0.601	0.760
	1.5	$\mu_{\phi,req}$	9.25	7.62	6.30	4.01	1
		$P/P_y=0$	0.44	0.468	0.497	0.584	0.800
		$P/P_y=0.2$	0.441	0.469	0.497	0.584	0.790
		$P/P_y=0.4$	0.448	0.475	0.503	0.587	0.780
		$P/P_y=0.6$	0.464	0.491	0.52	0.598	0.770
		$P/P_y=0.8$	0.490	0.517	0.545	0.617	0.760
	2	$\mu_{\phi,req}$	8.50	6.79	5.45	3.29	1
		$P/P_y=0$	0.452	0.485	0.522	0.639	0.800
		$P/P_y=0.2$	0.453	0.486	0.522	0.637	0.790
		$P/P_y=0.4$	0.459	0.492	0.528	0.636	0.780
		$P/P_y=0.6$	0.475	0.508	0.543	0.642	0.770
		$P/P_y=0.8$	0.502	0.534	0.568	0.654	0.760
0.03	1.1	$\mu_{\phi,req}$	6.47	5.45	4.60	3.11	1
		$P/P_y=0$	0.493	0.522	0.554	0.658	0.800
		$P/P_y=0.2$	0.493	0.522	0.554	0.655	0.790
		$P/P_y=0.4$	0.500	0.528	0.558	0.652	0.780
		$P/P_y=0.6$	0.516	0.543	0.572	0.655	0.770
		$P/P_y=0.8$	0.541	0.568	0.594	0.666	0.760

续表

α	r	类别	Ⅰ-E	Ⅱ-E	Ⅲ-E	Ⅳ-E	Ⅴ-E
0.03	1.5	$\mu_{\phi,\text{req}}$	6.09	5.09	4.27	2.86	1
		$P/P_y=0$	0.503	0.534	0.57	0.691	0.800
		$P/P_y=0.2$	0.503	0.534	0.57	0.687	0.790
		$P/P_y=0.4$	0.509	0.539	0.573	0.68	0.780
		$P/P_y=0.6$	0.525	0.554	0.586	0.678	0.770
		$P/P_y=0.8$	0.55	0.577	0.606	0.683	0.760
	2	$\mu_{\phi,\text{req}}$	5.67	4.60	3.77	2.42	1
		$P/P_y=0$	0.515	0.554	0.6	0.788	0.800
		$P/P_y=0.2$	0.516	0.554	0.599	0.777	0.790
		$P/P_y=0.4$	0.521	0.558	0.601	0.753	0.780
		$P/P_y=0.6$	0.537	0.572	0.611	0.733	0.770
		$P/P_y=0.8$	0.561	0.594	0.628	0.724	0.760

表 9 和表 10 中的正则化宽厚比限值 λ 不便于实际工程应用，需要将 λ 与板件宽厚比联系起来。文献 [15] 拟合的箱形截面屈曲系数 k 的表达式是 h/b, t_f/t_w, β 的函数。与工形截面的情况相同，给定应力状态参数 β，从表 9 和表 10 查出对应的宽厚比限值 λ，取任意的 h/b, t_f/t_w 得到 k 代入式 (11) 可算出 $\left(\dfrac{h}{t_w}\right)\varepsilon_k$，再由式 (10) 即可得到 $\dfrac{b}{t_f}\varepsilon_k$。

取 h/b 的变化范围为 $1\sim4$，t_f/t_w 的变化范围为 $1\sim2$，通过上述步骤可得到对应于给定 λ 的 $\dfrac{h}{t_w}\varepsilon_k - \dfrac{b}{t_f}\varepsilon_k$ 相关关系。图 8~图 12 分别给出了不考虑超强和考虑超强时的五组 $\dfrac{h}{t_w}\varepsilon_k - \dfrac{b}{t_f}\varepsilon_k$ 相关关系。

同工形截面的 $\dfrac{h}{t_w}\varepsilon_k - \dfrac{b}{t_f}\varepsilon_k$ 相关关系相类似，箱形截面的 $\dfrac{h}{t_w}\varepsilon_k - \dfrac{b}{t_f}\varepsilon_k$ 相关关系外凸也很明显。在受弯和压弯状态下，箱形截面的受压翼缘板处于均匀受压状态，所以当腹板宽厚比为 0 时，翼缘宽厚比限值也不受轴压比影响，其只随结构影响系数的增大而逐渐增大。当翼缘宽厚比为 0 时，结构影响系数和轴压比的变化均会影响到腹板宽厚比的取值。轴压比一定时，结构影响系数越大，腹板宽厚比限值越大。结构影响系数一定时，轴压比越大，腹板宽厚比限值越小。考虑超强系数之后，宽厚比限值比不考虑超强系数时放宽了很多。其他条件相同，增大 r 或 α 也可以使板件宽厚比限值得到放宽。由于同工形截面的相关关系很相似，经过数据拟合发现仍然可以用式 (13) 来近似计算箱形截面的板件宽厚比相关关系，对图 8 和图 9，公式中系数分别为式 (16) 和式 (17)。

不考虑超强系数 ($r=1.1$, $\alpha=0.01$):

$$\begin{cases} m = 5 + 5p \\ A = 20 + 20C \\ B = 65 + 55\tanh(C-0.25)^{1.1} - \\ \quad (38 + 86\tanh(C-0.25)^{1.9})\tanh(p^{1.1} + p^{0.75}) \end{cases} \quad (16)$$

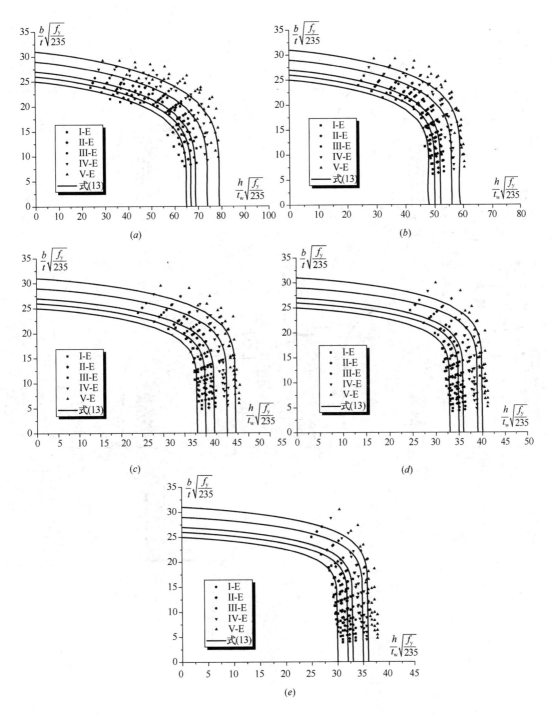

图 8 箱形截面 b/t_f—h/t_w 相关关系（不考虑超强，$r=1.1$，$\alpha=0.01$）
(a) $P/P_y=0$；(b) $P/P_y=0.2$；(c) $P/P_y=0.4$；(d) $P/P_y=0.6$；(e) $P/P_y=0.8$

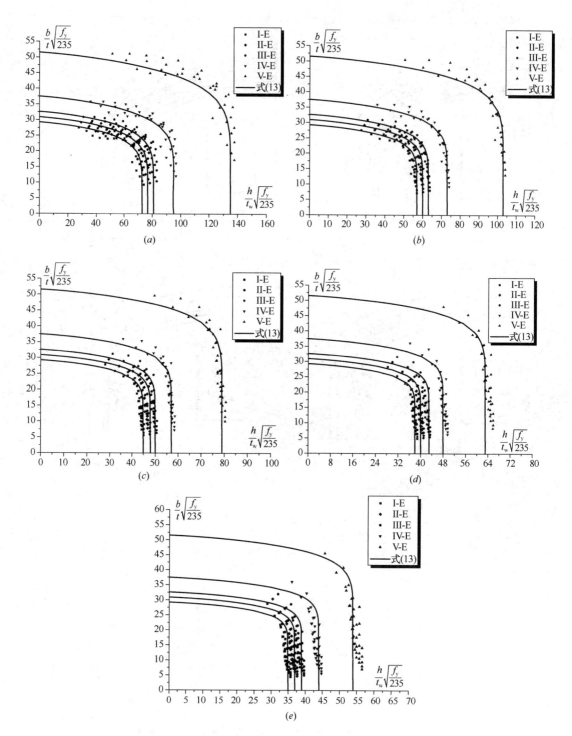

图9 箱形截面 b/t_f—h/t_w 相关关系（考虑超强，$r=1.1$，$\alpha=0.01$）
(a) $P/P_y=0$；(b) $P/P_y=0.2$；(c) $P/P_y=0.4$；(d) $P/P_y=0.6$；(e) $P/P_y=0.8$

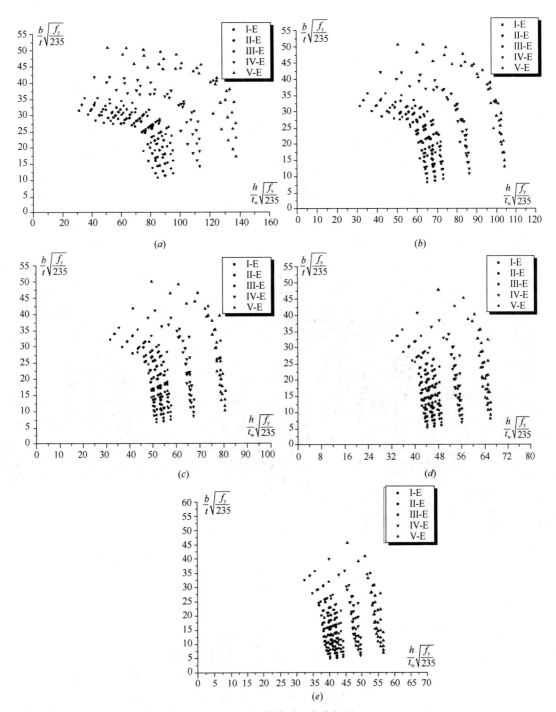

图 10 箱形截面 b/t_f—h/t_w 相关关系（考虑超强，$r=1.1$，$\alpha=0.03$）
(a) $P/P_y=0$；(b) $P/P_y=0.2$；(c) $P/P_y=0.4$；(d) $P/P_y=0.6$；(e) $P/P_y=0.8$

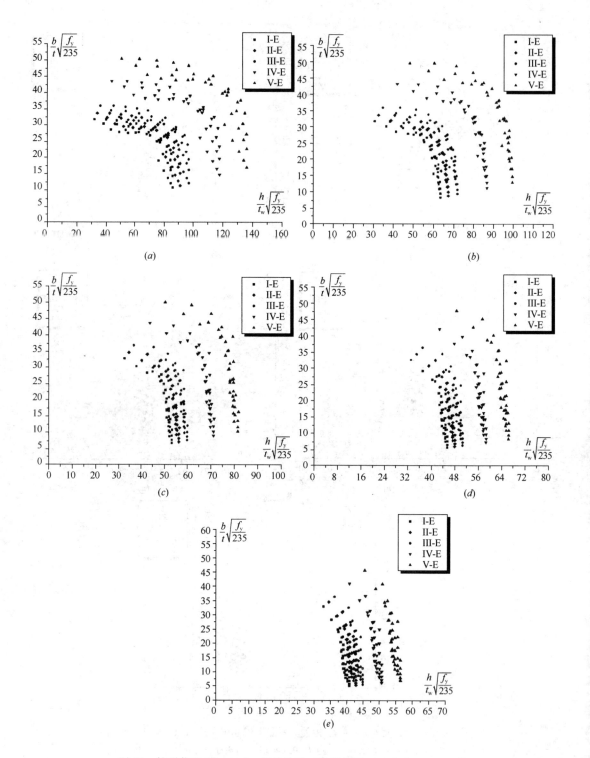

图 11 箱形截面 b/t_f—h/t_w 相关关系（考虑超强，$r=1.5$，$\alpha=0.03$）
(a) $P/P_y=0$; (b) $P/P_y=0.2$; (c) $P/P_y=0.4$; (d) $P/P_y=0.6$; (e) $P/P_y=0.8$

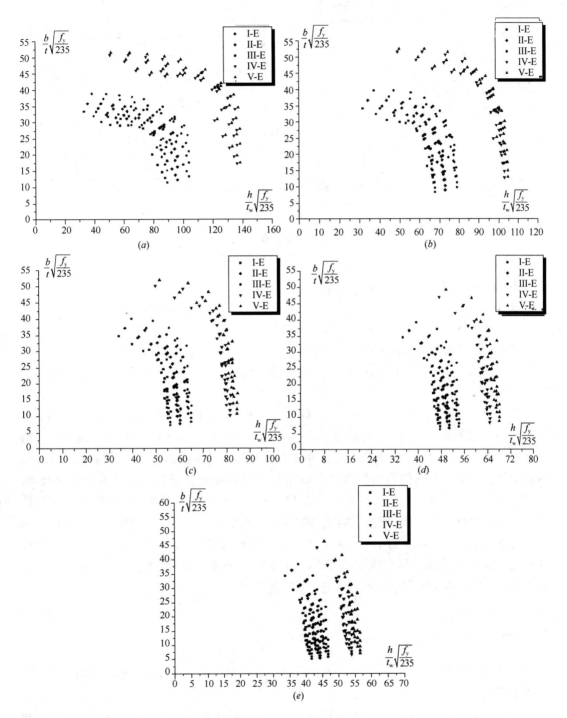

图 12 箱形截面 b/t_f—h/t_w 相关关系（考虑超强，$r=2$，$\alpha=0.03$）
(a) $P/P_y=0$；(b) $P/P_y=0.2$；(c) $P/P_y=0.4$；(d) $P/P_y=0.6$；(e) $P/P_y=0.8$

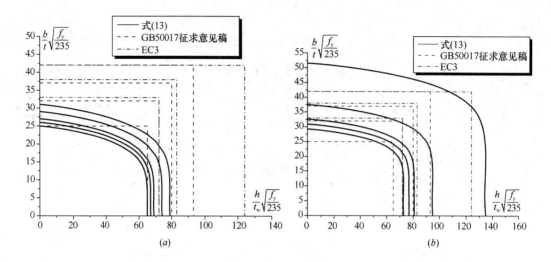

图 13 箱形梁截面分类比较
(a) 不考虑超强 ($r=1.1$, $\alpha=0.01$); (b) 考虑超强 ($r=1.1$, $\alpha=0.01$)

考虑超强系数 ($r=1.1$, $\alpha=0.01$):

$$\begin{cases} m = 7+5p \\ A = 21+33C+5100\tanh[(C-0.25)^5] \\ B = 53+80C+8500\tanh(C-0.25)^{4.5} - \\ \quad \{[31+40C+8500\tanh(C-0.25)^{4.5}]\tanh[(2.1-4500)\tanh(C-0.25)^8]p\} \end{cases}$$
(17)

图 13 分别给出了 EC3 和 GB 50017 征求意见稿(新 GB 50017)对箱形梁板件的四类和五类截面分类曲线,图上同时绘出了式(13)($r=1.1$, $\alpha=0.01$)的曲线。从宽厚比限值来看,新 GB 50017 的 A 类和 B 类截面相当于 EC3 的 I 类截面,C~E 类截面对应 II~IV 类截面,但 III 类截面的腹板宽厚比限值比 D 类截面要放宽很多。不考虑超强系数时,本文的 I-E~V-E 类截面宽厚比限值要比 EC3 和新 GB 50017 严格,大致相当于 EC3 的前二类截面和新 GB 50017 的前三类截面的水平,也即处在允许截面塑性开展的范围。考虑了超强系数之后,本文的 I-E~III-E 类截面分界和新 GB 50017 的 A~C 类截面分界比较接近,V-E 类截面分界大致相当于 EC3 的 III、IV 类截面分界。由图 10~图 12 可知,增大 r 或 α 值,本文的宽厚比限值可进一步得到放宽。

6. 结论

本文尝试提出了面向抗震设计的截面分类方法,根据地震力计算的结构影响系数大小对截面进行分类,主要结论如下:

(1) 从结构影响系数,区分考虑和不考虑结构的超强系数,得到结构的延性系数,利用文献[13]提供的公式反推出对截面的延性系数要求。计算发现考虑超强系数和增大后期刚度系数 α 或 r 值均可放宽对截面延性的需求。

(2) 根据文献[16]第 2~3 章拟合得到的受压四边简支板、三边简支一边自由板和受压弯荷载的四边简支板延性系数公式,按照各类截面延性需求得到了相应的宽厚比分界,并以结构影响系数和轴压比为变量,拟合了各类截面宽厚比限值的计算公式。由于没有考虑板件间的相互作用,按简支边界条件得到的各类板件宽厚比限值过于严格。

(3)根据文献[16]第4章得到的工字形截面延性系数公式,按照各类截面延性需求得到了相应的正则化宽厚比限值λ。利用文献[15]拟合的工形截面屈曲系数k和参数h/b,t_f/t_w,β的表达式,将各类截面的λ限值转化为翼缘宽厚比和腹板宽厚比的相关关系,并拟合了相关关系的计算式。

(4)同工字形截面相类似,根据文献[16]第5章得到的箱形截面延性系数公式,按照各类截面延性需求得到相应的正则化宽厚比限值λ。利用文献[15]拟合的箱形截面屈曲系数k和参数h/b,t_f/t_w,β的表达式,将各类截面的λ限值转化为翼缘宽厚比和腹板宽厚比的相关关系,并拟合了相关关系的计算式。

参考文献

[1] EN1993-1-1:2006. Eurocode 3:Design of steel structures Part 1-5:Plated structural elements[S]. Brussels:European Communities for Standardization,2006.

[2] EN1998-1-1:2004. Eurocode 8:Design of structures for earthquake resistance Part 1:general rules, seismic actions and rules for buildings.[S]. Brussels:European Communities for Standardization,2004.

[3] ANSI/AISC 341-2005:Seismic Provisions for Structural Steel Buildings[S]. Chicago,U. S. A:American Institute Steel Construction,2005.

[4] ANSI/AISC 360-2005:An American National Standard Specification for Structural Steel Building [S]. Chicago,U. S. A:American Institute Steel Construction,2005.

[5] 《建筑抗震设计规范》GB 50011—2010[S]. 北京:中国计划出版社,2010.

[6] 《钢结构设计规范》GB 50017—2003[S]. 北京:中国计划出版社,2003.

[7] 《钢结构设计规范(征求意见稿)》GB 50017—201X[S]. 北京:《钢结构设计规范》国家标准管理组,2012.

[8] AIJ2010 鋼構造限界狀態設計指針・同解說[M]. 日本建築學會,2010.

[9] C. Faella, F. M. Mazzolani, V. Piluso, and G. Rizzano. Local Buckling of Aluminum Members:Testing and Classification [J]. Journal of Structural Engineering,2000,126(3),353-360.

[10] Federico M. Mazzolani, Vincenzo Piluso, and Gianvittorio Rizzano. Local Buckling of Aluminum Alloy Angles under Uniform Compression [J]. Journal of Structural Engineering,2011,137(2),173-184.

[11] 童根树. 钢结构设计方法[M]. 北京:中国建筑工业出版社,2007.

[12] Kappos A J. Evaluation of behaviour factors on the basis of ductility and overstrength studies[J]. Engineering Structures,1999,21(9):823-835.

[13] 蔡志恒. 双周期标准化的弹塑性反应谱研究[D],浙江大学博士学位论文,2011.

[14] 陈绍蕃. 钢结构设计原理(第三版)[M]. 北京:科学出版社,2005.

[15] 彭国之. 薄壁截面的局部稳定性研究[D],浙江大学硕士学位论文,2012.

[16] 付波,板件延性系数和面向抗震设计的截面分类[D],浙江大学博士学位论文,2014

基于高性能计算的工程抗震与防灾：从单体到城市

陆新征[1]，卢啸[1,2]，许镇[1]，熊琛[1]，韩博[1]，叶列平[1]

（[1] 土木工程安全与耐久教育部重点试验室，清华大学土木工程系，北京，100084；
[2] 北京交通大学土木建筑工程学院，北京，100044）

摘　要：土木工程是人类生活和城市功能的主要载体。由于土木工程体量庞大、造价高昂、灾变行为复杂，数值模拟成为研究其灾变行为的重要手段。本文结合作者近年来在工程结构和城市区域震害数值模拟及防震减灾对策方面的科学研究和工程应用，介绍了重大工程结构的数值计算模型、基于GPU的高性能数值求解方法、基于精细化模型的城市区域建筑群震害模拟方法、基于物理引擎的城市建筑群地震倒塌模拟方法，以及超高层建筑、特大跨桥梁的倒塌模拟及其在工程设计领域的一些应用，为国内外相关领域的研究提供参考。

关键词：高性能计算；抗震减灾；灾变；重大工程；城市区域震害

Seismic Design and Disaster Mitigation Based on High Performance Computing: From Single Structure to Urban Area

LU Xinzheng[1], LU Xiao[1,2], XU Zhen[1], XIONG Chen[1], HAN Bo[1], YE Lieping[1]

([1] Key Laboratory of Civil Engineering Safety and Durability of China Education Ministry, Department of Civil Engineering, Tsinghua University, Beijing, P. R. China, 100084;
[2] School of Civil Engineering, Beijing Jiaotong University, Beijing, 100044, China)

Abstract: Civil engineering structures are the foundation of human life and city function. Because of the huge size, expensive cost and complicated disaster evolution behavior of civil engineering structures, numerical simulation is one of the most important methodologies to study their performance subjected to hazards. This work will review the research and application of the authors' group on the numerical simulation and disaster mitigation of engineering structures and urban areas, including the numerical models for mega engineering structures, the GPU-based high performance solution method, the seismic damage prediction of urban buildings based on high-fidelity model, the physical engine-driven collapse simulation for urban buildings, and the application of collapse simulation in real super tall buildings and super large-span bridges. These works will provide reference for the research in related fields.

Key words: High performance computing; seismic design and disaster mitigation; disaster evolution; mega structures; urban regional seismic damage

1. 引言

土木工程结构在其漫长的使用寿命中，会遇到各类极端灾害的作用。由于土木工程结构自身体量庞大、造价高昂、结构复杂，完全依赖物理试验手段研究其灾变过程难度很大。即使采用缩比模型，也依然存在尺寸效应、相似比设计等诸多困难。与此同时，土木工程也是城市功能的主要载体，当工程结构的灾变研究从单体发展到城市区域规模时，采用物理实验手段更是无能为力。因此，计算机数值模拟作为一种重要的科学研究手段，在工程结构抗震防灾领域得到日益广泛的应用。

工程结构计算机数值模拟的核心工作，是将工程结构的各种复杂行为（力学、热学等），建立相应的数学方程，而后通过计算机对这些方程进行求解，以预测相应的工程结构响应。工程结构数值模拟包括三个主要的构成部分：

(1) 工程结构的数值计算模型；
(2) 数学方程的求解算法；
(3) 完成工程结构数值模拟所需的计算机硬件平台。

其中高性能的硬件平台是基础，高效的求解算法是重要手段，而工程结构的数值计算模型是核心研究内容。

由于土木工程结构自身体量的庞大和行为的复杂，使得准确描述其复杂非线性受力行为的数值模型的计算量非常大。例如，虽然研究早已发现，实体单元是描述三维物体受力行为最为合适的单元类型，但是基于实体单元的建筑结构非线性计算，即便是利用当前最先进的计算平台，也只能完成一些简单的多、高层单体结构的非线性分析。例如，Yamashita 等[1]利用实体单元建立了一座高 129.7m 的规则高层钢结构的计算模型，单元总数超过了 1600 万，这样的计算量，即便是对于 E-Simulator 这样的超级计算机，也是一个非常有挑战性的工作。计算机有限的计算能力与工程结构数值模拟几乎无限的计算量需求，构成了工程结构数值模拟的一个主要矛盾，同时也成为工程结构数值模拟不断进步的一个重要原动力。

实际上，科学研究/工程应用的需求与试验能力（包括物理试验和数值模拟试验）的限制之间的矛盾，无论是对于物理试验还是对于数值模拟试验都同样存在。然而，计算机技术日新月异的发展，为突破数值模拟计算能力限制不断提供新的手段。与此相对的，物理试验能力的发展却遇到了巨大的困难。例如，以振动台试验为例，目前世界上最大的振动台为 1995 年落成的日本 E-Defense 振动台。此后近 20 年，振动台试验能

图 1.1 基于精细实体单元的结构非线性分析[1]

力都很难进一步得到提高。而世界上最快的超级计算机的头衔，几乎每年都在变化。甚至家家户户使用的桌面电脑的速度，已经可以和 15 年前世界上最快的超级计算机相媲美。日趋庞大而廉价的计算机数值计算能力，正沿着摩尔定律飞快发展，并不断为工程结构的数值模拟提供强有力的推动力。

世界各国的研究者，都从高性能计算技术的飞速发展中看到了工程结构数值模拟的美好前景。很多研究团队都做出了非常出色的研究成果。例如，日本东京大学在工程结构精细化数值模拟、城市区域震害数值模拟等方面做出了非常出色的研究成果，建立了 IES

等先进震害模拟系统。美国加州伯克利大学等领导开发的 OpenSees 计算程序结合 NEE-Shub 高性能计算平台，对促进工程结构数值模拟起到了重要的推动作用。其他例如 Carnegie Mellon 大学的 J. Bielak 教授等，在这一领域也都做出了非常出色的研究成果，极大的深化了人们对工程震害机理的理解及防灾减灾对策的研究。2009 年，中国住建部、美国国家科学基金会（NSF）和日本文部省组织中美日三国 50 余位地震工程和结构工程知名专家在广州举行"中美日建筑结构抗震减灾研讨会"，探讨未来结构工程和地震工程的发展方向。与会专家一致认定："基于超大规模计算的区域综合震害预测是未来地震工程领域具有重大价值的研究方向"。

本文将结合作者近年来在工程结构和城市区域震害数值模拟及防震减灾对策方面的科学研究和工程应用，介绍在计算模型、求解技术、硬件平台等方面的一些最新成果，为国内外相关领域的研究提供参考。

2. 重要工程地震灾变模拟

超高层建筑和超大跨桥梁等重大基础工程设施的建设，不仅反映了一个国家或地区的经济繁荣和社会进步，也是一个国家或地区建设科技发展水平的重要标志。随着我们城市化进程的不断推进，超高层建筑和超大跨桥梁等重大基础工程设施的建设势头强劲。然而，重大工程结构体系庞大，单元种类繁多，当面临强震作用时，其动力灾变效应无疑将更为复杂；同时，重大工程结构的损伤和破坏会给整个社会带来巨大的负面影响。因此，确保重大工程的地震安全是地震工程界的核心任务之一。而要保障重大工程的地震安全，首先要深入理解重大工程在强震作用下的损伤演化与倒塌灾变过程。故本节将通过建立高效的重大工程数值分析模型、提出倒塌灾变模拟方法，实现重大工程结构在强震下的损伤演化、灾变机理及倒塌机制的预测与评估，为重大工程抗震设计理论的进一步完善提供参考。

2.1 计算模型和算法

强大的软硬件计算平台是模拟工程结构地震灾变行为的基础，而合理的计算模型是高效准确的得到工程结构地震灾变行为的关键。目前工程结构灾变行为模拟的实现手段主要有两大类：

（1）基于成熟的商用有限元软件平台，利用其二次开发功能，开发针对性的构件或材料数值模型，以满足工程结构地震灾变模拟的需求。其优势在于，成熟的商用平台具有稳定且强大的求解能力以及方便的前后处理功能，研究人员可以专心于开发专用的构件和材料模型，进而提高研究工作的效率，特别是在解决一些工程问题时，利用成熟的商用有限元软件平台无疑是一个很好的选择。本课题组以通用有限元软件 MSC.Marc 为基础，在材料、构件及单元生死准则上开展了一系列的研究工作，并在多个重要工程结构分析中得到了成功的应用。采用商用有限元软件平台的一个主要不足之处在于，商用有限元软件是一个封闭的"黑盒子"，一些内部机理及深入的开发受到很多限制。

（2）利用具有针对性的开源有限元软件平台，并开发完善其特定功能，以满足工程结构地震灾变模拟的需求。典型的开源有限元软件平台即 OpenSees，其优势在于，其绝大部分源程序都是公开的，研究人员可以深入研究其程序执行机理并进行一些深层次的开发。当然，其缺点在于其稳定性、易用性与商用有限元软件相比还有一定的差距。本课题组在 OpenSees 开源有限元软件平台上，开发了相应的分层壳单元模型及混凝土本构模型，使其可以实现重大工程结构地震灾变的模拟。并进一步利用其可深层开发的特点，加入了基于 GPU 的高性能矩阵求解算法，从而使得其计算效率得到了极大的提高。

2.1.1 纤维梁和分层壳模型

在建筑结构的地震响应分析中，学者们提出了很多数值模型来模拟梁、柱和墙等构件。如利用集中塑性铰模拟梁、柱构件[2]；三垂杆[3]和多垂杆模型[4]模拟剪力墙构件等。为了能够准确把握结构和构件在地震作用下发生灾变前后整个过程的非线性响应，特别是柱中复杂的轴力-弯矩的耦合，剪力墙中平面内外弯曲、平面内弯矩、剪力耦合的非线性行为，本研究将采用纤维梁单元模拟梁柱构件，分层壳单元模拟剪力墙构件，两个模型的简要介绍如下。

纤维梁单元已经被广泛地运用到了弯曲破坏的梁柱构件地震响应模拟[5,6]。在纤维梁模型中，梁柱截面被离散成若干个纤维（如图 2.1a 所示），每个纤维可以赋予不同的单轴应力-应变关系，同一截面上的所有纤维满足平截面假定。纤维梁模型能够很好地模拟梁柱构件中轴力和弯矩的耦合作用，同时也能适用不同的截面形状。此外，通过赋予核心区混凝土纤维单轴约束本构，可以方便地考虑柱截面中箍筋的约束作用。而梁柱构件中的剪切破坏将以弹脆性破坏模型来近似考虑，即：当构件的剪力超过抗剪承载力时，构件发生剪切破坏，其抗剪强度和刚度设定为 0。

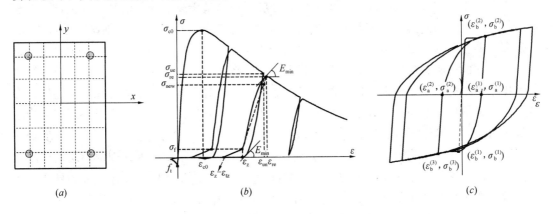

图 2.1 纤维梁截面及材料本构模型
(a) 截面纤维分布；(b) 混凝土本构曲线示意图；(c) 钢筋本构曲线示意图

纤维梁模型中，对于混凝土材料，本课题组在 Légeron & Paultre 模型[7]的基础上加以改进，考虑了约束效应、裂面效应、滞回效应等影响[8]，其单轴应力应变关系曲线如图 2.1b 所示，可以较好反映约束效应、软化行为，以及反复受力下的滞回和刚度退化的特性；对于钢筋材料，在 Esmaeily & Xiao 等提出的模型[9]基础上进行改进，可反映钢筋单调加载时的屈服、硬化和软化现象，并合理考虑了钢筋的 Bauschinger 效应[8]（如图 2.1c）。

基于以上的材料模型，通过 UBEAM 子程序的二次开发将纤维梁模型内嵌到了通用有限元软件MSC.Marc中，通常每个混凝土梁柱构件被分成6~8个单元来保证分析结果的精度。大量分析结果表明，该纤维梁模型的模拟结果与试验吻合良好，能较好把握梁柱构件的弹塑性行为[10-13]。为了进一步验证钢筋混凝土柱临近倒塌状态，强度和刚度迅速退化时纤维梁模型的准确性，对唐代远等[14]人进行的某一具有明显退化现象的钢筋混凝土压弯柱的滞回性能进行了模拟。构件的主要尺寸和配筋如图2.2所示，构件的纵向配筋率约为1.29%，轴压比为0.348。滞回曲线的比较如图2.2所示，表明该纤维梁模型能较好地考虑构件临近倒塌状态时的强度和刚度退化现象。

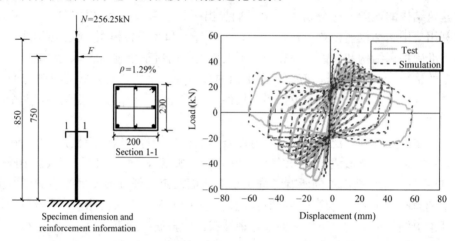

图2.2 纤维梁模型模拟结果与试验结果比较

本研究中采用分层壳模型模拟剪力墙。分层壳模型基于复合材料的基本理论，可以较好地模拟剪力墙面内、面外的弯曲耦合以及面内的弯曲剪力耦合。与纤维梁的原理类似，分层壳模型将剪力墙沿厚度方向分成若干不同厚度的层，每一层可以具有不同的材料属性，同一截面的所有层依然满足平截面假定。通过将剪力墙中的纵横向钢筋网弥散成等效钢筋层来模型剪力墙中的纵横向配筋，如图2.3所示。对于剪力墙中的边缘约束构件，将纵向配筋离散成一系列杆单元进行模拟，通过共节点的方式保证两者的变形协调。大量的计算分析表明该模型可以较好模拟剪力墙构件的受力特点[10,15]。同样，为了进一步验证分层壳模型在剪力墙临近倒塌阶段的适用性，对一具有明显承载力退化现象的矮墙滞回试验进行了模拟对比。该矮墙高1050mm，宽1100mm，厚度约为120mm，构件的具体尺寸和加载示意图如图2.4所示[16]。该构件滞回曲线的对比如图2.5所示，可见，分层壳模型仍能较好地反映剪力墙临界倒塌阶段的承载力迅速退化现象。

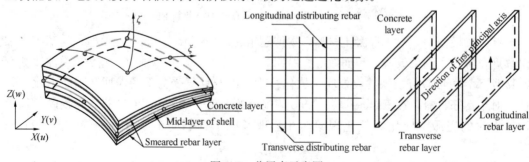

图2.3 分层壳示意图

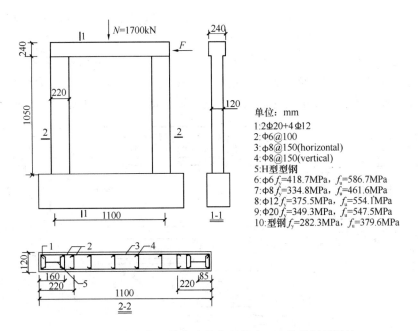

图 2.4 钢骨混凝土剪力墙体主要尺寸、配筋及材料属性

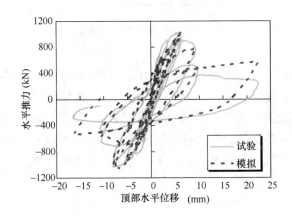

图 2.5 典型剪力墙滞回曲线模拟与试验比较

OpenSees 作为一款开源的有限元软件,在建筑结构抗震研究领域得到了广泛的运用。OpenSees 程序中的单元以纤维梁柱模型为主,该模型能较好地考虑双向弯矩和轴力的耦合作用,但不能方便准确的考虑剪力墙的受剪特性。本课题组基于分层壳理论,在 OpenSees 中开发了二维混凝土本构模型和剪力墙分层壳模型。本研究提出采用基于损伤力学和弥散裂缝模型来模拟分层壳单元中混凝土的二维受力行为。该模型具有形式简单、计算稳定性好等优点,适合作为基本模型集成在开源程序中供研究人员进一步发展和完善[17]。

混凝土的二维本构模型的基本方程可以表示为:

$$\sigma'_c = \begin{bmatrix} 1-D_1 & \\ & 1-D_2 \end{bmatrix} D_e \varepsilon'_c \tag{2-1}$$

式中,σ'_c、ε'_c 分别为主应力坐标系下混凝土的应力和应变,D_1、D_2 为混凝土在主应力坐标系下的损伤标量，受拉与受压损伤分开考虑。其中受压损伤参考 Løland 建议的受压损伤演化曲线[18]，受拉损伤参考 Mazars 建议的受拉损伤演化曲线[19]。对于剪力墙内的分布钢筋，本研究基于 OpenSees 中现有的钢筋模型开发了相应的多维钢筋模型。

结合各层厚度和各层所使用的多维钢筋、混凝土材料可生成 RC 剪力墙分层壳截面，进而将该截面赋予 4 节点壳单元完成分层壳剪力墙的定义，其流程如图 2.6 所示。

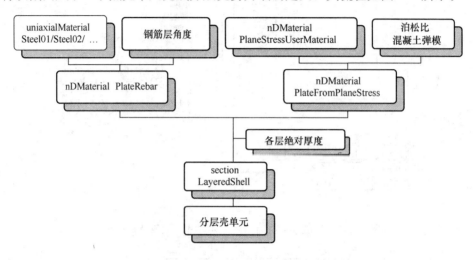

图 2.6 OpenSees 中分层壳模型框架

2.1.2 生死单元技术

建筑结构在特大地震作用下发生倒塌破坏的过程中，构件会逐渐破坏失效而退出工作。在有限元分析中，这一过程可以采用生死单元技术来模拟。生死单元技术是通过修改单元的刚度矩阵来实现的，当某一单元的响应超过某一设定的阈值时将会被"杀死"，通过调整该单元的刚度矩阵，赋予被"杀死"单元一个极小的刚度，避免直接归零带来的刚度矩阵奇异的问题。当该单元被杀死，其相应的应力和应变将释放为 0，在整个结构的质量和阻尼矩阵中该单元所对应的元素也将归为 0。在 MSC.Marc 通用有限元分析中，可以方便地采用 UACTIVE 子程序接口，通过编写相应的代码来实现这一过程。

由于纤维梁和分层壳模型都是基于材料本构层次的数值模型，可以建立基于材料层次的单元失效准则来监控整个地震灾变过程各构件的应力-应变行为。对于纤维梁单元，每个单元至少包含 36 个混凝土纤维和 4 个钢筋纤维，而每个纤维沿长度方向有 3 个高斯积分点；类似，对于分层壳模型，每个壳单元至少分成 10 层（具体层数根据剪力墙的实际配筋情况而定），每层具有 4 个高斯积分点。当每一个纤维或者层上任意积分点的应变值超过了预设的失效准则，该纤维和层的应力和刚度将被杀死，当某一单元所有的纤维或者层均被杀死，则认为该构件完全失效，将其从整个模型中杀死并移除。

2.1.3 基于 GPU 的高性能矩阵并行求解技术

随着有限元计算在土木工程领域的不断应用，越来越多的工程采用有限元计算进行抗震分析，其中不乏超高层建筑、大跨建筑等，从而导致对于有限元软件计算能力的需求也日渐提高。现如今，并行计算技术已经成为了有限元计算加速的主流，而日益发展的 GPU 技术也将并行计算加速提升到了一个新的高度。OpenSees 作为开源有限元软件，虽然拥有良好的可扩展性，但是计算效率一直以来都是它的瓶颈。通过分析 OpenSees 求解大型模型的时间分布，可以发现：影响 OpenSees 计算效率的主要因素是线性方程组求解模块。因此，本研究采用 GPU 加速技术，为 OpenSees 编写了线性方程组 GPU 加速求解器模块，将 OpenSees 的计算效率大幅提高，满足大型模型地震下动力时程分析的时效性要求。

在 OpenSees 中，求解线性方程组需要两个基本类：

（1）线性方程组类（LinearSOE）。该类用于存储、集成和读取线性方程组。对于一个特定的线性方程组 Ax＝b，LinearSOE 将 A、x、b 分别储存，并包含了相应的数据读取方法，使之可以被其他友元类调用。随着 A 的矩阵类型不同，LinearSOE 派生了许多子类，其基本类型分为一般矩阵、带状矩阵、分块矩阵和稀疏矩阵等等。

（2）线性方程组求解器类（LinearSOESolver 类）。该类用于特定的 LinearSOE 求解。根据求解方式的不同，分为直接法和迭代法两类。一种 LinearSOESolver 往往仅能求解一种 LinearSOE，然而一种 LinearSOE 可以采用多种 LinearSOESolver 进行求解。

在结构动力方程中，刚度阵、质量阵和阻尼阵往往都具有一定的稀疏矩阵的特征，因此，动力方程迭代求解所采用的线性方程组 $Ax=b$ 中，矩阵 A 也具有稀疏矩阵的特征，适合采用稀疏矩阵求解。同时，稀疏矩阵可以有效减小数据所需存储空间，适合超大规模模型的计算。在原版 OpenSees 中，针对稀疏矩阵，有 SparseSYM，SuperLU 等许多 LinearSOESolver 类可供选择。

图 2.7 为 GPU 加速 OpenSees 方程组求解流程。其求解模块与 OpenSees 原版主要有以下几个区别：

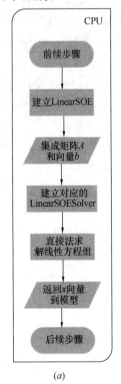
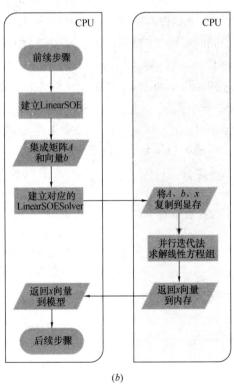

(a)　　　　　　　　　　　　(b)

图 2.7　GPU 加速 OpenSees 方程组求解流程

(a) OpenSees 原版；(b) GPU 加速 OpenSees

（1）在 CPU 线程中集成矩阵，拷贝到显存中，再并行计算。在 OpenSees 原版的 CPU 计算模块中，LinearSOE 集成方程组之后，LinearSOESolver 直接调用 LinearSOE 存储于内存中的数据进行计算。而在 GPU 加速求解模块下，矩阵集成将在 CPU 线程中完成，之后 LinearSOESolver 会先将方程组矩阵和向量拷贝至显存中，再调用显卡并行求解方程组。

（2）采用迭代法计算。原版 OpenSees 中，对于稀疏矩阵求解，多数采用直接求解法进行计算。直接求解法逻辑计算很多，并行度低，并不适合 GPU 计算。相对的，采用迭代法计算则可以很好地发挥 GPU 并行计算的能力，提升计算效率。因此，在 GPU 加速求解模块中采用迭代法进行稀疏矩阵线性方程组求解。

此外，本研究引入两个基于 GPU 加速的稀疏矩阵方程组求解库，用于 OpenSees 中稀疏矩阵方程组加速求解：

（1）CulaSparse。CulaSparse 是一个基于 GPU 加速的线性代数函数库[20]，用于迭代求解稀疏矩阵方程组。它由 EM Photonics 公司开发和维护，基于英伟达公司（NVIDIA）开发的 CUDA（Compute Unified Device Architecture，统一计算设备架构）编写。它支持许多常用的稀疏矩阵方程组求解器、预处理器以及存储格式等等，其加速效率最高可达到 10 倍以上；此外，CulaSparse 提供 C、C++和 Fortran 语言接口，支持 Linux、Windows、OSX 等操作系统。CulaSparse 为研究人员免费提供研究许可，目前最新的 CulaSparse 版本为 CulaSparseS5，支持 NVIDIA CUDA 5.0。

（2）CuSP。CuSP 是一个开源的 C++稀疏矩阵函数模板库[21]，可以进行多种稀疏矩阵运算。CuSP 同样基于 CUDA 编写，它有着方便、快捷的程序结构和函数调用接口，容易上手；同时，由于其以模板库的形式与 CUDA 合并，因此其编译较为方便，仅依赖 CUDA 库即可。目前最新的 CuSP 版本为 0.4.0，支持 NVIDIA CUDA 5.5。

通过调用以上两个 GPU 加速求解库，可以快速进行稀疏矩阵方程组的求解。本研究基于 OpenSees 架构，为 OpenSees 编写了 CulaSparseSolver 和 CuSPSolver 两个求解器（均为 SparseGenRowLinSolver 的派生类，对应于 SparseGenRowLinSOE）。图 2.8 为这两个求解器的 UML 类图。由于它们继承自 OpenSees 自身的求解器基类，因此调用简单方便，仅需要在 OpenSees 命令流中，替换掉 System 部分的命令流即可。

采用上述配置对 OpenSees 的两个 GPU 加速求解器性能进行了测试。测试平台参见表 2.1，二者价格基本相当。其中 Intel Core i7-3970X 是 2013 年市场上可以买到的最快的 CPU 型号。三种求解器的计算用时间比较见表 2.2，顶点时程曲线比较如图 2.9 所示。可以看出，采用两种 GPU 加速求解器进行弹塑性时程分析，可以使计算效率提升 10～15 倍；同时，计算结果几乎完全重合。由此可以证明，采用 GPU 对 OpenSees 弹塑性时程分析进行加速是可靠且高效的。

测试平台硬件配置及求解器　　　　　　　　　　表 2.1

平台	硬件	价格	求解器
CPU	Intel Core i7-3970X 3.5GHz	￥15,645（2013 年 6 月价格）	SparseSYM
GPU	Intel Core i7-4770X 3.4GHz & NVIDIA Geforce GTX Titan	￥15,005（2013 年 12 月价格）	CulaSparseSolver & CuSPSolver

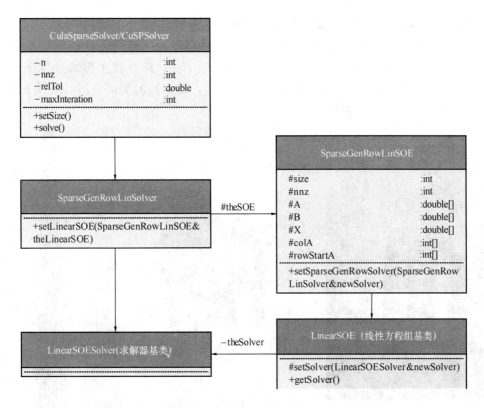

图 2.8 求解器 CulaSparseSolver/CuSPSolver UML 类图

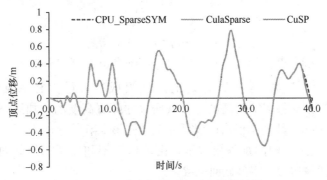

图 2.9 三种求解器顶点时程曲线结果

三种求解器计算用时　　　　　　　　表 2.2

平台	求解器	计算用时	相对于 CPU 加速比
CPU	SparseSYM	409h	—
GPU	CulaSparseSolver	38h	10.76
	CuSPSolver	27h 30m	14.87

2.2　超高层建筑地震灾变模拟

2.2.1　上海中心大厦简介

上海中心大厦位于上海陆家嘴，是一栋以甲级写字楼为主的综合性超高层建筑（图 2.10a），主体塔顶建筑高度 632m，结构屋顶高度约 580m，共 124 层，采用"巨柱一核心筒—伸臂桁架"的混合抗侧力体系（如图 2.10b），该体系的主要组成如下：

（1）核心筒主体为一个边长约 30m 的方形钢筋混凝土筒体，核心筒底部翼墙厚 1.2m，随高度增加核心筒墙厚逐渐减小，顶部厚 0.5m；核心筒内腹墙厚度由底部的 0.9m 逐渐减薄至顶部的 0.5m。由于建筑功能的要求，核心筒的角部在第五区以上被逐步切去，最终形成一个十字形核心筒[22]。

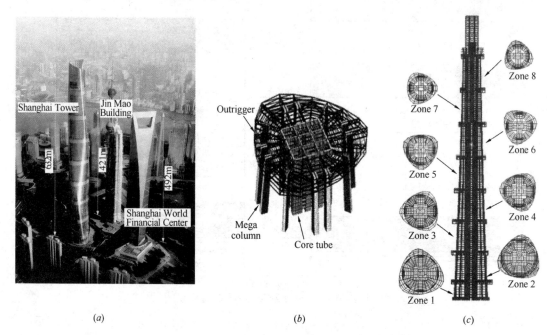

图 2.10　上海中心大厦模型图
(a) 上海中心位置示意图；(b) 抗侧力体系示意图；(c) 整体有限元模型

（2）巨柱系统由 12 根型钢—混凝土巨柱组成[22]，最大柱截面约为 5300mm×3700mm。8 根巨柱贯穿整个结构高度，柱截面尺寸随着高度的增加逐渐减小，最终减小为 2400mm×1900mm；其余 4 根角柱仅延伸至结构第 5 节段。

（3）桁架系统位于结构的加强层位置，由环形桁架和伸臂桁架共同组成，高度约为 9.9m，所有桁架杆件均为工字型截面钢梁。

对于上海中心大厦中的剪力墙，采用分层壳单元进行模拟，外框架、环向桁架、伸臂桁架以及塔顶均采用工字型钢梁，均采用纤维梁单元进行模拟，对于巨型钢骨混凝土柱，采用了分层壳和杆单元的组合式模型进行模拟，并以巨柱构件的精细实体有限元模型为基准，通过纯压、纯弯、单向压弯、双向压弯等多种工况的数值试验对组合式模型进行了验证和标定。更详细的模型信息及建模方法见文献 [23]，最终完成的上海中心大厦的有限元模型如图 2.10c 所示。

2.2.2　典型地震倒塌过程

上海中心大厦通过精心设计和反复论证，已经可以保证在设计罕遇地震下的安全性[22]，但是为了更好的研究此类结构的倒塌机理和倒塌过程，本研究中人为加大地震动强度，直至结构发生倒塌。虽然这样大的地震在上海地区基本不可能发生，但是由此得到的结构倒塌模式和破坏机理，对认识超高层结构体系受力特性有一定的科学意义。

上海中心大厦倒塌分析时,以科研中广泛采用的 1940 年的 El-Centro 地震动记录作为典型输入,采用 PGA 调幅方法将其 PGA 调至 19.6m/s²,约为实际地震动的 6.4 倍(实际地震动的 PGA=0.313g),且沿 x 方向单向输入。倒塌分析采用经典的 Rayleigh 阻尼,阻尼比取 5%,结构最终倒塌模式如图 2.11 所示。

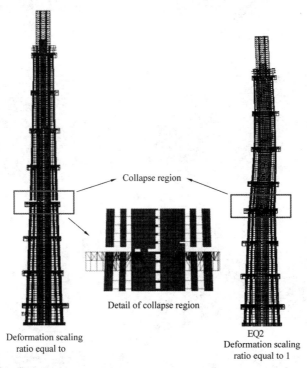

图 2.11 上海中心大厦典型倒塌模式

其倒塌过程如图 2.12 所示:当 $t=2.58$s 时(图 2.12a),核心筒的部分连梁发生破坏,且由于核心筒从第 6 节段到第 7 节段的转变中,洞口布局有改变,核心筒开洞从中间移至外缘,存在刚度的突变,极大削弱了核心筒的抗弯能力,因此,第 7 节段底部核心筒外缘的剪力墙被压溃;当 $t=3.90$s(图 2.12b)时,由于核心筒从第 4 节段到第 5 节段截面形状改变,四个角部被切除,因此第五区段底部剪力墙开始被压溃;当 $t=5.88$s(图 2.12c)时,由于第五区段部分剪力墙大量破坏,内力产生重分配,巨柱受到的侧向力和竖向力逐渐增大,巨柱开始压弯破坏;当 $t=6.18$s 时(图 2.12d),第五区段核心筒和巨柱完全破坏,结构倒塌开始发生。

El-Centro 地震动输入下结构顶点的位移时程曲线和楼层位移分布如图 2.13 所示。由于该高层结构的第一、二阶平动周期非常长,相应的地震力也比较小,因此,在地震作用下的破坏主要由高阶振型(水平向三阶振型)控制。故而结构临近倒塌时,结构变形模式呈高阶振型形状。破坏部位以上结构的重心并未有显著偏移(图 2.13b),导致结构倒塌以竖向倒塌模式为主,而非侧向倾覆倒塌。

显然,该超高层结构在 El-Centro 波作用下,第五区段以上部位破坏比较严重,主要集中在第五、六、七区段,而最终在第五区段发生折断,整个结构断成两截,可见第五区段是引起结构倒塌的潜在薄弱部位,在设计过程中应该给予更多关注。在上述倒塌过程

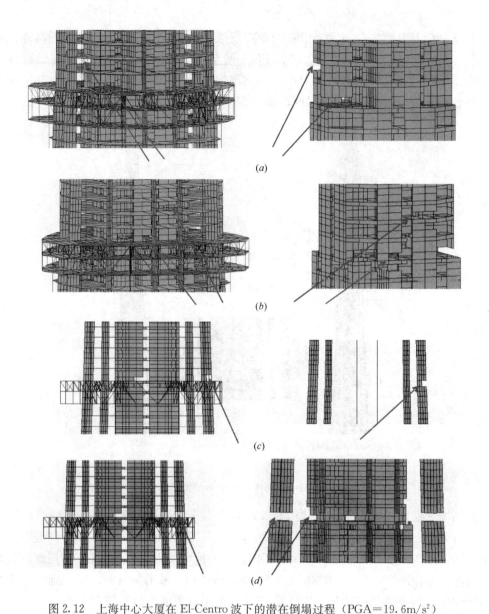

图 2.12 上海中心大厦在 El-Centro 波下的潜在倒塌过程 （PGA=19.6m/s²）
(a) t=2.58s, shear walls at the bottom of Zone 7 begin to fali; (b) t=3.90s, shear walls at the base of Zone 5 begin to fail; (c) t=5.88s, mega-columns at Zone 5 begin to fail; (d) t=6.18s, more than 50% shear walls and all mega-columns destroyed at Zone 5 the whole structure behins to collapse

中，连梁最先发生破坏，是结构的第一道防线，耗散部分地震能量，随后核心筒内的剪力墙和巨柱开始发生破坏，只有在同一区段的剪力墙和巨柱都发生破坏后，结构才可能发生倒塌。而通过常规弹塑性分析得到的初始屈服部位可能并非是引起结构倒塌的关键部位，如果设计不当，对初始屈服部位进行加强，可能使得耗能构件不能充分耗散能量，反而使结构变得不安全，这进一步说明了倒塌分析的重要性。

2.2.3 关键构件应力-应变曲线

结构倒塌过程是结构在地震作用下的宏观响应，为了进一步研究结构倒塌过程中关键构件力学行为，本节对关键构件的应力-应变行为进行了研究。在 El-Centro 地震动单向输

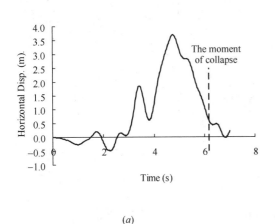 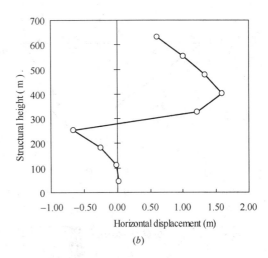

图 2.13　上海中心大厦在 El-Centro 波作用下灾变位移响应

(*a*) 顶点水平位移时间；(*b*) 倒塌时刻层间位移角分布

入下计算得到的结构典型破坏部位如图 2.11 所示。选取破坏部位的典型结构单元（图 2.14），得到相应的倒塌过程中混凝土和钢材应力-应变滞回关系如图 2.15 所示。

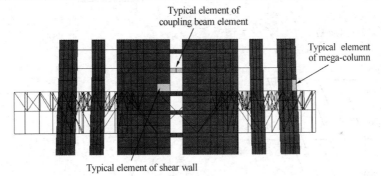

图 2.14　典型单元分布

对于巨柱单元，其在地震输入下受力过程的混凝土应力-应变滞回曲线如图 2.15*a* 所示。显然，巨柱在整个破坏过程中，主要承受压应力，仅在较少时刻出现了拉应力，最终混凝土被压溃，巨柱发生压弯破坏；对于连梁，其在地震输入下受力过程的剪力与剪应变的关系如图 2.15*b* 所示，最终发生剪切破坏；核心筒和伸臂桁架在地震输入下受力过程的应力-应变关系如图 2.15*c*～*f* 所示。

2.3　特大跨桥梁地震灾变模拟

2.3.1　琼州海峡大桥简介

琼州海峡位于广东雷州半岛和海南岛之间，东西长约 80km，南北宽约 30km，平均水深 44m，最大深度 114m。本研究的主要对象是琼州海峡大桥前期方案中的斜拉桥方案之一，该方案全桥总长 4304m，主跨 1500m，中塔标高 46m，边塔标高 386m，大桥两端各设置 4 个辅助墩，墩高 62m，航道净空为 1270m×81m，结构总体布置如图 2.16。

琼州海峡大桥主要包括桥塔(墩)、桥面板、拉索三个部分。调查表明，地震中斜拉桥的破坏主要集中在桥台、桥墩、支座、主梁、结构基础等部位，而主梁的破坏也主要由于

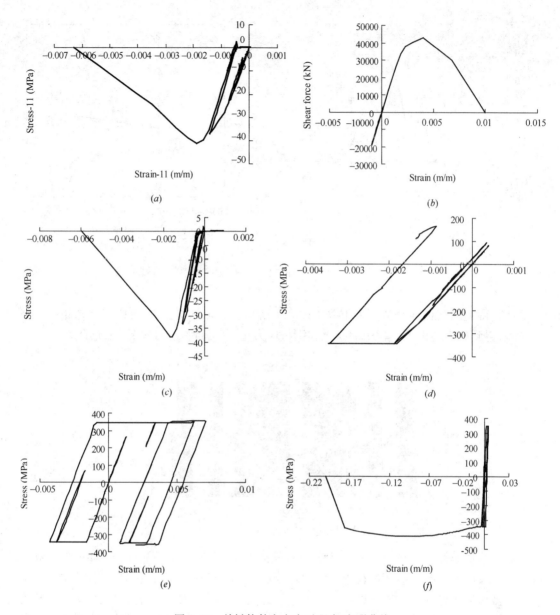

图 2.15 关键构件灾变全过程力-变形曲线

(a) 巨柱混凝土应力-应变曲线；(b) 连梁剪力-剪应变曲线；(c) 核心筒中混凝土应力-应变曲线；
(d) 核心筒中型钢约束构件应力-应变曲线；(e) 未破坏伸臂桁架应力-应变曲线；
(f) 破坏伸臂桁架应力-应变曲线

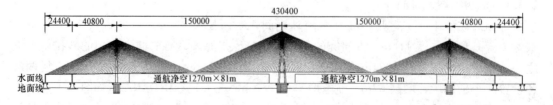

图 2.16 琼州海峡大桥斜拉桥设计方案立面布置图（单位：cm）

墩、台的滑移或者倾斜造成[24,25]，因此准确模拟桥塔（墩）在地震下的响应是大跨斜拉桥地震分析的关键。琼州海峡大桥设计方案采用空心桥塔，研究中可以采用分层壳单元模拟。为验证该单元的合理性，本研究对文献中的空心桥塔试验进行了分析[26]，构件沿高度方向划分 20 段，每一面沿长度方向划分 4 段，单元沿厚度方向划分 16 层，截面配筋及有限元模型如图 2.17 所示，计算结果如图 2.18 所示。结果表明，MSC.Marc 及 OpenSees 的分层壳单元计算结果均与试验均吻合良好，有较高的计算精度，可以采用该单元模拟琼州海峡大桥的桥塔构件在地震下的受力行为。

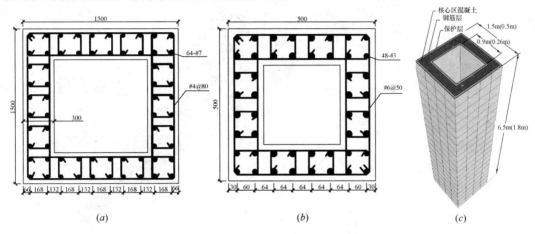

图 2.17　桥墩试验截面配筋及有限元模型示意图[26]

(a) 构件 PSI 截面配筋（6.5m）；(b) 构件 MSI 截面配筋（缩尺 1∶3 1.8m）；(c) 有限元模型示意图

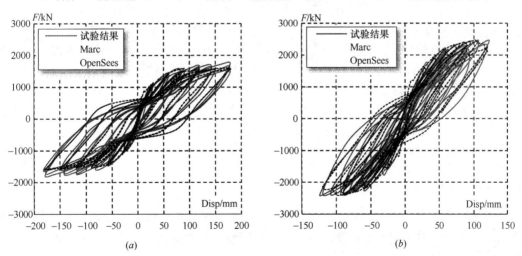

图 2.18　桥墩试验模拟结果（实验数据[26]）

(a) 试件 PSⅠ（足尺）顶点-力位移曲线；(b) 试件 MSⅠ（缩尺 1∶3）顶点-力位移曲线

由于 OpenSees 缺少友好的前后处理界面，在处理大型复杂结构时建模工作量极大，因此本研究利用已经发展相当成熟的有限元软件 MSC.Marc 对该斜拉桥进行建模，通过转换程序获得对应的 OpenSees 模型，并利用 MSC.Marc 的计算结果与 OpenSees 进行对比，检验 OpenSees 在计算大型桥梁时的可行性及可靠性。在此基础上，建立 MSC.Marc 与 OpenSees 中单元及连接的对应关系，编写转换程序完成 Marc 模型到 OpenSees 模型的转化（如图2.19所示）。最终的有限元模型如图2.19所示。

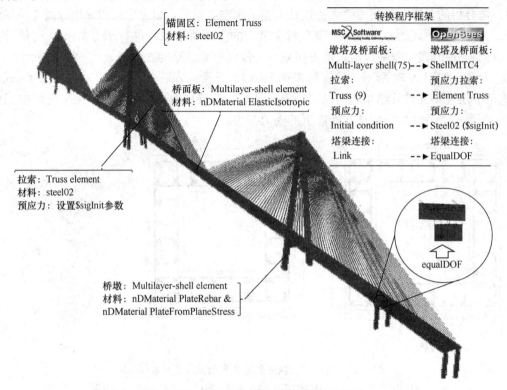

图2.19 琼州海峡大桥有限元模型示意图

2.3.2 典型地震倒塌模拟

琼州海峡大桥的模态分析表明，结构顺桥向和横桥向的一阶周期分别为15.7s和8.5s，因此，进行强震倒塌分析时，选择了科研中常用的长周期分量较高的 ChiChi 1999 及 Tōnankai 1944 进行地震波输入，计算结果如图2.20所示。

结果表明，在上述两条地震波作用下，琼州海峡大桥均在主塔的下横梁与桥墩交界处发生了混凝土的压坏，进而触发了结构的整体倒塌，从主塔顶点位移时程曲线也可以看出，塔顶沿着一个方向发生倾斜。

2.4 灾变模拟的工程价值

倒塌模拟作为一种工程结构地震响应分析的重要手段，不仅可以对重大工程在强震下的损伤演化和倒塌灾变全过程进行预测，判断结构的薄弱部位，还能有助于完善重大工程的抗震设计理论与设计方法。

2.4.1 超高层建筑地震动强度指标

地震动强度指标是基于性能设计方法中的重要组成部分，是连系地震动危险性与结构响应的桥梁。合理的地震动强度指标能够有效减小结构响应预测结果的离散性，众多学者已经对适用于常规结构的地震动强度指标展开了大量的研究[27-30]。但是这些研究和提出的地震动强度指标基本都是针对0～4s的常规结构而展开的，而对于大量涌现的超高层建筑，这些地震动强度指标的适用性待考证，进一步提出适用于超高层建筑地震动强度指标也是十分必要的。

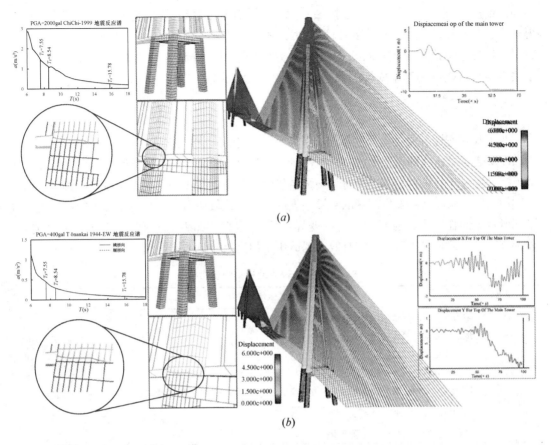

图 2.20 琼州海峡大桥地震倒塌模拟结果

(a) 横桥向 PGA=2000gal ChiChi 1999 波作用下琼州海峡大桥倒塌计算结果;
(b) 双向 1∶1 PGA=400gal Tōnankai 1944 波作用下琼州海峡大桥倒塌计算结果

与常规结构相比，超高层建筑结构的显著特点就是基本周期很长，在地震动作用下结构的响应中高阶振型参与明显，结构的破坏模式甚至受高阶振型控制[23]。因此，适用于超高层建筑结构的地震动强度指标必须能够反映结构响应中高阶振型的影响；其次，作为一个合理的地震动强度指标，还要形式简单，便于工程人员所接受，现行规范的抗震设计中都是基于加速度反应谱的，且地震危险性的衰减关系也是基于谱加速度的，故本研究中提出的地震动强度指标将以谱加速度 S_a 作为基本参数；最后，由于超高层建筑结构都采用基于性能设计，性能水准比较高，完全满足规范中所规定的大震需求，且具有较高的安全储备[22]。在大震下，结构的主要抗侧力构件基本保持弹性或者只有少量的屈服，整个结构进入非线性的程度并不深，因此，本研究在提出新的地震动强度指标时，初步不考虑结构进入非线性周期变长带来的影响。鉴于上述原因，本研究提出的适用于超高层建筑的地震动强度指标的一般形式如式（2-2）所示：

$$\overline{S}_a = \sqrt[n]{\prod_{i=1}^{n} S_a(T_i)} \tag{2-2}$$

式中，$S_a(T_i)$ 是第 i 阶周期对应的谱加速度；n 是所考虑的结构的平动周期数，与结构的基本周期相关的待定参数；$\overline{S}_a$ 表示新的地震动强度指标，其反映了前 n 阶周期对应的结构的谱加速度的几何平均值,对于一个给定的结构,该指标可以直接通过加速度反应谱得到,且与常用的 $S_a(T_1)$ 指标有较好的延续性。显然,当结构周期非常短时,即 $n=1$ 时,该指标退化成一阶周期对应的谱加速度 $S_a(T_1)$。卢啸等[31]通过大量参数研究,给出了关键参数 n 与结构基本周期的关系如式(2-3)所示。

$$n = \begin{cases} 1, & T_1 \leqslant 1\text{s} \\ 0.39T_1 + 1.15, & 1\text{s} < T_1 \leqslant 10\text{s} \end{cases} \quad (2\text{-}3)$$

为了验证所提出地震动强度指标的优越性,以上海中心大厦(即模型 A)和另一 580m 的超高层建筑结构(即模型 B)的三维有限元模型为基础,以 FEMA p695 中推荐的 22 组远场地震动记录作为基本输入,进行弹塑性增量时程分析,逐步增大地震动强度,直至结构发生倒塌,找到引起结构发生倒塌的临界地震动强度,比较该指标与其他较常用的地震动强度指标所对应的临界倒塌强度 $IM_{critical}$ 的变异性系数 COV,如表 2.3 所示。变异性系数越小,表示该指标对超高层建筑结构越有效,越适用于超高层建筑的抗震设计和分析。

本研究中所比较的地震动强度指标　　　　　　　　　　　表 2.3

指标名称	基本描述	单位
PGA	地震动记录的峰值加速度	g
PGV	地震动记录的峰值速度	cm/s
PGD	地震动记录的峰值位移	cm
$S_a(T_1)$	一阶周期对应的谱加速度	g
$IM_{1E\&2E}$	Luco & Cornell 提出的考虑近场冲击效应引起结构高阶振型参与的地震动强度指标[29] $IM_{1E\&2E}$ $=\sqrt{[PF_1^{[2]}S_d(T_1,\zeta_1)]^2 + [PF_2^{[2]}S_d(T_2,\zeta_2)]^2}$	
$\bar{S}_a$	本研究中所提出的地震动强度指标	g

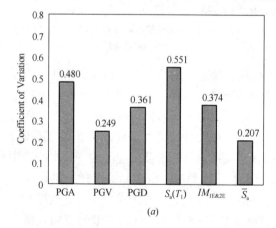 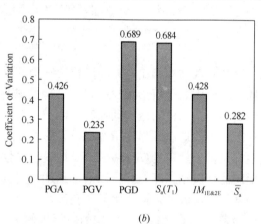

图 2.21　倒塌临界状态各地震动强度指标的变异系数
(a) 模型 A;(b) 模型 B

图 2.21 为两栋超高层建筑发生倒塌的临界地震动强度指标的变异系数。可见,在利用不同地震动强度指标表征模型 A 发生倒塌的临界值时,指标 $\bar{S}_a$ 对应的变异性系数最小,约为 0.207,仅为 $S_a(T_1)$ 对应变异性系数的 37.5%;PGV 指标所对应的变异性系数次之,约为 0.249,为 $S_a(T_1)$ 对应变异性系数的 45.1%;$S_a(T_1)$ 的变异性系数最大,说明仅考虑结构一阶模态是不能反映超高层结构地震响应的;$IM_{1E\&2E}$ 考虑了结构前两阶振动

模态，其变异性系数比 $S_a(T_1)$ 有所减小，但仍大于 $\overline{S}_a$ 的变异性系数，说明仅考虑前两阶模态依然不足以反映超高层建筑结构的地震响应。同理，图 2.21b 可以看出，在利用不同的地震动强度指标表征模型 B 发生倒塌的临界值时，指标 PGV 对应变异性系数最小，指标 $\overline{S}_a$ 的变异性系数次之，略大于 PGV 的变异性系数；PGD 和 $S_a(T_1)$ 的变异性系数最大。总的说来，PGV 和本研究提出的指标 $\overline{S}_a$ 对超高层结构的都具有较好的适用性。但是，根据叶列平等[32]的研究，对于短周期结构，PGV 指标精度较差，而 $S_a(T_1)$ 的精度最好，本研究提出的指标 $\overline{S}_a$ 在短周期情况下自动退化为 $S_a(T_1)$ 指标，因此比 PGV 指标有着更好的通用性。

另外，本研究提出的地震动强度指标是基于谱加速度的，能较好地利用规范的反应谱，也便于地震风险性分析中直接利用现有的地震动衰减关系，因此，本研究提出的地震动强度指标 $\overline{S}_a$ 也是适用于超高层建筑结构的抗震设计的地震动强度指标之一。

2.4.2 最小地震剪力系数对超高层建筑抗倒塌能力影响

目前在建筑结构抗震设计时，基本都是基于加速度反应谱确定地震作用力。长周期地震动记录中的速度或者位移特性对结构可能具有更大的破坏性，基于规范加速度反应谱的底部剪力法或振型分解法所确定的地震力，并不能很好地反映长周期地震动对结构的影响。对于超高层长周期结构，基于规范加速度反应谱确定的底部剪力会很小。出于对结构安全的考虑，国内外的抗震相关规范中都提出了最小地震剪力系数（或最小基底剪力）的要求。而近年来，我国大量的超高层建筑设计实例表明[33]对于基本周期很长的超高层建筑，按振型分解法计算的地震剪力系数并不满足我国现行规范的最小地震剪力系数的要求。显然最小地震剪力系数已经成为超高层建筑抗震设计中的关键难题。而通过倒塌分析方法，则可定量地研究最小地震剪力系数对超高层建筑抗倒塌能力的影响，进而完善超高层建筑的抗震设计理论。

为讨论抗震规范中最小地震剪力系数对超高层建筑结构抗震性能及抗特大震倒塌能力的影响，以我国 8 度区的某超高层为例，按照表 2.4 中两种不同设计地震剪力系数取值方法，设计了 2 个模型，其设计地震剪力的取值方法具体介绍如下：

两个方案设计地震剪力取值差异比较　　　　　表 2.4

楼层地震作用取值方法	模型 A	模型 B
$V_{d,i} = V_{EK,i} \geqslant \lambda G_i$	是	否
$V_{d,i} = \max(V_{EK,i}, G_i)$ 且 $V_b \geqslant 0.8\lambda G$	—	是

注：$V_{d,i}$ 表示第 i 层的用于结构设计的水平地震剪力标准值；$V_{EK,i}$ 表示用振型组合法计算得到的第 i 层的水平地震剪力；G_i 表示第 i 层以上结构的重力荷载代表值；G 表示结构总重力荷载代表值；V_b 表示结构的总基底剪力。

(1) 模型 A：振型分解法计算得到的楼层剪力完全满足我国现行《建筑抗震设计规范》GB 50011—2010 和《高层建筑混凝土结构技术规程》JGJ 3—2010 中最小地震剪力系数要求；

(2) 模型 B：振型分解法计算得到的楼层剪力不满足抗震规范最小地震剪力系数要求，但基底总剪力不小于最小基底剪力限值的 80%，对于不满足规范的楼层，按照最小地震剪力系数调整得到的楼层剪力进行内力组合和构件设计。

该超高层建筑位于我国抗震设防 8 度区 II 类场地第一分组，场地特征周期为 $T_g =$

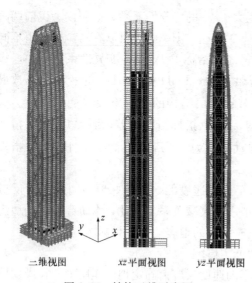

图 2.22 结构三维示意图

0.35s，主体结构总高度为 439m，共 97 层，顶部为 4 层观光层，地下 4 层，约 18m。该结构平面尺寸约为 36.1m×53.7m，外围为 16 根矩形钢管混凝土巨柱，内部为 21m×37m 的矩形核心筒，在结构的长边方向采用了"巨型钢管混凝土柱－核心筒－桁架"的混合抗侧力体系；为保证结构短边方向的刚度，在结构短边方向沿结构全高布置了"X"型巨型钢斜撑抵抗水平荷载。5 道腰桁架均匀分布在结构的高度方向，每道桁架体系高约 8.3m，结构整体示意图如图 2.22 所示。

模型 A 与模型 B 的结构布置完全一致。为使得模型 A 振型分解法计算的楼层剪力满足我国规范最小地震剪力系数的要求，必须尽可能增大结构刚度，减小结构周期。虽然采用了加大框架柱边长、增加剪力墙厚度、剪力墙内插钢板等一系列手段，但是仍无法满足周期要求。因此，从研究的角度出发，通过反复调整，最终模型 A 的核心筒采用 C100 钢纤维高强混凝土，使得结构周期可以降低到 6 s 以下。虽然超高强混凝土在实际工程中还较少采用，但已经有了较多的研究[34]。从表 2.5 的对比可以看出，毫无疑问，模型 A 的材料用量和建设难度要远大于模型 B。

模型 A 和模型 B 典型截面差异比较　　　　表 2.5

		模型 A	模型 B
剪力墙	x 向外墙厚	2900 内置两块 124mm 钢板	1900
	y 向外墙厚	1800 内置两块 76mm 钢板	1200
	混凝土等级	C100 钢纤维高强混凝土	C80
	典型边缘构件[注1]	650×1300×50×70	650×1300×50×70
巨柱	x 向中柱[注2]	3700×3150×45	3200×2650×45
	y 向中柱[注2]	2850×2850×40	2350×2350×40
	角柱[注2]	4950×3300×50	3950×2700×50

注：表中截面尺寸单位均为 mm；边缘构件均为 H 型钢；外围巨柱为矩形钢管柱，内填充 C80 混凝土。

模型 A 和模型 B 的基本动力特性比较如表 2.6 所示。显然，对于采用刚度调整方案的模型 A，其基本自振周期明显小于模型 B，结构整体抗侧刚度明显大于模型 B；由于模型 A 外围钢管混凝土柱和剪力墙的尺寸要明显大于模型 B，因此，模型 A 在增大结构抗侧刚度的同时，其重力荷载代表值也有了明显增加，是模型 B 重力荷载代表值的 1.24 倍，也就意味着模型 A 的材料用量要明显多于模型 B。显然，要使得振型分解法计算的地震剪力完全满足规范最小地震剪力系数的要求，必须要付出很多的代价，且这样的设计在实际工程中实现难度也非常大，经济性也较差。

模型 A 和模型 B 基本动力特性比较　　　　　　　　　　表 2.6

	T_1(s) x 向平动	T_2(s) y 向平动	T_3(s) 扭转	重力荷载代表值 （ton）	基底总设计剪力 （N）
模型 A	5.51	5.37	2.62	7.12×10^5	1.67×10^8
模型 B	6.91	6.60	3.66	5.76×10^5	1.13×10^8

为了进一步比较两种不同剪重比控制方案模型的抗倒塌能力差异，利用倒塌分析方法对上述两个模型（模型 A 和 B）进行了典型地震波下的倒塌模拟。从 FEAM p695 推荐的 22 组远场地震动记录中选择 Kocaeli_DZC180 为基本地震动输入，其 5% 阻尼比的弹性反应谱与规范反应谱的比较如图 2.23 所示。倒塌分析方法为，沿模型 x 轴单向输入逐步增大的地震力，直至结构发生倒塌，从而得到引起结构倒塌的临界地震动强度，并以此来分析结构的抗倒塌能力。

对于模型 A，当 Kocaeli_DZC180 地震动记录的 PGA 增大到 1.4g 时，结构发生倒塌；对于模型 B，当 Kocaeli_DZC180 地震动记录的 PGA 增大到 2.0g 时，结构发生倒塌。在建筑结构抗倒塌研究中，一般采用抗倒塌安全储备 CMR 来量化反映结构的抗倒塌能力[35]，可按式（2-4）计算。

$$CMR = \frac{PGA_{\text{Collapse}}}{PGA_{\text{MCE}}} \tag{2-4}$$

其中，PGA_{Collapse} 为引起结构倒塌的临界地震动强度，PGA_{MCE} 为结构大震所对应的地震动强度。从而可得到模型 A 和模型 B 在 Kocaeli_DZC180 地震动记录输入下抗倒塌安全储备系数（CMR）的比较如图 2.24 所示。模型 A 在 Kocaeli_DZC180 地震动记录输入的抗倒塌安全储备系数 CMR 约为 3.5（1.4g/0.4g），而模型 B 在 Kocaeli_DZC180 地震动记录输入的抗倒塌安全储备系数 CMR 约为 5.0（2.0g/0.4g）。可见，虽然模型 A 振型分解法计算的地震剪力完全满足抗震规范中最小地震剪力系数的要求，结构抗侧刚度、构件和结构整体的绝对承载能力均高于模型 B，但是，由于结构抗侧刚度大同时也使得结构在地震作用下的水平地震剪力大幅增大，地震力需求明显提高，导致模型 A 的抗震能力－需求比并没有明显提高。反而模型 B 在振型分解法计算的地震剪力不满足规范最小地震剪力系数的要求时，通过直接增大楼层地震剪力的方法进行结构设计，在不改变结构地震力

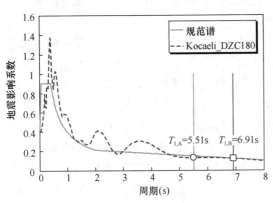

图 2.23　Kocaeli_DZC180 地震动记录 5% 阻尼比反应谱与规范谱比较

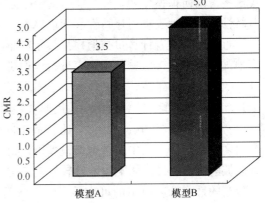

图 2.24　模型 A 和模型 B 的 CMR 比较

需求的同时增大了结构的抗震能力,使其能力-需求比明显提高,因此,在 Kocaeli_DZC180 地震动记录输入下,模型 B 体现出了更高的抗倒塌安全储备。故而在相关超高层建筑结构的抗震设计中,当最小地震剪力系数不能满足规范要求时,建议采用强度调整的办法,直接增大设计地震剪力的方法进行设计。

3. 城市区域的地震灾变模拟

近年来,全球范围内地震频发,从 2008 年到 2012 年,全球上 7 级以上大地震共发生了 111 起[36],其中有两起大地震发生在我国内陆地区,分别是 2008 年的汶川地震和 2010 年的玉树地震。不仅如此,中、小型地震也频频发生。从 2008 年到 2012 年,我国共发生了 5 级以上地震 206 起[36],小震更是不计其数。城市人口密度和财产密度都非常高,尤其是特大型城市,一旦发生强烈地震,必然会造成严重的人员伤亡和财产损失。中小地震引起的非结构构件破坏和室内设备损坏,也可能导致非常严重的损失。因此,科学、合理地进行城市区域震害预测是目前我们国家乃至世界范围内的一项迫切需求。

3.1 城市区域建筑震害模拟的精细化建模

3.1.1 精细化建模需求

在过去的 30 年,基于易损性矩阵的区域建筑震害预测方法得到广泛应用[37],该方法通过给出不同场地烈度下某类型建筑达到不同破坏状态的概率(此概率一般可通过历史震害经验调查得到)来对震害损失进行预测。但是易损性矩阵方法只能给出宏观和统计意义上的预测结果,无法反映具体建筑物的详细损伤情况,也难以反映某一具体地震事件的特点。因此,基于能力-需求分析的建筑震害预测方法也得到了大量研究[38-40]。例如,Hazus 99 的 Advanced Engineering Building Modules(AEBM)[39]通过计算建筑弹塑性能力谱和地震需求谱的交点得到性能点,进而根据性能点来判断建筑物的损失情况。该方法相对易损性矩阵方法有了很大的进步,它能够计算出每栋建筑的位移和最大绝对加速需求,因此可以更加准确地对结构构件和非结构构件进行损失预测,也可以考虑不同地震动之间的差异。然而,该方法仍存在以下两方面的问题:(1)将实际建筑物简化为单自由度体系,无法考虑高阶振型影响;(2)采用基于静力推覆的能力-需求分析,无法考虑地震动的一些特性(例如地震动时域特性和速度脉冲)的影响[41]。

近年来,随着计算机数值模拟技术的不断发展,精细化模型和非线性时程分析越来越多地被应用于城市区域结构震害分析,但带来的问题是计算量过大。为了解决计算精度和计算效率之间的矛盾,本文作者[42]提出了采用 GPU 作为计算平台,以剪切层模型(如图 3.1 和图 3.2)作为计算模型,采用非线性时程计算作为分析手段,进行高效城市区域震害分析的方法。剪切层模型有着比较适中的简化度以及相对较低的计算量[43],相对于单自由度体系,采用集中质量剪切模型可以得到结构每一层的破坏状态,结构高阶振型以及地震波的速度脉冲的影响也能够较好地考虑。国内外许多研究都证明了采用该模型进行多层建筑的地震响应模拟的可靠性[44-47]。然而,如何确定层模型中的层间滞回行为对保障预测结果的可靠性至关重要。对于城市区域,想要精确得到其中每座建筑每一层的侧移刚度和滞回参数是十分困难的,多数情况下仅能得到一些结构的宏观信息。国内外学者在此方面做了大量的研究[48-52]，但是该问题尚未得到圆满解决。

本研究以 Hazus 软件作为基础，提出了相应的城市区域建筑物震害预测剪切层模型，并基于 Hazus 软件中通过大量调查统计得到的建筑抗震性能参数，提出了剪切层模型参数的确定方法，为精细化的城市区域建筑震害预测提供参考。

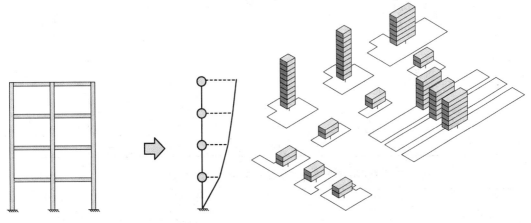

图 3.1　单体建筑集中质量剪切模型　　　图 3.2　群体建筑集中质量剪切模型

3.1.2　结构类型和设防等级

Hazus 将城市区域中一般建筑按照结构类型和高度分为 36 个建筑类型[53]。本研究考虑我国工程实际情况，选择了其中 19 种建筑类型（见表 3.1）。对于同一建筑类型，Hazus 考虑到不同建造年代设计规范的差异，其抗震性能参数也有所不同。本研究采用了 Lin 等[54]提出的 Hazus 抗震设计等级与我国建筑的对照关系（表 3.2），用来确定我国建筑与 Hazus 所提供参数的对应关系。

本研究使用的建筑类型[53]　　　　表 3.1

标签	描述	高度			
		包含范围		Hazus 中典型建筑	
		命名	层数	层数（N_0）	高度（m）
W1	木制框架		1-2	1	4.27
S1L	钢框架	低层	1-3	2	7.32
S1M		中层	4-7	5	18.3
S1H		高层	8+	13	47.58
S3	轻钢框架	所有		1	4.575
C1L	钢筋混凝土框架	低层	1-3	2	6.1
C1M		中层	4-7	5	15.25
C1H		高层	8+	12	36.6
C2L	钢筋混凝土剪力墙	低层	1-3	2	6.1
C2M		中层	4-7	5	15.25
C2H		高层	8+	12	36.6

续表

标签	描述	高度			
		包含范围		Hazus 中典型建筑	
		命名	层数	层数（N_0）	高度（m）
C3L	带砌体填充墙的钢筋混凝土框架	低层	1-3	2	6.1
C3M		中层	4-7	5	15.25
C3H		高层	8+	12	36.6
RM2L	带预制混凝土板的配筋砌体	低层	1-3	2	6.1
RM2M		中层	4-7	5	15.25
RM2H		高层	8+	12	36.6
URML	无筋砌体	低层	1-2	1	4.575
URMM		中层	3+	3	10.675

Hazus 抗震设计等级与我国建筑对应关系[54]　　　　　表 3.2

设防烈度（基本加速度）	建筑年代		
	1978 年以前	1979 年至 1989 年	1989 年之后
9 度(0.40g)	Pre-Code	Moderate-Code	High-Code
8 度(0.30g)	Pre-Code	Moderate-Code	Moderate-Code
8 度(0.20g)	Pre-Code	Low-Code	Moderate-Code
7 度(0.15g)	Pre-Code	Low-Code	Low-Code
7 度(0.10g)	Pre-Code	Pre-Code	Low-Code
6 度(0.05g)	Pre-Code	Pre-Code	Pre-Code

3.1.3 层间恢复力模型

Hazus 建议结构的骨架曲线可采用三线性模型（图 3.3），本研究层间滞回模型的骨架线也采用此模型。该模型可以分别描述结构层间的弹性阶段，屈服阶段和塑性阶段。整个骨架曲线需要 5 个参数来标定：k_0（初始侧向刚度），V_y（层间剪切屈服强度），η（强化系数），β（极限强度和屈服强度比）和 Δ_c（结构的层间极限位移）。

不同结构在地震下往复受力行为有较大差异。根据结构类型的不同，本研究采用以下三种不同的层间滞回模型来描述不同类型的结构。

(1) 修正的 Clough 模型

修正的 Clough 模型[55]（图 3.4 a）被广泛应用于模拟弯曲破坏的钢筋混凝土框架结构，因此本研究采用该模型模拟钢筋混凝土框架结构的层间滞回关系（也就是 C1 类型）。

(2) 理想弹塑性模型

理想弹塑性模型（图 3.4b）可以用于模拟钢框架结构。因此本研究采用该模型模拟钢框架结构的层

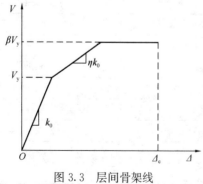

图 3.3　层间骨架线

间滞回关系(也就是 S1 和 S3 类型)。

（3）捏拢模型

对于剪切破坏起主导作用的结构，采用捏拢模型(图 3.4c)进行模拟是一个可行的方法。本研究采用 Steelman 和 Hajjar[56] 提出的捏拢模型。在该模型中，卸载刚度等于初始

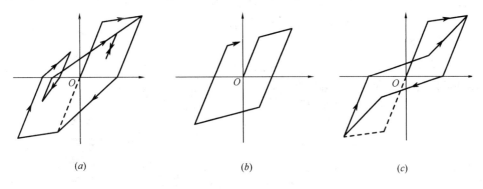

图 3.4　三种滞回模型
(a) 修正的 Clough 模型；(b) 理想弹塑性模型；(c) 捏拢模型

加载刚度，加载捏拢的转折点将出现在固定的转折线上，如图 3.5 所示。因此，本模型仅需 1 个描述结构退化程度的参数即可确定其滞回行为，如式所示：

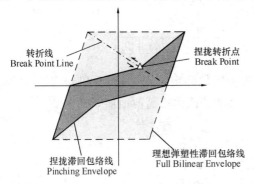

图 3.5　捏拢模型滞回行为

$$\tau = \frac{A_p}{A_b} \tag{3-1}$$

其中 A_p 和 A_b 分别是捏拢滞回包络面积和理想弹塑性滞回包络面积。选用该捏拢模型是因为其参数 τ 可以通过 Hazus 技术手册[53]提供的退化系数 κ 比较方便的确定[56]。如果采用其他多参数的捏拢模型(例如 Ibarra 捏拢模型[57])，参数标定的工作将非常复杂，不适用于区域震害预测的实现。在本研究中，除 C1、S1、S3 之外的所有建筑类型均采用此模型。

3.1.4　一般建筑剪切层模型参数确定

在 Hazus 中，在每一个抗震等级下，每一种结构类型都有一个典型建筑，可通过该典型建筑的已知的参数来标定该类型结构的抗震性能。这些参数包括：①结构初始弹性一阶周期；②结构性能曲线；③结构振型质量系数；④结构高度、楼层数；⑤结构材料弹性阻尼；⑥结构退化系数；⑦结构各个破坏状态所对应的层间位移角限值。

在本研究中，对于区域中的某一个目标建筑，仅需知道其结构类型、层高、层数、建筑年代、面积，就能通过 Hazus 已知的参数来确定其集中质量剪切模型的参数，具体方法如下：

（1）结构周期

对于区域中的一般多层建筑，其地震响应主要受前两阶振型控制[58]。对于城市区域中的一般建筑，前两阶周期通过式进行计算[59]：

$$T_1 = \frac{N}{N_0} \cdot T_0$$
$$T_2 = \frac{1}{3} T_1 \tag{3-2}$$

其中 T_1 和 T_2 分别是目标建筑的一阶和二阶振动周期，N 是目标建筑的层数。N_0 和 T_0 分别是 Hazus 给出的典型建筑的层数和一阶周期。

(2) 结构骨架线参数

对于一般多层建筑，其层质量和层初始刚度可认为是各层基本相同的[39]。因此，剪切层模型的刚度矩阵和质量矩阵可写为：

$$[K] = k_0 \begin{bmatrix} 2 & -1 & & & \\ -1 & 2 & -1 & & \\ & -1 & \ddots & \ddots & \\ & & \ddots & 2 & -1 \\ & & & -1 & 1 \end{bmatrix} = k_0 [A] \tag{3-3}$$

$$[M] = m \begin{bmatrix} 1 & & & & \\ & 1 & & & \\ & & 1 & & \\ & & & \ddots & \\ & & & & 1 \end{bmatrix} = m[I] \tag{3-4}$$

其一阶圆频率可由式(3-5)计算[58]：

$$\omega_1 = \sqrt{\frac{[\varPhi_1]^T [K][\varPhi_1]}{[\varPhi_1]^T [M][\varPhi_1]}} = \sqrt{\frac{k_0}{m}} \sqrt{\frac{[\varPhi_1]^T [A][\varPhi_1]}{[\varPhi_1]^T [I][\varPhi_1]}} \tag{3-5}$$

其中 $[\varPhi_1]$ 是结构的一阶振型向量。如果结构的楼层数确定，则无论 k_0 和 m 如何取值，$[\varPhi_1]$ 的各自由度比例是确定的，且其值可以用广义特征值法很简便地加以确定[58]。

于是结构的层间初始刚度可由式(3-6)计算

$$\begin{aligned} k_0 &= m\omega_1^2 \left(\frac{[\varPhi_1]^T [A][\varPhi_1]}{[\varPhi_1]^T [I][\varPhi_1]} \right) \\ &= \frac{4\pi^2 m}{T_1} \left(\frac{[\varPhi_1]^T [A][\varPhi_1]}{[\varPhi_1]^T [I][\varPhi_1]} \right) \\ &= \lambda \frac{4\pi^2 m}{T_1} \end{aligned} \tag{3-6}$$

$$\lambda = \left(\frac{[\varPhi_1]^T [A][\varPhi_1]}{[\varPhi_1]^T [I][\varPhi_1]} \right) \tag{3-7}$$

λ 取值与层数的关系见表 3.3(为计算方便，表中给出的数值为 $4\pi^2 \lambda$ 的数值)。

λ 取值与层数之间的关系(1 至 12 层)　　　表 3.3

楼层数	$4\pi^2 \lambda$	楼层数	$4\pi^2 \lambda$
1	39.5 ($4\pi^2$)	7	903.3
2	103.4	8	1159.3
3	199.3	9	1447.3
4	327.3	10	1767.3
5	487.3	11	2119.3
6	679.3	12	2503.3

于是，目标建筑第 i 层的层间骨架线参数（图 3.3）可由式（3-8）确定：

$$k_{0,i} = \lambda \cdot \frac{4\pi^2 m}{T_1^2} \quad (3\text{-}8)(a)$$

$$V_{y,i} = SA_y \cdot \alpha_1 \cdot m \cdot g \cdot N \cdot \Gamma_i \quad (3\text{-}8)(b)$$

$$\eta_i = \frac{SA_u - SA_y}{SD_u - SD_y} \cdot \frac{SD_y}{SA_y} \quad (3\text{-}8)(c)$$

$$\beta_i = \frac{SA_u}{SA_y} \quad (3\text{-}8)(d)$$

其中 m 是结构每层的质量，可以根据结构层面积和建筑用途确定[60]；g 是重力加速度；δ_{C0} 是 Hazus 建议的结构完全破坏所对应的层间位移角限值[53]；h 是结构层高；(SD_y, SA_y)，(SD_u, SA_u) 是 Hazus 中给出的典型结构性能曲线的屈服点和极限点[53]；α_1 是 Hazus 给出的典型建筑的振型质量系数[57]；Γ_i 是结构第 i 层的设计抗剪强度 $V_{y,i}$ 与基底设计抗剪强度 $V_{y,1}$ 的比值：

$$\Gamma_i = \frac{V_{y,i}}{V_{y,1}} \quad (3\text{-}9)$$

在我国规范中，设计地震力随结构高度基本呈倒三角形[61]。在本研究中，Γ_i 由式（3-10）确定：

$$\Gamma_i = \frac{\sum_{j=i}^{N} W_j H_j}{\sum_{k=1}^{N} W_k H_k} = 1 - \frac{i(i-1)}{(N+1)N} \quad (3\text{-}10)$$

其中 W_j，W_k 是结构第 j，k 层的重量，H_j，H_k 则是结构第 j，k 层所在平面距离地面的高程。

（3）结构滞回模型参数

对于修正的 Clough 模型和理想弹塑性模型，其行为仅需要骨架线就可以确定。对于本研究给出的捏拢模型，则需要一个参数 τ 来确定其行为。参数 τ 的取值可以参照 Steelman 和 Hajjar[56] 提出的方法，取所对应结构类型在 Hazus 中的退化参数 κ_m 加以确定[53]。

（4）结构阻尼

本研究中，结构阻尼采用瑞雷阻尼进行计算[58]，其阻尼比确定参照表 3.4。

结构阻尼比取值[53] 表 3.4

结构类型	$\zeta = \zeta_1 = \zeta_2$
钢结构（S1，S3）	0.05
钢筋混凝土结构（C1，C2，C3）	0.07
砌体结构（RM2，URM）	0.10
木结构（W1）	0.10

（5）结构破坏状态限值

对于城市区域中一般建筑，其结构破坏状态采用 Hazus 定义的 4 种破坏状态：轻微破坏，中等破坏，严重破坏，完全破坏。而确定结构破坏状态的指标则采用位移角进行确定，其对应的取值可在 Hazus 手册[53] 表 5.9 内查到。

3.1.5 特殊建筑剪切层模型的参数确定

对于一些特殊建筑（如新型结构类型，混合结构，不规则结构），其抗震性能与一般多层建筑差别很大，Hazus 软件对此类建筑也没有相应的规定。对于这类建筑，可采用更加精细的模型（如纤维梁模型或分层壳模型[11, 62-65]）来确定其参数，如图 3.6 所示。

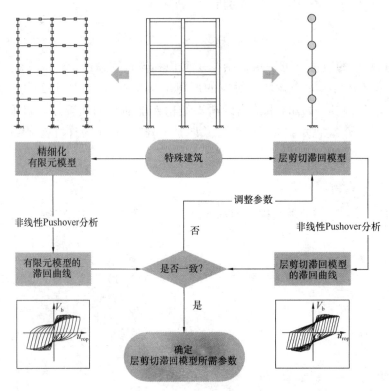

图 3.6 特殊建筑剪切层模型的参数确定流程

3.1.6 震害分析模型的数据获取

本研究中仅需获取结构类型、层高、层数、建筑年代、面积等 5 个参数，就能通过 Hazus 已知的参数来确定其建筑震害分析模型。然而，城市建筑数量庞大，每一栋建筑都需要采集相应的参数，这就导致城市区域建筑震害分析工作需要依赖详尽的城市基础数据库。当城市基础数据库的信息详尽程度难以满足震害分析需求时，如何高效、快速、准确地获取城市建筑的基础资料成了一个重要问题。

图 3.7 Google Earth 城市三维数字模型

图 3.8 "都市圈"的城市三维数字模型

近年来 Google Earth 的城市区域 3D 建筑模型在全球各大城市依次上线，如图 3.7。与此同时，国内的一些商业公司也逐渐完成了国内多个城市的区域 3D 数字模型，如图 3.8。城市区域建筑的 3D 数字虚拟化正在成为一种趋势。城市区域 3D 建筑模型提供了精细的城市建筑尺寸信息。充分利用该数据，为城市区域建筑震害的模拟提供了一种新的数据获取思路。

为此，本研究提出了一种基于 Google Earth 城市区域 3D 建筑模型的建筑参数获取方法。该方法可以通过 Google 的 Dae 三维数字模型格式，获得分析区域内每一栋建筑的三维模型参数，如层高、楼面形状等，进而可以获得层高、层数、楼面面积等震害分析所需数据。

（1）方法介绍

在 Google Earth 城市区域 3D 建筑模型进行城市区域 3D 建筑几何参数的抓取，面临如下问题：

1）城市三维数字模型本质上是由多边形或者三角形组成的。Dae 模型文件中仅包含所有多边形的端点坐标信息以及连接关系信息。事先通常不知道某栋建筑是由哪些多边形构成的，如图 3.9 所示。

2）城市三维数字模型中除了建筑模型，通常还包括地表，道路，植被等模型。因此需要剔除掉其他的模型，单独识别出构成建筑的多边形，如图 3.9 所示。

3）为了实现建筑的地震响应模拟，在获得到了组成某栋建筑的所有多边形之后，需要根据这些多边形数据，抽取出结构外形的参数。包括结构高度，结构各层的平面布置情况。

图 3.9　城市区域 3D 建筑模型示意

面对以上问题，本研究采用了基于城市切片的建筑参数抽取方法，具体实现步骤如下所示：

1）对整个城市区域 3D 模型中的所有多边形进行切片。对于每个多边形，得到对应该高程的一条直线。

2）提取出能围成闭合多边形的直线，将该多边形作为某栋建筑的层外形多边形，如图 3.10 所示。重复步骤 2，得到各个高程的建筑层外形多边形，如图 3.11 所示。

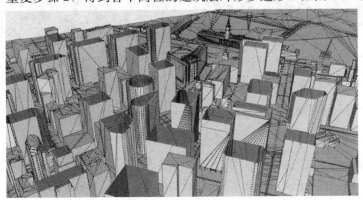

图 3.10　对模型的所有多边形进行切片

3) 判断各个多边形的竖向重叠关系，对于竖向重叠的多边形认为这些多边形组成一栋建筑，这样就获取到了区域中建筑的各层外形数据。

4) 对于已经识别的建筑，计算其多边形的高程差，从而估算得到建筑的高度。

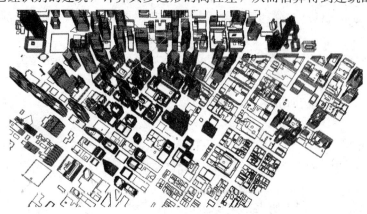

图 3.11　得到各个高程的建筑层外形多边形

(2) 实例验证

通过上述的步骤，就能基本获得区域中建筑的各层外形参数以及各建筑的高度参数。为了校验本方法的可用性，首先可以对单栋建筑进行识别验证，如图 3.12 所示。图中半透明的模型为原建筑 3D 模型。橙色的部分为识别得到的各层建筑外形多边形。

此外对于整个区域，为了校验识别的效果，也可以与原模型进行对比，如图 3.13 所示。图中黄色部分为识别得到的建筑层外形多边形，可以看出虽然原模型中包含地形等无关的多边形数据。但是该方法依然能较好地识别出每一栋建筑。

通过识别出的建筑以及楼层多边形，可以快速计算楼层高度、楼层数量、楼层面积等建筑外形参数，为城市震害分析提供了重要的数据支持。

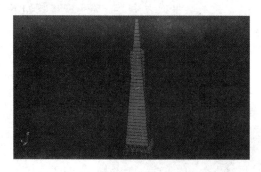
图 3.12　单栋建筑层外形多边形识别效果

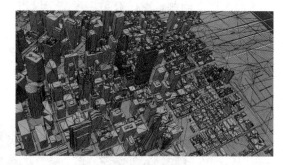
图 3.13　区域建筑层外形多边形识别效果

3.2　城市区域震害模拟的高性能计算

3.2.1　基于 GPU 粗粒度并行的建筑震害计算

虽然基于精细化模型的非线性时程分析在单体建筑中已经得到了广泛的应用[66]，但是城市区域拥有海量的建筑，该方法所需的计算量巨大。2008 年 Hori 和 Ichimura[67]基于精细化结构模型和非线性时程分析开发了 Integrated Earthquake Simulation（IES）平台，实现了整个东京的震害预测。但是该平台必须依赖超级计算机，这意味着高昂的运营费用和维护成本。因此，为更好地开展城市区域建筑震害的精细化模拟，迫切需要一种低成本、高性能的计算平台。

最近几年，图形处理单元（GPU）技术飞速发展并在通用计算领域得到广泛应用。虽然 GPU 单核的计算能力相对 CPU 较弱，但是 GPU 的处理核心数远多于相同价格的 CPU，因此相同价格的 GPU 相对 CPU 具有更高的计算性能[68]。随着 NVIDIA 公司发布了 CUDA（Compute Unified Device Architecture），GPU 通用计算的编程难度被大大降低。目前 GPU 计算在生物、电磁场、地理等领域得到了大量的应用[69-71]。

最能发挥 GPU 性能的是细粒度的并行计算[72]，即将每个子任务划分成许多更细小的操作步，然后 GPU 在操作步层面对任务进行并行计算。这种并行方式在神经网络以及有限元领域得到了很广泛的应用[73,74]。但是这种并行计算方式需要对 GPU 程序进行细致的调试以实现对大量的操作步高效的运算，程序实现难度较大。而相对于细粒度的并行，粗粒度的并行方式则更容易实现。粗粒度的并行计算方式采用的是基于相对独立的子任务的并行而不是操作步的并行。这种并行方式在优化和控制领域都有应用[75]。当满足以下情况时，粗粒度和细粒度的并行效率相当：

(1) 子任务的数量远多于 GPU 计算的核心数；
(2) 每个子任务的计算量比较适中，能够独立在一个 GPU 核心上完成；
(3) 各个子任务之间不需要太多的数据交换。并且计算过程中不需要全局同步，每个 GPU 核心上的子任务可以按顺序逐个的进行计算。

幸运的是区域城市建筑震害预测分析可以满足以上三个特性。虽然整个城市有成千上万栋的建筑，但是如果通过采用适当的计算模型使每栋建筑结构的计算量不超过 GPU 单核的计算能力，则每栋单体建筑的计算就能被看作一个相对独立的计算子任务。此外不同建筑的地震响应相对独立，不同 GPU 线程之间几乎不需要进行数据交换。因此 GPU 数百计的计算核心可以实现几百栋建筑同时计算，这样计算效率将非常高。

本研究提出适合多自由度剪切层模型的 GPU/CPU 协同粗粒度并行计算的区域建筑震害预测方法，对并行计算的效率进行了详细的讨论，并对一个中等大小的城市进行了精细化的震害模拟。

3.2.2 CPU/GPU 协同的程序构架

整个程序包括 3 个模块：前处理模块、结构分析模块和后处理模块。前处理模块的主要任务是获取计算模型的主要参数，并且选择地震场景进行非线性时程分析。后处理模块主要展示模拟结果并预测损失。

计算分析模块是程序的核心，由 CPU 和 GPU 协同完成计算任务。其中，CPU 负责文件读写、任务分配等工作；GPU 开启大量线程，每一条线程负责完成每一栋单体建筑的结构非线性时程计算，如图 3.14 所示。

本研究采用多自由度剪切层模型来模拟每个单体建筑物[42]，它可以很好地平衡计算量和计算精度之间的关系。为了避免隐式动力计算的收敛性问题，并提高 GPU 并行计算效率，本研究采用中心差分法求解动力方程[58]。动力方程中的阻尼采用经典瑞利阻尼。

3.2.3 GPU 计算性能分析

为了对比本研究提出的基于 GPU/CPU 协同粗粒度并行计算的结构分析模块的性能，

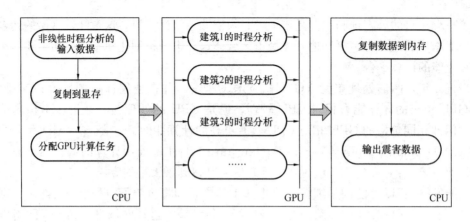

图 3.14　计算分析模块的工作流程

在传统 CPU 平台上也开发了一个功能相同计算程序。本研究采用的对比算例和对比平台如下所示：

(1) 根据 Hazus 中的建筑类型，随机生成不同结构类型的 1024 栋建筑，如表 3.5 所示。

(2) 采用广泛使用的 El Centro 地震动时程记录[76]进行对比分析。地震动时程记录的峰值加速度按照 200cm/s^2 进行调幅。

(3) 计算时长为 40s，时间步为 8000 步。

(4) 为了避免读写速度的影响，下文中的计算时间不包括硬盘的读写时间。

计算平台如表 3.6 所示。所采用的两套计算平台均为 2011 年采购，在当时两者价格基本相同。因此采用以上两套平台可以用于比较相同价格情况下的性能差异。

性能对比分析中使用的建筑结构类型分布情况　　表 3.5

结构类型	类型描述	楼层范围	建筑数量
W1	木结构	1～2	51
S1L	低层钢框架结构	1～3	43
S1M	多层钢框架结构	4～7	55
S1H	高层钢框架结构	8～10	83
S3	轻钢结构	任意层数	54
C1M	低层 RC 框架结构	1～3	55
C1M	多层 RC 框架结构	4～7	58
C1H	高层 RC 框架结构	8～10	52
C2L	低层 RC 剪力墙结构	1～3	68
C2M	多层 RC 剪力墙结构	4～7	52
C2H	高层 RC 剪力墙结构	8～10	69
C3L	低层无砌体填充墙 RC 框架结构	1～3	51
C3M	多层无砌体填充墙 RC 框架结构	4～7	34
C3H	高层无砌体填充墙 RC 框架结构	8～10	49
RM2L	低层配筋砌体结构	1～3	51
RM2M	多层配筋砌体结构	4～7	57
RM2H	高层配筋砌体结构	8～10	47
URML	低层无配筋砌体结构	1～2	50
URMM	多层配筋砌体结构	3～7	45
总计			1024

表 3.6 对比采用的计算平台

平台	硬件部分	编译器
CPU 计算平台	Intel Core i3 530 @2.93GHz & DDR3 4G 1333MHz.	Microsoft Visual C++ 2008 SP1.
GPU/CPU 协同计算平台	Intel Celeron E3200 @ 2.4GHz & NVIDIA GeForce GTX 460 1GB.	Microsoft Visual C++ 2008 SP1 & CUDA 4.2

首先采用单栋建筑对比 CPU 平台和 CPU/GPU 协同计算平台的计算效率。楼层数和单栋建筑平均计算时间的关系如图 3.15 所示。从图中可以看出对于一栋建筑 GPU/CPU 协同计算的时间要长于 CPU 计算的时间，即单颗 GPU 核心的计算能力要明显弱于 CPU 的计算能力。此外还可以注意到对于一栋 10 层楼建筑，采用单精度计算 GPU/CPU 协同计算平台需要 5 秒，如果采用双精度计算则需要 8 秒。然而 CPU 平台的单精度和双精度计算时间非常接近（双精度还稍快），这是因为使用的 CPU 是基于 64 位构架，其默认计算精度就是双精度。

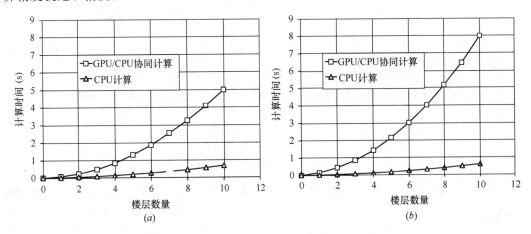

图 3.15 单体建筑的层数和平均计算时间的关系
(a) 单精度；(b) 双精度

对于 1024 栋建筑物，如果改变 CUDA 的 block size（每个计算块的线程数量），得到的计算时间如图 3.16 所示，可见当程序的 block size 为 32 时程序能达到最高的计算能力。这是因为，当采用 CUDA 进行并行计算的时候每 32 个线程组成一个 wrap，而 GPU 计算任务的最小单位是一个 wrap[77]。因此，如果 block size 小于 32 时，GPU 的性能不能充分发挥。而当单精度的 block size 大于 520 或者双精度的 block size 大于 260 时，将超出 GPU 中速度最快的寄存器的容量（本平台GPU 每个 block 只能使用 32K 的 32 位寄存器，可存储 32,768 个单精度数或 16,384 个双精度数），因此也会导致并行效率下降。

在 1024 栋建筑中，随机选取不同

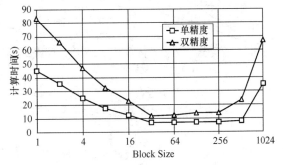

图 3.16 在 GPU/CPU 平台上对 1024 栋建筑进行分析的计算时间和 block size 之间的关系

的建筑数量在CPU平台和GPU/CPU平台上进行性能对比（GPU/CPU平台的block size是32），结果如图3.17所示。CPU的计算时间与建筑的数量基本上是线性的关系。然而GPU/CPU协同计算平台的计算时间则主要取决于计算时间最长的那个线程（图3.17）。此外GPU/CPU协同计算平台双精度计算的曲线上有几个曲线的突跃，这是由于GTX460的双精度计算是在特殊函数单元中完成的（Special Function Unit），其数量是CUDA核心数的1/6。这个问题可望在Fermi架构的Tesla GPU上得到修正，因为其双精度计算是直接在CUDA核心上完成的[77]。

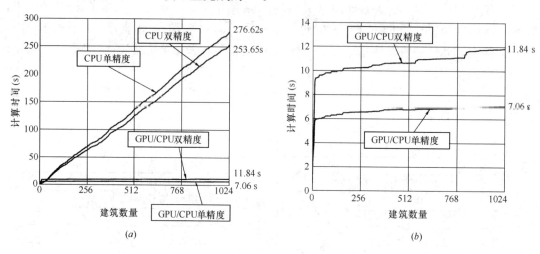

图3.17 CPU平台和GPU/CPU平台的计算效率对比
(a) CPU平台和GPU/CPU平台的结果；(b) GPU/CPU平台的结果

由GPU/CPU平台与CPU平台的计算时间对比可以看出GPU对繁重的非线性时程计算具有巨大的速度优势。如图3.18所示，当对1024栋建筑进行单精度计算时，采用GPU/CPU协同计算的时间仅是CPU计算的1/39，如果采用双精度则是1/21。此外，本文作者[42]对比了GPU单精度和双精度计算结果，发现差异不足0.1%，因此对于区域建筑震害模拟而言，单精度计算已经可以满足要求。

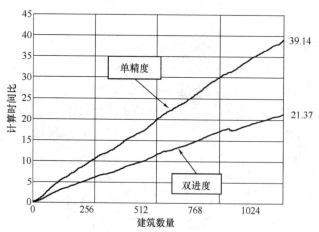

图3.18 CPU计算平台和GPU/CPU协同计算平台的计算时间比与建筑数量之间的关系

3.2.4 中等城市的建筑震害模拟

以某中等城市为例应用本研究中提出的方法对该城市进行了建筑震害模拟，并讨论本研究方法的优势。该中等城市总共有4255栋建筑。将该地区的建筑按照表3.7对不同年代的建筑进行了抗震等级分类[54]。对不同的抗震设计等级将根据Hazus中提出的方法确定建筑结构的层间滞回参数[53]。不同结构类型建筑的统计信息如表3.8所示。

结构抗震标准的划分　　　　　　　　　　　　　　　　　表 3.7

抗震标准	建造时间		
	1978 之前	1979—1989	1989 之后
	Pre-Code	Low-Code	Moderate-Code

该区域建筑的统计情况　　　　　　　　　　　　　　　　　表 3.8

结构类型	楼层范围	Moderate-Code	Low-Code	汇总
C1L	1~3	6	0	6
C1M	4~7	2421	428	2849
C2H	8+	14	2	16
RM2H	8+	5	1	6
S1H	1~3	4	0	4
S1M	4~7	6	21	27
URML	1~2	0	1	1
URMM	3+	0	1308	1308
W1	1~2	34	4	38
总计		2490	1765	4255

分析采用了 3 类地震动数据：远场地震动，近场无速度脉冲地震动，近场有速度脉冲地震动。对于每一类地震动，从 FEMA p695 推荐的地震动数据库中分别选取 5 条（表 3.9）[78]。根据该城市的设防烈度，对于小震、中震、大震分别采用 70, 200 和 400cm/s^2 的地面峰值加速度对选取的地震动记录进行调幅。由于本研究主要讨论建筑震害模拟方法，因此没有考虑场地对地震动的影响，每栋建筑都采用相同的地震动记录。

从 PEER-NGA 数据库中选择的地震动时程记录　　　　　　表 3.9

编号	远场地震动记录	近场有脉冲地震动记录	近场无脉冲地震动记录
1	NORTHR/MUL009	IMPVALL/H-E06_233	GAZLI/GAZ_177
2	DUZCE/BOL000	ITALY/A-STU_223	IMPVALL/H-BCR_233
3	HECTOR/HEC000	SUPERST/B-PTS_037	NAHANNI/S1_070
4	IMPVALL/H-DLT262	LOMAP/STG_038	LOMAP/BRN_038
5	KOBE/NIS000	ERZIKAN/ERZ_032	CAPEMEND/CPM_260

为了体现本研究采用的多自由度剪切层模型的优势，将分别使用本研究的多自由度模型方法和传统的单自由度模型方法对该城市进行模拟，单自由度模型的参数根据 Steelman 和 Hajjor 建议的方法[56]确定。对于这样一个中等城市，多自由度剪切层模型和单自由度模型完成一次地震模拟的计算时间完全可以满足应急震害预测的速度要求。

计算得到的最大楼层损伤状态的分析结果如图 3.19 所示。图中结果显示采用多自由度剪切层模型得到的楼层损伤结果普遍比单自由度模型的预测结果大一些。这是由于多自由度剪切层模型可以得到各个楼层的响应，从而能够考虑损伤集中现象。对于 200gal 远场地震波记录和近场无速度脉冲地震动记录，单自由度模型和多自由度剪切层模型的结果差异较小。但是对于近场有脉冲地震动记录，单自由度模型和多自由度剪切层模型的结果具有较大的差别。这是由于速度脉冲能够激励出高阶阵型从而导致结构更加严重的破坏，而单自由度模型无法考虑高阶阵型的影响。

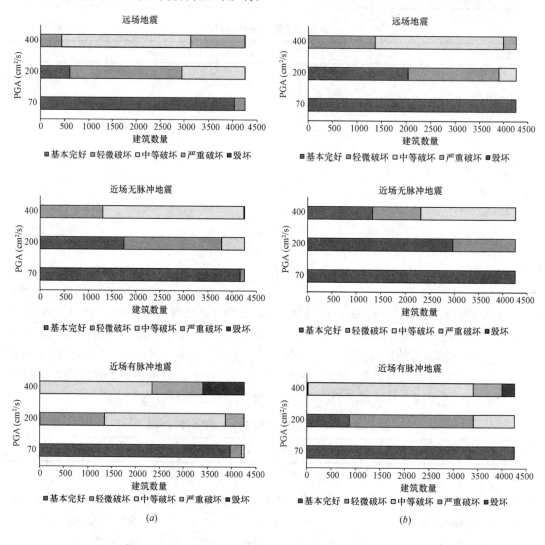

图 3.19　不同模型计算结果的损伤分布情况（5 条地震波的平均结果）
(a) 多自由度剪切层模型；(b) 单自由度模型

以峰值加速度为 200 cm/s^2 的 IMPVALL/H-BCR_233 地震动为例，建筑破坏的预测结果如图 3.20、图 3.21 所示。结果显示采用多自由度剪切层模型，能够准确获得建筑损伤的楼层位置和最大加速度位置，显著优于单自由度模型的计算结果。

本研究的结果可以证明 GPU/CPU 协同粗粒度并行计算具有很高的计算效率，其性能价格比能够达到传统 CPU 计算的 39 倍。并且，采用的多自由度剪切层模型可以考虑地震动中速度脉冲的影响，也能确定不同楼层的损伤情况，从而能够更加准确的进行损失预测。

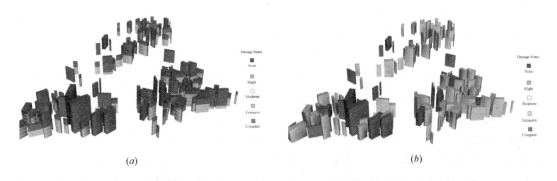

图3.20 区域中部分建筑各楼层损伤情况(地震记录:IMPVALL/H-BCR_233,PGA:200 cm/s²)
(a) 多自由度剪切层模型;(b) 单自由度模型

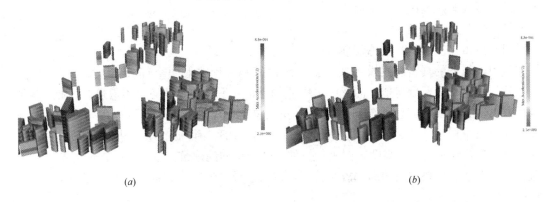

图3.21 区域中部分建筑各楼层的峰值加速度
(地震记录:IMPVALL/H-BCR_233,PGA:200 cm/s²)
(a) 多自由度剪切层模型;(b) 单自由度模型

3.3 城市区域震害的高真实度展示

3.3.1 震害真实感展示的需求

城市区域震害模拟的高真实度展示对防灾规划、虚拟应急训练、成果交流等都具有重要意义。在这方面,国内外很多研究者已经开展了基于三维城市模型的震害展示研究,如加拿大约克大学[79]基于分布式GIS建立了温哥华的三维地形和城市模型,并对城市整体震害进行评估;土耳其中东技术大学[80]、清华大学[81]利用CAD和纹理技术建立了真实感的3D城市,通过静态方式展现地震易损性或震害效果。但是这些研究所采用的震害预测模型往往过于简单,不能得到建筑物破坏的细节信息,因而其震害展示效果真实感不够。而近年来得到迅速发展的城市建筑群精细预测模型[60,67],可以得到更加清晰准确的震害计算结果(例如整个结构的震害时程以及楼层的局部损伤等),为提高震害模拟的真实感提供了关键的数据支持。

但是,精细化的区域建筑震害预测模型一般也难以包含建筑倒塌过程、残骸运动等倒塌细节。缺乏倒塌细节不仅影响震害模拟的真实感,也无法评价建筑倒塌对交通、救援等的影响。物理引擎[82]是近些年计算机图形学的新技术,专门用于计算场景中物体的复杂物理行为,可以模拟震害场景中建筑物的开裂、破碎、碰撞、堆积等一系列破坏过程,在科学仿真、电影制作上有相关应用[83-85]。应用物理引擎技术，可以弥补数值模拟缺失的细节，如清华大学[86]曾用物理引擎技术模拟桥梁垮塌全过程中残骸的运动，弥补有限元"生死单元"技术的不足。因此，物理引擎技术也可以应用在城市震害模拟中，为城市工程震害的动态破坏细节过程提供了技术支持。然而，国内外尚缺乏基于物理引擎等技术的城市工程震害视景仿真方面的深入研究，因此，在震灾真实感展示中，垮塌过程的真实感特效问题面临较大挑战。

本研究基于城市区域建筑震害精细化数值分析结果，提出真实感的城市区域建筑三维建模方法以及震害过程的动画实现方法，以准确而又真实感的表现震害分析结果；利用物理引擎，提出建筑震害过程中的垮塌效果的实现算法，以弥补数值模拟缺乏的细节，增加模拟完整性和真实感。并且，对我国一个中等规模城市的建筑震害过程进行综合模拟，为城市防灾减灾策略提供参考意见。

3.3.2 真实感建模与震害动态演示

（1）城市建筑群的真实感建模

城市建筑群的真实感建模主要包括三维建模和纹理映射两个部分，如图 3.22 所示。区域建筑震害预测模型将楼层简化为集中质量点，并不包含完整的三维信息。因此，需要利用城市区域的地理信息系统 GIS 数据，获取位置、建筑外形、楼层高度等扩充信息来建立建筑三维模型。本研究采用徐峰等人的基于 GIS 三维城市建筑建模方法[81]，利用 GIS 中建筑轮廓多边形在高度方向进行拉伸来建立不同的建筑三维模型。同时，为了使建筑三维模型能够反映层间的局部震害特征，建筑三维模型应与震害预测模型保持一致，同样以楼层为基本单元。

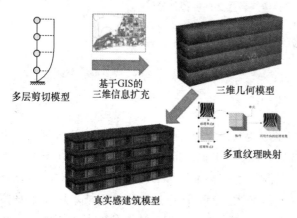

图 3.22 城市建筑群真实感建模方法

为增加建筑模型的真实感，需要对三维模型进行纹理映射。本研究基于 Tsai 和 Lin 的方法[87]，采用多重纹理技术，对建筑物的不同表面分别进行贴图，以更加细致地体现建筑物的特征。根据图 3.22 所示方法，可以建立城市区域的真实感模型，为震害动态演示奠定基础。

（2）基于精细化预测的震害动态演示

如何准确、高效地表现大规模的震害预测的时变数据是本研究城市区域建筑震害动态演示的主要问题。为了对图形表现过程进行充分控制，本研究选择开源图形引擎 OSG（Open Scene Graph）作为视景模拟的主要平台[88]。在 OSG 中，一般会采用回调机制（Callback）来完成每一帧渲染前需要的工作，如空间变化、顶点更新、视点变化等。利用 OSG 的回调机制，本研究直接用震害预测数据在每一帧更新建筑模型的顶点坐标，以动态地表现建筑震害过程，具体算法如图 3.23 所示。

图 3.23 所示算法包含三个重要步骤：

1）更新器的创建与初始化

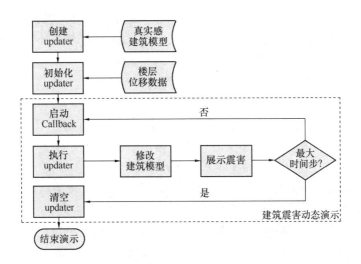

图 3.23 建筑震害的动态可视化方法

为了实现对图形顶点的操作，需要创建更新器 updater。它需要继承 OSG 中图形的更新回调类 UpdateCallback，以实现对回调过程的全面控制。在 updater 初始化阶段，更新器需要在动画渲染前加载对应建筑震害的节点位移数据。

2）更新器的动态执行

在回调过程中，需要将震害预测中的时间步与动态演示中的帧数相对应。在动态演示的每一帧中，更新器将利用对应时间步的震害数据来更新建筑的顶点坐标。重载更新器中的 Update（）函数可以实现对图形顶点的动态修改功能。通过回调，不断修改建筑模型，可形成区域建筑震害的动态演示过程。

3）清空更新器

当动态演示达到最大的时间步时，清空更新器，建筑模型不再更新，震害过程的动态演示结束。

3.3.3 基于物理引擎的倒塌模拟

区域震害预测模型只能给出建筑倒塌的判断，缺乏细节的倒塌过程。细节的倒塌过程可以增强震害模拟的真实感，也为地震疏散、救援调度等提供重要的参考，因此，倒塌模拟在地震模拟中是不可缺少的。而且，倒塌过程需要符合物理规律，并不是纯粹的图形表现。

为匹配上述需求，本研究选择物理引擎模拟建筑倒塌过程。由于在多刚体动力学和碰撞检测上的优势，物理引擎善于模拟倒塌过程中大量残骸的复杂运动和相互碰撞过程。本研究对比了三大物理引擎 Havok，PhysX 和 Bullet[89-91]，发现由于 PhysX 可以采用 GPU 来加速计算过程，更适合大量物体的计算，而在区域震害模拟中，建筑的规模是庞大的，因此，PhysX 更适合用于区域建筑的倒塌过程模拟。

为了保证物理计算与图形显示的准确对应，需要根据震害预测数据在物理空间中建立与图形一致的建筑倒塌模型。在本研究建筑模型场景层次中，楼层图形由 OSG 中的 Geode 节点储存和管理[88]。在 PhysX 中，物理角色 Actor 类是基本的计算单元，代表着物理世界中所有的物体[91]。因此，对应于 Geode 节点，在物理引擎中楼层由 Actor 建立。物理引擎的楼层模型，需要楼层的材料属性（如弹性系数、线阻尼等），也需要楼层倒塌时的状态参数（如位移、速度等）。材料属性需要根据不同建筑结构类型，查询相关实验结果确定，而状态参数通过数值分析数据获得，以保证模拟的准确性。在物理引擎中，楼层的外形参数通过对应的楼层的 Geode 节点获得。为减少震害模拟过程中的耗时，楼层的物理模型可以在渲染前建立，在需要物理计算时被激活。

物理引擎与图形引擎是两个独立的程序，需要构建两者的动态联系才能模拟建筑的倒塌过程。目前，关于 OSG 与 PhysX 的结合工作研究还非常有限。本研究中基于 OSG 回调机制，在每一帧中，利用 PhysX 实时计算数据不断更新楼层的顶点坐标，以形成楼层倒塌过程的动态演示[88]。为了建立 OSG 与 PhysX 的对应关系，创建了一个专门的类 Stories。在 Stories 类中，同时包含楼层对应的 Actor 对象和 Geode 节点，以建立楼层物理模型与图形模型的联系。为了实现建筑倒塌的动态过程，需要控制 Stories 类的回调过程，因此，设计了一个新类 Move-callback。由于继承了 Callback 机制的控制类，Move-callback 类可以控制 Stories 类的渲染主循环。Move-callback 类从 PhysX 的 Actor 中获取运动信息，然后应用运动信息来更新楼层图形。通过 Stories 类和 Move-callback 类实现的渲染循环过程形成了 PhysX 计算支持的建筑倒塌模拟，如图 3.24 所示。

建立 OSG 与 PhysX 的动态联系后，可以通过物理引擎模拟建筑倒塌过程。但是，为表现完整的建筑震害过程，震害动态演示与倒塌过程必须形成协同模拟的整体。根据震害预测给出的倒塌判断，将整个建筑震害过程分为两个阶段：倒塌前和倒塌后。首先，建立区域建筑群的真实感模型。在倒塌前，利用位移时程数据来动态地演示建筑的位移和变形。发生垮塌后，清空动态演示的更新器，根据前述方法建立楼层图形 Geode 节点与物理角色 Actor 的动态联系。然后，将 Geode 的运动状态信息传递给 Actor，并激活 Actor。物理引擎将不断计算 Actor 的倒塌过程，计算数据也会同步传递给 Geode，以实时地楼表现建筑的倒塌过程。上述过程（如图 3.25 所示）实现了完整的城市区域建筑震害高真实度模拟。

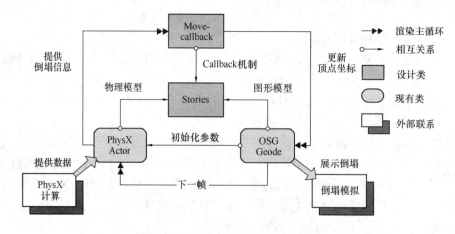

图 3.24 OSG 与 PhysX 的动态联系

3.3.4 中等城市的震害真实感展示

本研究以前文中的中型城市震害模拟作为算例，基于精细化震害预测结果和 GIS 数据建立了真实感的三维城市模型（图 3.26a），并展示了震害过程中不同楼层间的位移和变形（图 3.26b）。由于城市区域的建筑数量众多，所有建筑倒塌过程的非线性时程分析

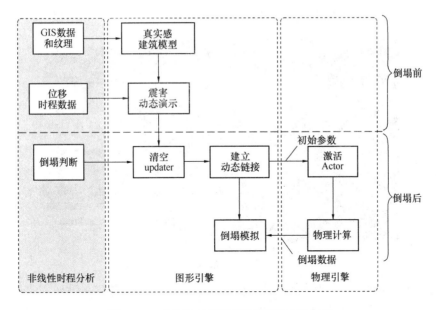

图 3.25 建筑震害高真实度模拟的完整过程

是不可能通过普通计算机模拟的。然而，通过使用 PhysX，可以在桌面型计算机中快速模拟建筑倒塌过程（图 3.26c 和 图 3.26d）。倒塌过程模拟很好地弥补了非线性时程分析的不足，使整个震害模拟过程更加真实、完整。

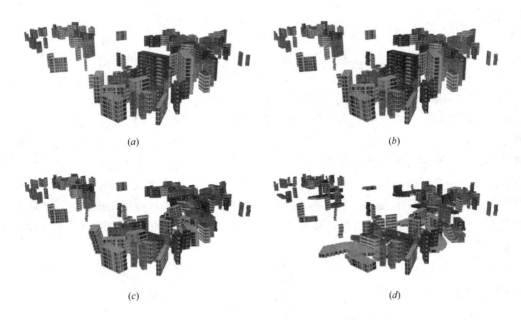

图 3.26 城市区域建筑震害高真实度模拟结果
(a) 真实感建筑模型；(b) 放大的建筑位移；(c) 倒塌过程；(d) 倒塌结束

为进一步增加震害场景的完整性和真实感，本研究将震害场景加入了地形、天空等环境模型，并且利用立体投影设备进行了区域建筑震害过程的立体感展示，进一步加强了震害过程模拟的沉浸感，如图 3.27 所示。图 3.27 的结果可以用于虚拟环境下地震疏散、应急调度等应用，综合提升防震减灾能力。

数字光处理投影仪

双屏监视器

图 3.27　城市区域建筑震害的立体展示效果

本研究基于精细化的结构模型和物理引擎技术，实现了城市区域建筑震害的真实感显示，对我国一个中型城市进行了区域建筑震害模拟。本研究不仅准确、真实感地表现了区域建筑震害过程，同时，也给出符合物理规律的倒塌过程，弥补了震害模拟的细节缺失。研究结果可以用于震害预测、灾后应急、救援训练等，为城市防震减灾提供了重要技术支持。

4. 结论

Reitherman 指出，地震工程学所面临的三个最大困难在于"风险、非弹性和动力学 (Risk, inelasticity and dynamics)"[92]。虽然地震工程很多重要思想，在 20 世纪中期、甚至 19 世纪末期，就已经提出来了。但是离开了计算机的帮助，定量且精确的研究上述三个问题中的任何一个都是非常困难的。而自从计算机出现后，结构工程和地震工程的研究取得了翻天覆地的巨大变化。Roesset and Yao 评价："20 世纪及未来若干年结构工程学最主要的改变是因为数字计算机的发展，包括将计算机作为一个强大的工具进行烦琐的计算以及作为一个新的通信工具供设计团队、教授和学生，及其他人员开展交流"[93]。充分利用计算机科学和技术方面取得的最新成果，加深对工程结构和城市在强烈地震下的灾变演化规律和机理的理解，进而提出高效可靠的工程对策，对工程防灾减灾有着重大而深远的意义。

致谢

感谢国家自然科学基金（No.51222804，51178249，51378299，51308321，90815025，51261120377），北京市自然科学基金（No. 8142024），国家科技支撑计划课题（No. 2013BAJ08B02），交通运输部交通运输建设重大科技专项（No. 2011-318-223-170），霍英东教育基金（No. 131071）和教育部新世纪优秀人才支持计划（No. NCET-10-0528）等项目的资助。并感谢清华大学钱稼茹教授、任爱珠教授，澳大利亚 Griffith 大学 H Guan

博士,日本东京大学 M Hori 教授,美国斯坦福大学 KH Law 教授,美国加州大学伯克利分校 F McKenna 博士,ARUP 工程咨询公司 YL Huang 博士等对本文研究的贡献。参与本文研究的还有清华大学博士研究生汪训流、解琳琳、曾翔,硕士研究生唐代远、张万开等。

参考文献

[1] Yamashita T, Kajiwara K, Hori M. Petascale Computation for Earthquake Engineering. Computing in Science & Engineering, 2011, 13 (4): 44-49.

[2] Ibarra LF, Krawinkler H. Global Collapse of Frame Structures Under Seismic Excitations. Pacific Earthquake Engineering Research Center, 2005.

[3] Kabeyasawa T, Shiohara H, Otani S, Aoyama H. Analysis of the Full-Scale Seven-Story Reinforced Concrete Test Structure: Proc. 3rd JTCC, US-Japan Cooperative Earthquake Research Program, BRI, Tsukuba, Japan, 1982.

[4] 李国强,周向明,丁翔. 钢筋混凝土剪力墙非线性动力分析模型. 世界地震工程, 2000(02): 13-18.

[5] Taucer FF, Spacone E, Filippou FC. A Fiber Beam-Column Element for Seismic Response Analysis of Reinforced Concrete Structures. Report No. UCB/EERC-91/17, UC Berkeley, America, 1991.

[6] Spacone E, Filippou FC, Taucer FF. Fibre Beam-Column Model for Non-Linear Analysis of R/C Frames. 1. Formulation. Earthquake Engineering & Structural Dynamics, 1996, 25 (7): 711-725.

[7] Legeron F, Paultre P. Uniaxial Confinement Model for Normal- and High-Strength Concrete Columns. Journal of Structural Engineering-ASCE, 2003, 129 (2): 241-252.

[8] 汪训流,陆新征,叶列平. 往复荷载下钢筋混凝土柱受力性能的数值模拟. 工程力学, 2007(12): 76-81.

[9] Esmaeily A, Xiao Y. Behavior of Reinforced Concrete Columns Under Variable Axial Loads: Analysis. ACI Structural Journal, 2005, 102 (5): 736-744.

[10] 叶列平,陆新征,马千里,汪训流,缪志伟. 混凝土结构抗震非线性分析模型、方法及算例. 工程力学, 2006(S2): 131-140.

[11] Li Y, Lu X, Guan H, Ye L. An Improved Tie Force Method for Progressive Collapse Resistance Design of Reinforced Concrete Frame Structures. Engineering Structures, 2011, 33 (10): 2931-2942.

[12] Lu X, Lu X, Guan H, Ye L. Collapse Simulation of Reinforced Concrete High-Rise Building Induced by Extreme Earthquakes. Earthquake Engineering & Structural Dynamics, 2013, 42 (5): 705-723.

[13] Ren P, Li Y, Guan H, Lu X. Progressive Collapse Resistance of Two Typical High-Rise RC Frame Shear Wall Structures. Journal of Performance of Constructed Facilities-ASCE, 2014.

[14] 唐代远. 等跨 RC 框架结构抗地震倒塌性能的试验研究及理论分析. 北京: 清华大学硕士学位论文, 2011.

[15] 门俊,陆新征,宋二祥,陈肇元. 分层壳模型在剪力墙结构计算中的应用. 防护工程, 2006, 28 (3): 9-13.

[16] 魏勇,钱稼茹,赵作周,蔡益燕,郁银泉,申林. 高轴压比钢骨混凝土矮墙水平加载试验. 工业建筑, 2007(06): 76-79.

[17] 解琳琳，黄羽立，陆新征，林楷奇，叶列平. 基于 OpenSees 的 RC 框架-核心筒超高层建筑抗震弹塑性分析. 工程力学，2014(01)：64-71.

[18] Loland KE. Continuous Damage Model for Load-Response Estimation of Concrete. Cement and Concrete Research，1980，10（3）：395-402.

[19] Mazars J. A Description of Microscale and Macroscale Damage of Concrete Structures. Engineering Fracture Mechanics，1986，25 (5-6)：729-737.

[20] Humphrey JR, Price DK, Spagnoli KE, Paolini AL, Kelmelis EJ. CULA：Hybrid GPU Accelerated Linear Algebra Routines，1000 20TH ST，PO BOX 10，BELLINGHAM，WA 98227-0010 USA，2010.

[21] CuSP Home Page. http：//cusplibrary.github.io/.

[22] 蒋欢军，和留生，吕西林，丁洁民，赵昕. 上海中心大厦抗震性能分析和振动台试验研究. 建筑结构学报，2011(11)：55-63.

[23] Lu X, Lu X, Zhang W, Ye L. Collapse Simulation of a Super High-Rise Building Subjected to Extremely Strong Earthquakes. Science China-Technological Sciences，2011，54 (10)：2549—2560.

[24] 范立础，胡世德，叶爱君. 大跨度桥梁抗震设计. 人民交通出版社，2001.

[25] 陈幼平，周宏业. 斜拉桥地震破坏的计算研究. 地震工程与工程振动，1995(03)：127-134.

[26] Yeh Y, Mo YL, Yang CY. Seismic Performance of Rectangular Hollow Bridge Columns. Journal of Structural Engineering-ASCE，2002，128 (1)：60-68.

[27] Shome N, Cornell CA, Bazzurro P, Carballo JE. Earthquakes，Records，and Nonlinear Responses. Earthquake Spectra，1998，14 (3)：469-500.

[28] Tothong P, Luco N. Probabilistic Seismic Demand Analysis Using Advanced Ground Motion Intensity Measures. Earthquake Engineering & Structural Dynamics，2007，36 (13)：1837-1860.

[29] Luco N, Cornell CA. Structure-Specific Scalar Intensity Measures for Near-Source and Ordinary Earthquake Ground Motions. Earthquake Spectra，2007，23 (2)：357-392.

[30] Baker JW, Cornell CA. A Vector-Valued Ground Motion Intensity Measure Consisting of Spectral Acceleration and Epsilon. Earthquake Engineering & Structural Dynamics，2005，34（10）：1193-1217.

[31] Lu X, Ye L, Lu X, Li M, Ma X. An Improved Ground Motion Intensity Measure for Super High-Rise Buildings. Science China-Technological Sciences，2013，56 (6)：1525-1533.

[32] Ye L, Ma Q, Miao Z, Guan H, Yan Z. Numerical and Comparative Study of Earthquake Intensity Indices in Seismic Analysis. Structural Design of Tall and Special Buildings，2013，22（4）：362-381.

[33] 汪大绥，周建龙，姜文伟，王建，江晓峰. 超高层结构地震剪力系数限值研究. 建筑结构，2012(05)：24-27.

[34] 李俊，王震宇. 超高强混凝土单轴受压应力-应变关系的试验研究. 混凝土，2008(10)：11-14.

[35] 陆新征，叶列平. 基于 IDA 分析的结构抗地震倒塌能力研究. 工程抗震与加固改造，2010(01)：13-18.

[36] 中国地震局. 历史地震目录. http：//www.cea.gov.cn/publish/dizhenj/468/496/index.html.

[37] Onur T, Ventura CE, Finn WDL. A Comparison of Two Regional Seismic Damage Estimation Methodologies. Canadian Journal of Civil Engineering，2006，33 (11)：1401-1409.

[38] Earthquake Loss Estimation Methodology - HAZUS97. Technical Manual. Washington, D. C.：Federal Emergency Management Agency - National Institute of Building Sciences，1997.

[39] Earthquake Loss Estimation Methodology - HAZUS99. Technical Manual. Washington, D. C.: Federal Emergency Management Agency - National Institute of Building Sciences, 1999.

[40] Sextos AG, Kappos AJ, Stylianidis KC. Computer-Aided Pre- and Post-Earthquake Assessment of Buildings Involving Database Compilation, GIS Visualization, and Mobile Data Transmission. Computer-Aided Civil and Infrastructure Engineering, 2008, 23 (1): 59-73.

[41] Iwan WD. Drift Spectrum: Measure of Demand for Earthquake Ground Motions. Journal of Structural Engineering-ASCE, 1997, 123 (4): 397-404.

[42] Lu X, Han B, Hori M, Xiong C, Xu Z. A Coarse-Grained Parallel Approach for Seismic Damage Simulations of Urban Areas Based On Refined Models and GPU/CPU Cooperative Computing. Advances in Engineering Software, 2014, 70 (0): 90-103.

[43] Moghaddam H, Mohammadi RK. Ductility Reduction Factor of MDOF Shear-Building Structures. Journal of Earthquake Engineering, 2001, 5 (3): 425-440.

[44] Veletsos AS, Vann WP. Response of Ground - Excited Elasto-Plastic Systems. Journal of Structural Division-ASCE, 1971, 97 (4): 1257-1281.

[45] Diaz O, Mendoza E, Esteva L. Seismic Ductility Demands Predicted by Alternate Models of Building Frames. Earthquake Spectra, 1994, 10 (3): 465-487.

[46] Pampanin S, Priestley M, Sritharan S. Analytical Modelling of the Seismic Behaviour of Precast Concrete Frames Designed with Ductile Connections. Journal of Earthquake Engineering, 2001, 5 (3): 329-367.

[47] 经杰, 叶列平, 钱稼茹. 基于能量概念的剪切型多自由度体系弹塑性地震位移反应分析. 工程力学, 2003(03): 31-37.

[48] Aoyama H. A Method for the Evaluation of the Seismic Capacity of Existing Reinforced Concrete Buildings in Japan. Bulletin of the New Zealand National Society for Earthquake Engineering, 1981, 14 (3): 105-130.

[49] Hajirasouliha I, Doostan A. A Simplified Model for Seismic Response Prediction of Concentrically Braced Frames. Advances in Engineering Software, 2010, 41 (3): 497-505.

[50] Muto K. Dynamic Design for Structures. Tokyo: Maruzen Publishing, 1977.

[51] 何广乾, 魏琏, 戴国莹. 论地震作用下多层剪切型结构的弹塑性变形计算. 土木工程学报, 1982 (03): 10-19.

[52] 陈光华. 地震作用下多层剪切型结构弹塑性位移反应的简化计算. 建筑结构学报, 1984(02): 45-57.

[53] FEMA. Multi-Hazard Loss Estimation Methodology Hazus - MH 2.1 Advanced Engineering Building Module (AEBM) Technical and User'S Manual. Washington, D. C.: Federal Emergency Management Agency, 2012.

[54] Lin S, Xie L, Gong M, Li M. Performance-Based Methodology for Assessing Seismic Vulnerability and Capacity of Buildings. Earthquake Engineering and Engineering Vibration, 2010, 9 (2): 157-165.

[55] Mahin SA, Lin J. Construction of Inelastic Response Spectra for Single-Degree-Of-Freedom Systems, Computer Program and Applications, Report No. UCB/EERC-83/17. Berkeley: University of California, Earthquake Engineering Research Center, 1984.

[56] Steelman JS, Hajjar JF. Influence of Inelastic Seismic Response Modeling On Regional Loss Estimation. Engineering Structures, 2009, 31 (12): 2976-2987.

[57] Ibarra LF, Medina RA, Krawinkler H. Hysteretic Models that Incorporate Strength and Stiffness

Deterioration. Earthquake Engineering & Structural Dynamics, 2005, 34 (12): 1489-1511.

[58] Chopra AK. Dynamics of Structures. Prentice Hall New Jersey, 1995.

[59] 《建筑结构荷载规范》GB 50009—2001. 北京: 中国建筑工业出版社, 2002.

[60] Sobhaninejad G, Hori M, Kabeyasawa T. Enhancing Integrated Earthquake Simulation with High Performance Computing. Advances in Engineering Software, 2011, 42 (5SI): 286-292.

[61] 《建筑结构抗震设计规范》GB 50011—2010. 北京: 中国建筑工业出版社, 2010.

[62] 马玉虎, 陆新征, 叶列平, 唐代远, 李易. 漩口中学典型框架结构震害模拟与分析. 工程力学, 2011(05): 71-77.

[63] Miao Z, Ye L, Guan H, Lu X. Evaluation of Modal and Traditional Pushover Analyses in Frame-Shear-Wall Structures. Advances in Structural Engineering, 2011, 14 (5): 815-836.

[64] Tang B, Lu X, Ye L, Shi W. Evaluation of Collapse Resistance of RC Frame Structures for Chinese Schools in Seismic Design Categories B and C. Earthquake Engineering and Engineering Vibration, 2011, 10 (3): 369-377.

[65] 施炜, 叶列平, 陆新征, 唐代远. 不同抗震设防RC框架结构抗倒塌能力的研究. 工程力学, 2011(03): 41-48.

[66] Krawinkler H. Van Nuys Hotel Building Testbed Report: Exercising Seismic Performance Assessment. Pacific Earthquake Engineering Research Center, College of Engineering, University of California, Berkeley, 2005.

[67] Hori M, Ichimura T. Current State of Integrated Earthquake Simulation for Earthquake Hazard and Disaster. Journal of Seismology, 2008, 12 (2): 307-321.

[68] Owens JD, Luebke D, Govindaraju N, Harris M, Krüger J, Lefohn AE, Purcell TJ. A Survey of General - Purpose Computation On Graphics Hardware, 2007.

[69] Stone JE, Phillips JC, Freddolino PL, Hardy DJ, Trabuco LG, Schulten K. Accelerating Molecular Modeling Applications with Graphics Processors. Journal of Computational Chemistry, 2007, 28 (16): 2618-2640.

[70] Weldon M, Maxwell L, Cyca D, Hughes M, Whelan C, Okoniewski M. A Practical Look at GPU-Accelerated FDTD Performance. Applied Computational Electromagnetics Society Journal, 2010, 25 (4SI): 315-322.

[71] Lv MH, Wei X, Lei C. A GPU-Based Parallel Processing Method for Slope Analysis in Geographical Computation. Advanced Materials Research, 2012, 538: 625-631.

[72] Che S, Boyer M, Meng J, Tarjan D, Sheaffer JW, Skadron K. A Performance Study of General-Purpose Applications On Graphics Processors Using CUDA. Journal of Parallel and Distributed Computing, 2008, 68 (10): 1370-1380.

[73] Hung SL, ADELI H. A Parallel Genetic/Neural Network Learning Algorithm for Mimd Shared-Memory Machines. IEEE Transactions On Neural Networks, 1994, 5 (6): 900-909.

[74] Mackerle J. FEM and BEM Parallel Processing: Theory and Applications - a Bibliography (1996—2002). Engineering Computations, 2003, 20 (3-4): 436-484.

[75] Adeli H. High-Performance Computing for Large-Scale Analysis, Optimization, and Control. Journal of Aerospace Engineering, 2000, 13 (1): 1-10.

[76] Cundumi O, Suarez LE. Numerical Investigation of a Variable Damping Semiactive Device for the Mitigation of the Seismic Response of Adjacent Structures. Computer-Aided Civil and Infrastructure Engineering, 2008, 23 (4): 291-308.

[77] NVIDIA CUDA Programming Guide. http://docs.nvidia.com/cuda/pdf/CUDA_C_Program-

ming_Guide. pdf.

[78] FEMA P695 - Quantification of Building Seismic Performance Factors. Washington, D. C.: Federal Emergency Management Agency, 2009.

[79] Abdalla R, Tao V. Integrated Distributed GIS Approach for Earthquake Disaster Modeling and Visualization//Berlin: Springer, 2005: 1183-1192.

[80] Duzgun HSB, Yucemen MS, Kalaycioglu HS, Celik K, Kemec S, Ertugay K, Deniz A. An Integrated Earthquake Vulnerability Assessment Framework for Urban Areas. Natural Hazards, 2011, 59 (2): 917-947.

[81] Xu Z, Lu X, Guan H, Han B, Ren A. Seismic Damage Simulation in Urban Areas Based On a High-Fidelity Structural Model and a Physics Engine. Natural Hazards, 2014, 71 (3): 1679-1693.

[82] Millington I. Game Physics Engine Development (The Morgan Kaufmann Series in Interactive 3D Technology). San Francisco: CRC Press, 2007.

[83] Boeing A, Br unl T. Evaluation of Real-Time Physics Simulation Systems, 2007.

[84] Joselli M, Clua E, Montenegro A, Conci A, Pagliosa P. A New Physics Engine with Automatic Process Distribution Between CPU-GPU, 2008.

[85] Maciel A, Halic T, Lu Z, Nedel LP, De S. Using the PhysX Engine for Physics-Based Virtual Surgery with Force Feedback. International Journal of Medical Robotics and Computer Assisted Surgery, 2009, 5 (3): 341-353.

[86] Xu Z, Lu X, Guan H, Ren A. Physics Engine-Driven Visualization of Deactivated Elements and its Application in Bridge Collapse Simulation. Automation in Construction, 2013, 35: 471-481.

[87] Tsai F, Lin HC. Polygon-Based Texture Mapping for Cyber City 3D Building Models. International Journal of Geographical Information Science, 2007, 21 (9): 965-981.

[88] OSG Community, OpenSceneGraph. [2012.05.26]. http://www.openscenegraph.org/projects/osg/.

[89] Havok. Havok Physics. [2012.07.24]. http://www.havok.com/products/physics.

[90] Game Physics Simulation, Bullet Physics Library. [2012.07.13]. http://www.bulletphysics.com.

[91] NVIDIA Corporation, PhysX. [2012.08.11]. http://www.nvidia.cn/object/physx_new_cn.html.

[92] Reitherman R. Earthquakes and Engineers: An International History. American Society of Civil Engineers, 2012.

[93] Roesset JM, Yao J. State of the Art of Structural Engineering. Journal of Structural Engineering-ASCE, 2002, 128 (8): 965-975.

PERFORMANCE-BASED EARTHQUAKE ENGINEERING APPLIED TO HIGH-VOLTAGE SUBSTATIONS USING REAL-TIME HYBRID SIMULATION AND PEER METHODOLOGY

Khalid M. Mosalam[1], Selim Günay[1] and Qiang Xie[2]

[1] Department of Civil and Environmental Engineering, University of California, Berkeley, CA 94564-1710, USA

[2] Deptartment of Building Engineering, Tongji University, Shanghai, P. R. China

Emails: mosalam@berkeley.edu, selimgunay@berkeley.edu, and qxie@tongji.edu.cn

Abstract: The objective of this paper is to utilize the results of a previously conducted experimental parametric study in the context of performance-based earthquake engineering (PBEE). The experimental parametric study, which utilized real-time hybrid simulation (RTHS), was conducted to investigate the effect of support structure stiffness and damping on the seismic response of electrical disconnect switches with different insulator post types. Because the seismic performance of the disconnect switches is of significant interest to various stakeholders, including electrical engineers, utility companies, manufacturers and policy makers, it is essential to transform the results from an engineering demand parameter format, which is meaningful and useful to structural engineers, to a format equally meaningful to all interested stakeholders. The PBEE methodology developed by the Pacific Earthquake Engineering Research (PEER) Center, that provides a probabilistic end-result in terms of fatalities, economical losses or downtime, is employed for this transformation. The paper also presents a direct quantitative evaluation of the influence of various parameters on the performance of disconnect switches, e.g. support structure configuration, insulator type, geographical location of substation, and amount of slack in connecting cables.

Key words: Damage analysis; electrical substation equipment; hazard analysis; loss analysis; PEER PBEE methodology; real-time hybrid simulation; structural analysis; uncertainty

1. INTRODUCTION

Disconnect switches constitute a crucial component of power transmission and distribution systems. Electrical insulator posts and a metallic tube (blade) that connects the top parts of the insulator posts comprise a typical vertical-break disconnect switch (Figure 1). In the field, disconnect switches are generally mounted on support structures, which are typically steel frames or columns with well-defined geometry (Figure 2).

Disconnect switches are used to control the flow of electricity between all types of substation equipment, e.g. interruption of power during maintenance. They are also used to manage the power distribution network, e.g. shifting electrical loads across the network or turning off an area of the network if a safety threat arises. Hence, proper functioning of disconnect switches is vital for power regulation in the aftermath of an earthquake. However, disconnect switches and their electrical insulator posts continue to experience severe damage during earthquakes. Figure 3 shows damaged disconnect switches, where the insulator posts failed in a brittle manner and separated from the switch and the metallic tubes (blades) were distorted.

Figure 1　A typical disconnect switch consisting of insulator posts and a metallic tube (blade)

Figure 2　245-kV vertical-break disconnect switch in field installation

Figure 3　Earthquake damage of disconnect switches:
Complete failure at Ertaishan Switchyard (left) [Photo by Q. Xie]; Damage to porcelain insulator posts and top blades (right) [Photo by E. Fujisaki]

A real time hybrid simulation (RTHS) system was developed for cost effective and efficient dynamic testing of electrical vertical-break disconnect switches as an alternative to conventional shaking table testing (Mosalam and Günay 2014). Using this system, an experimental parametric study was conducted to investigate the effect of support structure stiffness and damping on the response of disconnect switches of two insulator post types,

i. e. polymer and porcelain (Günay and Mosalam 2014). As a result of the parametric study, engineering demand parameters (EDP) of the insulator posts, including accelerations, forces, displacements, and strains, were obtained for various support structure stiffness and damping. Although the results of the parametric study provided useful information about the effect of support structures on the response of disconnect switches, such information is mainly meaningful to structural engineers. However, the seismic performance of disconnect switches is in direct interest to other stakeholders, e. g. electrical engineers, utility companies, manufacturers and policy makers.

The objective of this paper is to transform the results obtained from RTHS parametric study conducted in (Günay and Mosalam 2014), which are in the form of EDPs, to a format meaningful to all stakeholders. The performance-based earthquake engineering (PBEE) methodology developed by the Pacific Earthquake Engineering Research (PEER) Center, which provides a probabilistic end-result in terms of fatalities, economical losses or downtime, is the approach adopted for this purpose.

2. BACKGROUND

2.1 RTHS parametric study

In the developed RTHS system (Mosalam and Günay 2014) and the conducted parametric study (Günay and Mosalam 2014), a single insulator post used in 245-kV vertical-break disconnect switches (Takhirov 2008) is tested as the experimental substructure on a shaking table and a single degree-of-freedom (SDOF) system representing the support structure is employed as the analytical substructure (Figure 4b). This configuration, dictated by the dimensions of the shaking table used in RTHS, represents the testing of the disconnect switch in an open blade configuration, where the experimental element is the jaw post identified by an ellipse in Figure 4a.

The RTHS parametric study was conducted using 13 different stiffness values, 3

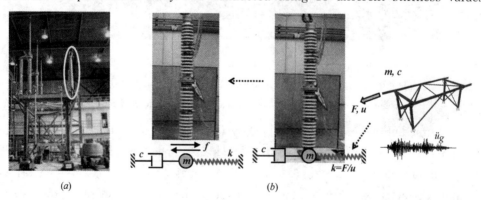

Figure 4 245-kV disconnect switch tests (Mosalam and Günay 2014; Günay and Mosalam 2014)
(a) Conventional shaking table test; (b) RTHS system used for parametric study

damping ratios and 2 insulator types. The considered stiffness range represented different configurations of the support structure, where the smallest stiffness, i.e. 385 kN/m, represents a steel frame with no braces and the largest stiffness, i.e. 10,508 kN/m, corresponds to the frame with full bracing. Employed damping ratios were 1%, 3%, and 5% of the critical damping value, representing typical damping range for steel structures. Porcelain and hollow core polymer insulator posts were physically tested. Examples of obtained results are presented in Figure 5, showing variations of bottom strains and top displacements of the insulator posts with the support structure stiffness.

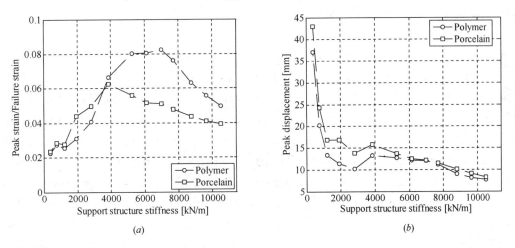

Figure 5 Sample results from the experimental parametric study (Günay and Mosalam 2014)
(a) Normalized peak strain; (b) Displacement with support structure stiffness

2.2 PEER performance-based earthquake engineering methodology

The PEER PBEE methodology is a second generation PBEE methodology developed to improve earlier PBEE methods (Moehle 2003). The main features of the PEER methodology are as follows:
- Performance of a structure is determined in a rigorous probabilistic manner by considering all sources of uncertainty that affect the performance.
- Performance is defined with decision variables (DV) which reflect global system performance.
- Performance is defined with DVs in terms of the direct interest of various stakeholders.

PEER performance assessment methodology has been summarized in various publications (Moehle 2003; Krawinkler 2002; Krawinkler and Miranda 2004; Moehle and Deierlein 2004; Porter 2003) and various benchmark studies have been conducted (Comerio 2005; Goulet et al. 2006; Krawinkler 2005; Kunnath 2006; Mitrani-Reiser wt al., 2006; Lee and Mosalam 2005, 2006). Recently, Günay and Mosalam (2013) summarized and explained the methodology in a simplified manner for its adoption by the broad engineering community. It is noted that the application of the methodology presented herein is mostly based on the explanations given in (Günay and Mosalam 2013).

PEER PBEE methodology consists of the four analysis stages shown in Figure 6, i. e. hazard, structural, damage, and loss. The methodology focuses on the probabilistic calculation of system performance measures meaningful to facility stakeholders by considering the four stages of analysis in an integrated manner, where uncertainties are explicitly considered in all stages. As shown in Figure 6, the outcome of each of the four stages is either a probability (or probability of exceedance, POE) distribution. The probabilities determined in each stage are combined using Equation 1.

The damageable parts of a facility are divided into damageable groups consisting of components affected by the same EDP in a similar manner, e. g. structural or non-structural components. Global collapse of a structure is treated separately in this methodology since its probability does not change from one damageable group to another. Equation 1, which resemble the well-known triple integration, referred to as the PEER PBEE framework equation in (Moehle 2003; Krawinkler 2005), is applicable only for the case of a single damageable group and no global collapse. A fourth summation is included to consider the presence of different damageable groups in Equation 2. The most general format of the formulation is given in Equation 3 for the case of multiple damageable groups and global collapse.

$$P(DV^n) = \sum_m \sum_i \sum_k P(DV^n | DM_k) \, p(DM_k | EDP^i) \, p(EDP^i | IM_m) \, p(IM_m) \tag{1}$$

$$P(DV^n) = \sum_m \sum_j \sum_i \sum_k P(DV_j^n | DM_k) \, p(DM_k | EDP_j^i) \, p(EDP_j^i | IM_m) \, p(IM_m) \tag{2}$$

$$P(DV^n) = \sum_m \left(\sum_j \sum_i \sum_k P(DV_j^n | DM_k) \, p(DM_k | EDP_j^i) \, p(EDP_j^i | IM_m) p(NC | IM_m) \right.$$
$$\left. + P(DV^n | C) \, p(C | IM_m) \right) p(IM_m) \tag{3}$$

where $p(IM_m)$ is the probability of the mth value of the earthquake intensity measure (IM), determined as an outcome of hazard analysis, $p(EDP_j^i | IM_m)$ is the probability of the ith value of the EDP utilized for the jth damageable group, when the mth value of IM occurs (outcome of structural analysis), $p(DM_k | EDP_j^i)$ is the probability of the kth Damage Measure (DM) when subjected to the ith value of the EDP utilized for the jth damageable group (outcome of damage analysis) and $P(DV_j^n | DM_k)$ is the POE of the nth value of the DV for the jth damageable group when the kth DM occurs (outcome of loss analysis). Moreover, $p(C|IM_m)$ and $p(NC|IM_m)$ are the probabilities of having and not having global collapse, respectively, under ground motion intensity IM_m. Finally, $P(DV^n|C)$ is the POE of the nth value of DV in the case of global collapse. Index j is dropped in Equation 1 since this equation represents the case of a single damageable group.

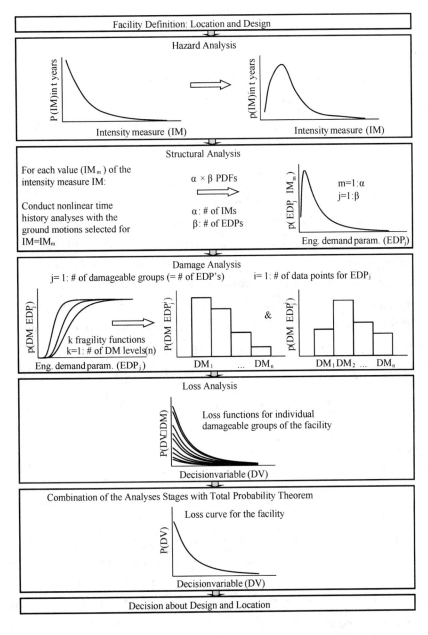

P(X|Y): Probability of exceedance of X given Y, P(X): probability of exceedance of X, p(X): probability of X

Figure 6　Outline of PEER PBEE Methodology

As mentioned before, the complete disconnect switch structure consists of the disconnect switch itself and its support structure. It has been observed that the support structures were not damaged during past earthquakes or shaking table tests (Mosalam et al. 2012). Therefore, global collapse of the complete disconnect switch structure is unlikely. Accordingly, Equation 2 is utilized in this paper.

The POE of the nth value of a DV in Equation 2, $P(DV^n)$, is interpreted as a weighted average of the POE of DVs from all possible cases, $P(DV_j^n|DM_k)$, where each case is defined by the probability of specific IM, EDP and DM, i.e. $p(DM_k|EDP_j^i)p(EDP_j^i|IM_m)p(IM_m)$. Similarly, the weighted average of any other quantity, such as the expected value of DV, E(DV), a valuable indicator (Der Kiureghian 2005), is calculated by replacing the POE with this quantity, Equation 4.

$$E(DV) = \sum_m \sum_j \sum_i \sum_k E(DV_j|DM_k) \, p(DM_k|EDP_j^i) \, p(EDP_j^i|IM_m) \, p(IM_m) \quad (4)$$

Equations 1-4 consider all possible scenarios of hazard, where each hazard level has a specific probability of occurrence during a considered time span, e.g. 50 years, as calculated from hazard analysis. However, in some situations, it may be useful to determine the loss in the case of a specific hazard level certainly taking place during the considered time span. These situations where the probability of the considered hazard level is 1.0 may arise if the considered structure is an important public facility, or if the return period of the considered hazard level is likely to be completed within the considered time span. In such cases, the POE and expected value of DV are represented with Equations 5 and 6, respectively, where IM_m is the considered intensity level.

$$P(DV^n) = \sum_j \sum_i \sum_k P(DV_j^n|DM_k) \, p(DM_k|EDP_j^i) \, p(EDP_j^i|IM_m) \quad (5)$$

$$E(DV) = \sum_j \sum_i \sum_k E(DV_j|DM_k) \, p(DM_k|EDP_j^i) \, p(EDP_j^i|IM_m) \quad (6)$$

3. ANALYSIS

3.1 Hazard analysis

In the PEER PBEE methodology, hazard analysis is conducted to determine the probability of different levels of earthquake hazard, by considering the uncertainty of all sources that define the earthquake hazard, such as the locations of nearby faults, their magnitude-recurrence rates, fault mechanism, source-site distance, site conditions, etc. Two sites are considered in hazard analysis, which correspond to the locations of two substations in Northern California. The first substation is Metcalf substation located in San Jose with 37.226° latitude and -121.744° longitude. The disconnect switches in this substation were damaged during 1989 Loma Prieta earthquake (EERI 1990). The second substation is Martin substation located in San Francisco with 37.705° latitude and -122.407° longitude. Both substations are located in highly active seismic zones, Figure 7.

OpenSHA (Field et al. 2003), an open-source software to conduct probabilistic seismic hazard analysis, is used in this study. Figure 8 shows a screenshot of OpenSHA graphical user interface, where main input parameters are identified with solid rectangles. These parameters are latitude and longitude of the site, site class parameters, type of intensity measure (IM), attenuation model and duration, in years, for which the hazard curve is calculated. Peak ground velocity (PGV) is chosen as the intensity measure because of its good correlation with the earthquake magnitude, effective duration and frequency content of the ground motion (Akkar S, Ozen 2005). The hazard curve is calculated using the input parameters as marked with a dashed rectangle in Figure 8. It represents the POE of each value of the IM. Numerical differentiation of the hazard curve gives the probability of each value of IM in 50 years as shown in Figure 9 for the two considered sites. It is observed that the probabilities of PGVs smaller than 15 cm/sec are larger in Martin substation, while those greater than 15 cm/sec are larger in Metcalf substation. Since the larger PGV levels are expected to induce damage, los-

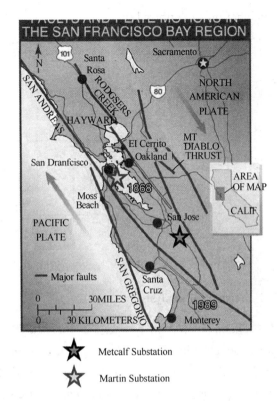

Figure 7 Locations of considered substations and nearby faults (source USGS)

ses due to earthquake hazard in Metcalf substation are expected to be larger than those in Martin substation. The plots in Figure 9 represent the term $p(IM_m)$ in the PEER PBEE

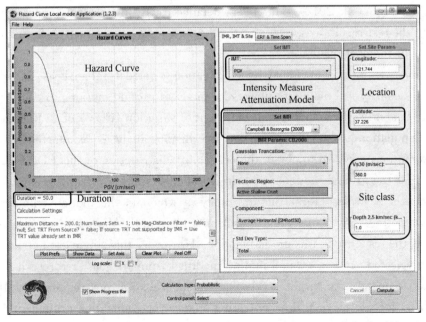

Figure 8 OpenSHA main window showing the input parameters and the hazard curve

combination Equations 1-4.

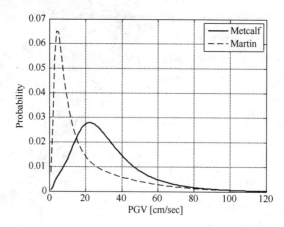

Figure 9 Probability of different levels of PGV for Metcalf and Martin substations

3.2 Structural analysis

The structural analysis determines the response to different levels and characteristics of earthquakes in a probabilistic manner. Conventionally, analytical simulation results are used for this purpose. Since the objective of this paper is to transform results obtained from previously conducted RTHS parametric study (Günay and Mosalam 2014) in the context of PBEE, hybrid simulation results are used herein. Furthermore, preference of RTHS over conventional analytical simulation is justified from modeling perspective. Analytical modeling of the complex geometry of the insulator posts and boundary conditions is a formidable task, even in the elastic range, requiring detailed 3D finite element (FE) models (Mosalam et al. 2012). Moreover, detailed FE results are only approximate in representing the test results, especially at the local response level, such as computed strains at the base of the post insulator.

In the conventional case of analytical simulation for structural analysis, for each intensity level, the analytical model is subjected to a set of ground motions to consider record-to-record variability. The median and coefficient of variation (COV) of an EDP determined from this simulation, commonly nonlinear time history analyses, conducted with this set of ground motions are used to obtain the probability distribution of this EDP for the intensity level that the ground motions are selected and/or scaled for. In this study, a single ground motion is utilized for structural analysis which is the motion employed in the RTHS parametric study (Günay and Mosalam 2014) and the median of an EDP is directly obtained as the response to this ground motion. However, this single ground motion is not arbitrarily selected. It is a ground motion modified from Joshua Tree record of Landers earthquake to match the target acceleration response spectrum required by IEEE 693 (IEEE 2006) for seismic qualification testing of electrical substation equipment (Takhirov et al. 2005). In recent years, this IEEE 693 compatible ground motion has become the standard one to be used in shaking table tests for seismic qualification of electrical substation equipment. Target acceleration response spectrum and that of the IEEE-compatible ground motion used in the RTHS parametric study (Günay and Mosalam 2014) are shown in Figure 10.

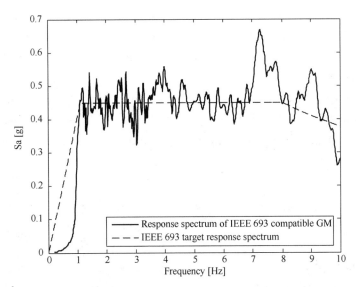

Figure 10 IEEE 693 target response spectrum and that of the ground motion used in the RTHS

One of the commonly used methods for ground motion selection used in structural analysis is to pick motions whose response spectra deviate the least from a target response spectrum (Beyer and Bommer 2007; Shantz 2006; Watson-Lamprey and Abrahamson 2006). Selected ground motions generally lead to a diverse response of the considered structure when the response is beyond the elastic range, even though the ground motions have similar response spectrum. However, response of the hybrid structure under investigation in this study, i. e. the disconnect switch and its support structure, is in the elastic range until failure occurs in a brittle manner. Hence, the peak value of any response quantity in each dynamic mode depends on the associated mode shape and spectral acceleration at the natural frequency of this mode. Combination of the peak responses with a modal combination rule (Chopra 2007) leads to conclude that any ground motion that matches the target response spectrum produces the same peak response, i. e. same EDP, in the elastic response. It is known that the modal combination is an approximation of the correct time history response, where the response from each mode is added in every time step and the peak response is determined. Moreover, a ground motion that is considered to match the target spectrum can still have acceptably small deviations from the target, refer to Figure 10. In this study, use of a single ground motion for the structural response of the disconnect switch is considered acceptable.

Probabilities of EDPs are determined in structural analysis where each damageable group has a specific EDP associated with it. Therefore, the damageable groups and corresponding EDPs are defined as a part of the structural analysis stage. Two damageable groups are considered for the investigated disconnect switches: (1) the insulator post and (2) the bus connection, for which strain and displacement are chosen as the corresponding EDPs, respectively. The insulator posts are damaged, lose functionality or fail in a brittle manner when the axial strain demand due to bending reaches its capacity. Therefore, strain is a suitable EDP to identify the response of an insulator post in relation to a damage state, particularly the failure state (Günay and Mosalam 2014). Insulator posts of the disconnect switches are connected to other equipment in an electrical substation with bus connections that possess slack, i. e. connecting cables are loose and not strained, in the un-deformed configuration. The relative displacement between an insulator post and another equipment should be limited to avoid bus damage and to prevent impact forces due to sudden tension that would otherwise develop in the connecting cables. Accordingly, displacement is a suitable EDP for the bus connection damageable group. The displacements of the insulator posts were determined from the RTHS parametric study (Günay and Mosalam 2014). However, displacements of the connected equipment can only be speculated. Relative displacement can be different from the displacement of an insulator post, varying widely depending on the frequency of the connected equipment compared to those of the support structure and the insulator post. It is impractical to consider all possible cases of connected equipment for the investigated 13 support structure configurations. Therefore, for simplicity and illustration of the use of RTHS results in the context of PEER PBEE methodology, it is assumed that the insulators are connected to very stiff equipment, i. e. the relative displacement is equal to the displacement of the insulator post. However, effect of the finite stiffness of connected equipment on the relative displacement is implicitly considered through the COV, accounting for the possibility of different configurations of connected equipment other than a very stiff one.

Besides the advantage of transforming the RTHS parametric study results to a format that is meaningful to all stakeholders, the choice of the selected EDPs highlights another benefit obtained from the PEER PBEE methodology. As observed in Figure 5, the most flexible support structures introduce the least strains on the insulator posts, while they result in the largest displacements. On the contrary, the stiffest support structures lead to the smallest displacements, while they tend to induce larger strains. Due to these two contradictory observations, the decision on the most suitable support structure can only be qualitative when the EDP values obtained from the RTHS parametric study are considered directly. Combination of the effect of the EDPs with PEER PBEE formulation allows a direct quantitative evaluation of the suitability of different support structures.

Tests of the RTHS parametric study were conducted using a single scale, namely 10% of IEEE 693 compatible input excitation, to avoid any nonlinearity such as grout cracking or loosening of bolts and to test the insulator in the same condition for all considered analytical damping and spring stiffness values of the support structures. Accordingly, from the test results, median EDPs are directly available for a single value of PGV. These values of EDPs are indicated with filled circles in Figure 11 for the RTHS test of the disconnect switch with a porcelain insulator mounted on a steel frame with full bracing and 1% damping ratio (ζ), Figure 12. Because the response of the disconnect switch and its support structure is elastic until sudden brittle failure, both the strain and displacement are accepted to be linearly proportional to PGV, Figure 11. Accordingly, median EDP values corresponding to all PGV values of interest are obtained with these linear functions. It is noted that the relationship between the EDPs and PGV were shown to be linear in (Mosalam and Günay 2014) with RTHS tests conducted using different scales of the input excitation.

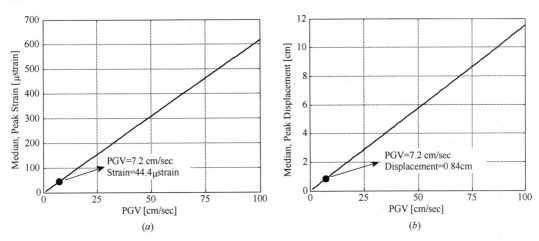

Figure 11 Median values for different PGV levels
(a) Peak strain; (b) Peak displacement

Due to the use of a single ground motion, it was not possible to determine the COV of EDPs from the test results. However, in the application of PEER PBEE methodology on a shear-wall building (Lee and Mosalam 2005, 2006; Günay and Mosalam 2014), COV of EDPs were determined to be in the range of 0.2 to 0.5 for linear elastic response levels. In this study, the COV for the strain is adopted as 0.4, which is closer to the upper bound of this range, to reflect the additional uncertainty due to the use of a single representative ground motion. Further uncertainty due to the effect of connected equipment on the displacement is considered by setting COV equal to 0.5 for this EDP. For each PGV value in the range of interest, the probability of each value of both EDPs are obtained using the determined median

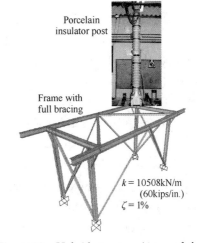

Figure 12 Hybrid structure in one of the RTHS tests for a disconnect switch

and adopted COV as the parameters of a lognormal distribution. The EDP probabilities obtained for the PGV values of 29, 59 and 94 cm/sec, corresponding to POE of 50%, 10% and 2%, respectively, in 50 years for the location of Metcalf substation are plotted in Figure 13. These plots represent the term $p(EDP_j^i | IM_m)$ in the PEER PBEE combination Equations 1-6.

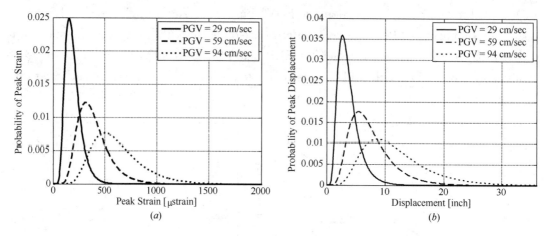

Figure 13　Probabilities of EDPs for different hazard levels on terms of PGV
(a) Peak strain; (b) Peak displacement

3.3　Damage analysis

The purpose of damage analysis is to probabilistically determine the damage level, i.e. damage measure (DM), of each damageable group for different values of the EDP associated with the damageable group. The damage levels in this study are defined as failure and bus connection damage (due to impact) for the insulator post and bus connection damageable groups, respectively. The insulator post behavior is linear elastic until failure occurs in a brittle manner, leaving failure as the only damage level to be considered for the insulator post group. The bus connection damage occurs due to the sudden tensile impact force applied by the connecting cables when the displacement is large enough to overcome the initial slack of the cable. Hence, the damage due to this sudden impact is the only damage level considered for the bus connection group.

Damage level of a damageable group shows variance, even for the same value of EDP. One of the reasons for this variance is the uncertainty in the material properties causing uncertainty in the corresponding EDP capacity term that defines the damage level. Therefore, one can only state that a specific value of strain leads to failure with a certain probability. In order to calculate this probability, the strain corresponding to failure is assumed to be lognormally distributed. Median failure strain is accepted to be 4800 and 1130 μstrain for the polymer and porcelain insulators, respectively, based on static pull-over tests conducted on these insulator posts (Mosalam et al. 2012). Considering the uncertainty in failure strain, both due to (a) variability of material properties from one insulator post to another with the same material and (b) effect of uncertain loading rate on the strain, a COV value of 0.3 is chosen. The probability of failure corresponding to different strain values are plotted in Figure 14(a) for the porcelain insulator post. The probabilities plotted in Figure 14 are commonly referred to as "fragility curves" and correspond to the term p $(DM_k | EDP_j^i)$ in the PEER PBEE combination Equations 1-6.

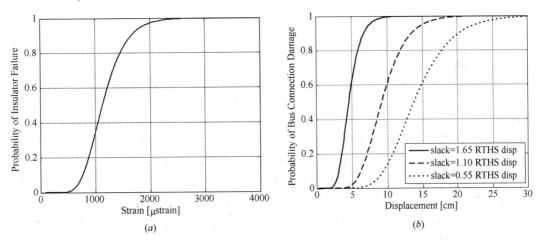

Figure 14　Probabilities of failure or damage corresponding to different EDPs
(a) Insulator failure; (b) Bus connection damage

A bus connection is damaged when the displacement at the top of an insulator post exceeds the slack in the connecting cables. Different median slack values are considered to determine the effect of slack on the economic loss for different support structures. Considered median slack values are multiples (1.65, 1.1 and 0.55) of the displacements obtained from the RTHS tests under the full scale IEEE 693 compatible ground motion. Here, the multiplier 1.1 corresponds to proper designs of the disconnect switches while the multiplier 1.1/2=0.55 corresponds to improper designs. The factors that determine the allowable maximum slack are (a) the mandatory electrical clearance requirements and (b) the instability or excessive displacements of the conductors under transverse wind loading (Dastous 2007; Fujisaki et al. 2014). It is generally beneficial to have more slack than required, as long as the slack is below the allowable maximum value because (i) the net additional cost of added conductor length is negligible and (ii) the conductors with more slack are easier to install (Dastous 2007). The multiplier 1.1×1.5=1.65 is considered to represent the case of more slack than required. The median slack values corresponding to 1.65, 1.1 and 0.55 of the displacements from the RTHS tests are therefore referred to as ample, required, and inadequate slack, respectively.

The RTHS tests were not conducted at the full scale of the IEEE 693 compatible ground motion for all the considered support structure configurations (Mosalam and Günay 2014; Günay and Mosalam 2014). Therefore, the displacements from the RTHS parametric study are linearly scaled up to full scale. This linear scaling is valid because of the linear elastic response of the switch and support structure and because the strain at full scale is smaller than the failure strain for the two considered insulator types (Günay and Mosalam 2014). The probabilities of bus connection damage are calculated from a lognormal distribution for the considered median values and a COV of 0.3, Figure 14(b). Variance in slack, represented with this COV value, is considered because (a) there is a rapid increase of stiffness and corresponding tension force when a cable is close to stretch completely (Dastous 2007) and (b) there may be slight variations in the field installations.

3.4 Loss analysis

The final analysis stage of the PEER PBEE methodology is the loss analysis, where the POE of each value of the chosen DV, such as fatalities, economic loss and downtime, is determined for all the damage levels of all damageable groups. Economic loss normalized by the replacement cost of a disconnect switch (referred to as the normalized economic loss in this paper) is used as the DV in this study. Both of the considered damage levels, i.e. insulator post failure and bus connection damage, are accepted to lead to the replacement of the disconnect switch. Accordingly, for both damage levels, the losses are defined by a lognormal distribution with a median of 1.0 and a COV of 0.3. The resulting probability and POE values are plotted in Figure 15, where the POE plot is called the loss function and corresponds to the term $P(DV_j^n | DM_k)$ in the PEER PBEE combination Equations 1-6. It is noted that the loss functions for the two damage levels overlap since they have the same median and COV values.

HAZUS (FEMA 2003) reports the replacement cost of an anchored component in a 245-kV substation as \$20 million. Although this value may seem excessive, it should be noted that it does not only include the cost of the disconnect switch itself, which is reported as \$32,500 including installation in (WAPA 2011), but it also includes the cost of removal, cost of demolition (if necessary), accompanying labor costs and most importantly the loss due to the interruption of the operation of the full substation during the replacement. It should be noted that the replacement cost of a disconnect switch may show variations depending on several factors, e.g. local manufacturing market, labor cost, the local cost and operation of electricity, among others. Detailed discussion of replacement cost determination is beyond the scope of the study.

There may be indirect economic losses which arise as a consequence of the replacement of a disconnect switch, such as the interruption of businesses due to the interruption of the substation operation. However, these indirect economic losses are difficult to quantify because of the complexity due to other aspects, e.g. socio-economics, and lack of systematic data that define these aspects. Furthermore, the indirect economic losses are likely to be similar in the chosen substation sites in the present study. Therefore, indirect losses are not considered for the seismic performance evaluation of the disconnect switches in this paper.

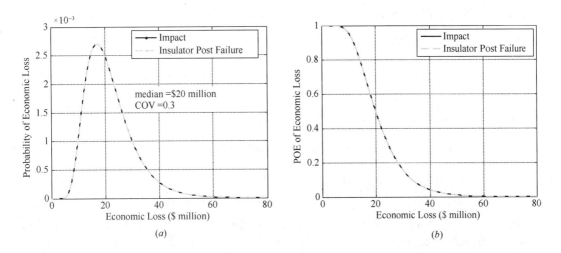

Figure 15　Economic loss for the considered damage levels of the disconnect switches
(a) Probability; (b) POE

4. PARAMETRIC STUDY ON THE ECONOMIC LOSSES

The outcome of the four PBEE analysis stages are combined using Equations 2, 4, 5 or 6 to obtain the loss curve or the expected value of the normalized economic loss. The loss curve represents the POE of different normalized economic loss values due to the effect of earthquake hazard on the 245-kV disconnect switches in 50 years, while the expected value defines the same effect with a single representative number. Determination of the loss curves and expected values using the RTHS parametric study results, together

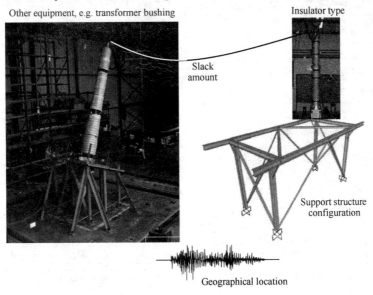

Figure 16　Parameters affecting the earthquake losses of the
245-kV disconnect switches

with the PEER PBEE formulation, allows the direct comparison of different values of various parameters on the seismic performance of these switches. The considered parameters are the support structure configuration, the amount of slack in the connecting cables (conductors), the insulator post type, and the geographical location of the substation where the disconnect switch is located, Figure 16. The effect of these parameters on the seismic performance of the disconnect switches are evaluated in the following sections.

4.1 Support structure configuration, slack amount and insulator post type

The most flexible support structures introduce the least amount of strain on the insulator posts, while they result in the largest displacement response, Figure 5. On the contrary, the stiffest support structures lead to the smallest displacement response, while they tend to induce large strains. Combination of the effect of these two EDPs with PEER PBEE formulation allows a direct quantitative evaluation of the suitability of different support structure configurations. However, for a specific support structure, the performance of a disconnect switch also depends on the amount of slack in the connecting cables. The expected normalized economic losses of the disconnect switches mounted on different sup-

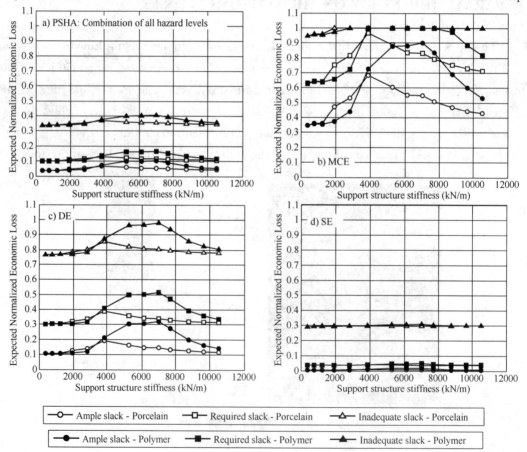

Figure 17　Expected normalized losses of disconnect switches mounted on different support structures with 1% damping and located in Metcalf substation for different slacks

port structures with 1% damping ratio and located in Metcalf substation, are plotted in Figure 17 for three slack values. The expected losses corresponding to a combination of all hazard levels from Equation 4 are plotted in Figure 17(a) and referred to as Probabilistic Seismic Hazard Analysis (PSHA) in the following discussions. Expected losses for three scenario-based hazard levels, corresponding to 2%, 10% and 50% POE in 50 years, are plotted in Figures 17(b), (c) and (d), respectively. The expected losses corresponding to these three hazard levels are determined with Equation 6 and referred to as Maximum Considered Earthquake (MCE), Design Earthquake (DE) and Serviceability Earthquake (SE), respectively. It is recalled that PGV values corresponding to POE of 2%, 10% and 50% in 50 years are 94, 59 and 29 cm/sec, respectively.

It is observed that the expected normalized losses are the smallest for the most flexible support structure for all considered cases, while a higher expected normalized loss is observed for the stiffest support structure. The difference between the economic losses for these two extreme support structure configurations is small for PSHA, DE and SE and increases for MCE. Loss curves of the disconnect switches with porcelain insulators determined for PSHA and MCE cases, Figure 18, lead to similar observations. The moderately stiff support structures are subject to higher expected losses compared to the most flexible and the stiffest support structures. Based on the expected losses and the loss curves, it is concluded that the most flexible support structures are the most suitable ones for disconnect switches with both porcelain and polymer insulator posts, provided that the required slack of the conductors is furnished.

In Figure 17, the expected losses for MCE and DE cases are much higher than those for PSHA. This is mainly because of the consideration of the probability of each hazard level in PSHA, Figure 9, as opposed to the probability of a specific hazard level being 1.0 and all others zero, in MCE, DE and SE. In that regard, PSHA determines the weighted average of the loss resulting from each hazard level, including MCE, DE and SE levels, where the weight multiplier for each hazard level is its probability. Obviously, the expected losses for MCE and DE levels are high. Therefore, if it is expected that DE or MCE level will occur for sure during the operational life of a disconnect switch, or if the considered substation is of high importance that it is not feasible to only consider a probable occurrence of DE or MCE levels, then the expected losses are high. However, if there is confidence in the probability of occurrence of each intensity level, then lesser losses are expected due to the finite probability of each intensity level. Similar discussion holds for Figure 18 where the POE of normalized economic losses for MCE are about 6.7 times higher than those for PSHA.

The slack amount has a significant effect on the expected loss for all considered hazard cases. Inadequate slack, half of the required in this case, increases the expected loss (up to 3.5 times) compared to the case when the required slack is provided. The disconnect switches with inadequate slack mounted on all support structure configurations are expec-

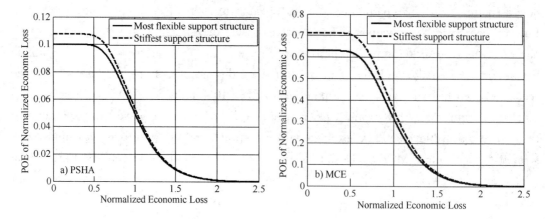

Figure 18　Loss curves of the disconnect switches with porcelain insulators, having the required slack, mounted on the stiffest and most flexible support structures and located in Metcalf substation

ted to be replaced almost certainly after the MCE level and consequently the economic loss equals the replacement value. It is observed that having ample slack, 1.5 times the required in this case, decreases the economic loss considerably (up to one-third). The mandatory electrical requirements are more critical for the flexible structures because of larger displacements. Therefore, it may not always be possible to provide more than the required slack for flexible support structures.

Expected normalized economic losses are almost the same for disconnect switches with porcelain and polymer insulator posts mounted on flexible support structures (those with less than 1300 kN/m stiffness). For stiffer support structures, disconnect switches with polymer insulator posts are expected to have higher economic loss where the differences are highest for the MCE level and the moderately stiff support structures. Because of some specific advantages of the polymer insulator posts over the porcelain, such as significantly lighter weight and compactness, lower installation cost, and improved power frequency insulation and contamination performance (Barrett 2003), polymer insulators are recently gaining wider spread usage (Koch 2014). It is possible to make use of such advantages of polymer insulators by mounting them on flexible support structures.

The effect of damping on the expected normalized loss is investigated in Figure 19. It is observed that damping does not have a significant effect on the expected losses. The effect of damping is minimal for the most flexible and the stiffest support structures. While this figure contains only the MCE level, similar observations hold for the other hazard cases. The expected normalized loss in Figures 17 and 19 can be replaced with "expected $\pm$ standard deviation" losses. This corresponds to the multiplication of the expected losses with 1.3 and 0.7 for the respective "+" and "-" standard deviation cases, since the COV of the used loss functions is 0.3.

4.2　Geographical Location

The geographical location of the substation where the disconnect switch is located has

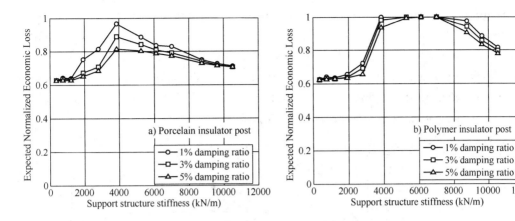

Figure 19 MCE level expected normalized losses of disconnect switches in Metcalf substation, having the required slack and mounted on different support structures with varied damping

a significant effect on the economic losses, since the level of hazard depends on the location. The expected normalized losses of disconnect switches with porcelain insulators, having the required slack and mounted on different support structures with 1% damping are compared for different hazard cases at Metcalf and Martin substations in Figure 20. Note that the different sub-figures have different vertical scale for better visibility of the results. As anticipated, expected economic losses are smaller in Martin substation for PSHA case. This is mainly because the probabilities of PGVs greater than 15 cm/s are larger in Metcalf substation, Figure 9. Expected losses for SE, DE and MCE levels directly depend on the values of PGV with 2, 10 and 50% POE in the considered sites. These values are 94 and 92 cm/sec for MCE, 59 and 51 cm/sec for DE and 29 and 12 cm/sec for SE in Metcalf and Martin substations, respectively. Accordingly, similar to the PSHA case, expected losses are smaller in Martin substations for the MCE, DE and SE hazard scenarios.

5. SUMMARY AND CONCLUDING REMARKS

Results of a real time hybrid simulation (RTHS) parametric study, previously conducted on electrical disconnect switches, were utilized in the context of performance based earthquake engineering (PBEE). Using the PBEE methodology developed by the Pacific Earthquake Engineering Research (PEER) Center, the RTHS results were transformed from an engineering demand parameters (EDPs) format to a probabilistic decision variable (DV) format to provide the seismic performance information to a broader range of stakeholders.

The study presented herein focused on the seismic performance evaluation of high voltage disconnect switches of electric substations. Information obtained from this PBEE, especially in terms of the transformation of the EDP information to an expected loss format demonstrate the intended objective of making the seismic performance information of the disconnect switches accessible to the broader stakeholders of electrical engineers, utility companies, manufacturers and policy makers. Moreover, the stakeholders can use the methodology not only to assess the seismic performance of disconnect switches, but also to design them, e. g. to determine suitable insulator type, support structure configuration, geographical location of substation and conductor slack.

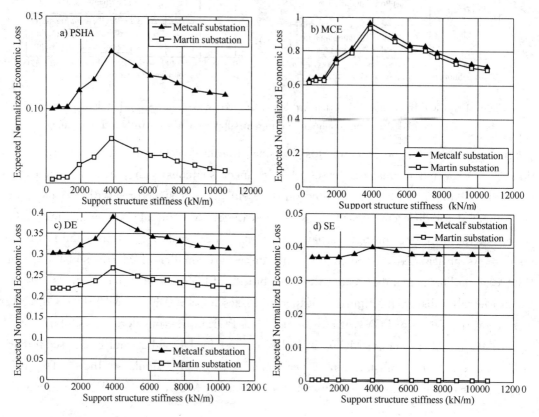

Figure 20　Expected normalized losses of disconnect switches with porcelain insulators, having required slack and mounted on different support structures (1% damping) at two substations

The PEER PBEE methodology provides a versatile framework. Accordingly, if a similar set of analytical or experimental data is obtained in the future, the structural analysis stage can be updated by replacing the RTHS parametric study results with the new data set. Similarly, different material properties or replacement cost values, e. g. in other parts of the world, can be used to update the damage and loss analyses stages. Hazard analysis can be modified in a straightforward manner to consider the substations located in other sites. Specific concluding remarks are listed as follows:

- Combination of the effect of different EDPs of a disconnect switch with PEER PBEE formulation allows a direct quantitative evaluation of the suitability of different support structures in a holistic manner.

- According to the obtained expected losses and loss curves, the most flexible support structures, i.e. those without braces, constitute the most suitable support structure configuration for the investigated 245-kV disconnect switches.
- The slack amount has a significant effect on the expected loss for all considered hazard cases. Inadequate slack can increase the expected loss significantly, while having ample slack can decrease the economic loss considerably. Therefore, it is advantageous to utilize the maximum allowable slack of the conductors between the switch and other equipment.
- Expected economic losses are almost the same for disconnect switches with porcelain and polymer insulator posts mounted on flexible support structures. For other support structures, disconnect switches with polymer insulator posts are expected to have higher economic loss. It is possible to make use of the benefits of polymer insulators, e.g. significantly lighter weight and compactness, lower installation cost, and improved power frequency insulation and contamination performance, by mounting them on flexible support structures.
- Damping of the support structure, ranging between 1% and 5%, is observed to have insignificant effect on the expected normalized losses.
- A disconnect switch in Metcalf substation located in San Jose is more vulnerable to earthquake hazard than the same type of disconnect switch in Martin substation located in San Francisco.

ACKNOWLEDGMENTS

This research was supported partly by Subcontract TRP-08-02 from California Institute for Energy and Environment and partly by the National Science Foundation (Award Number: CMMI-1153665). The authors thank S. Takhirov, M. Moustafa, and E. Fujisaki for their significant technical inputs.

REFERENCES

Mosalam, KM, Günay S. Seismic performance evaluation of high voltage disconnect switches using real-time hybrid simulation: I. System development and validation. *Earthquake Engineering and Structural Dynamics* 2014; 43(8):1205-1222.

Günay S, Mosalam, KM. Seismic performance evaluation of high voltage disconnect switches using real-time hybrid simulation: II. Parametric Study. *Earthquake Engineering and Structural Dynamics* 2014; 43(8):1223-1237.

Takhirov S. *Seismic Qualification Report on EC-1, P and EV-1 Types of 245-kV Disconnect Switches with EVG-1 Grounding Switch Installed*. Report No. SS-SCE-2008, Earthquake Engineering Research Center, University of California, Berkeley, 2008.

Moehle JP. A framework for performance-based earthquake engineering. *Proceedings of ATC-15-9 Workshop on the Improvement of Building Structural Design and Construction Practices*, 2003;

Maui, HI.

Krawinkler H. A general approach to seismic performance assessment. *Proceedings of International Conference on Advances and New Challenges in Earthquake Engineering Research*, ICANCEER. 2002; Hong Kong.

Krawinkler H, Miranda E. *Performance-based Earthquake Engineering*. Chapter 9 of *Earthquake Engineering: from Engineering Seismology to Performance-based Engineering*. Bozorgnia, Y. and Bertero, V. V., Editors, CRC Press, 2004.

Moehle JP, Deierlein GG. A framework for performance-based earthquake engineering. *Proceedings of 13th World Conference on Earthquake Engineering*. Paper No 679. 2004; Vancouver, Canada.

Porter KA. An overview of PEER's Performance-based earthquake engineering methodology. *Conference on Applications of Statistics and Probability in Civil Engineering* (ICASP9). Civil Engineering Risk and Reliability Association (CERRA), 2003; San Francisco, CA.

Comerio MC. Editor, *PEER Testbed Study on a Laboratory Building: Exercising Seismic Performance Assessment*.

Report No: PEER 2005/12, Pacific Earthquake Engineering Research Center, PEER, 2005.

Goulet C, Haselton CB, Mitrani-Reiser J, Deierlein GG, Stewart JP, Taciroglu E. Evaluation of the seismic performance of a code-conforming reinforced-concrete frame building - part I: ground motion selection and structural collapse simulation. *8th National Conference on Earthquake Engineering* (100th Anniversary Earthquake Conference), 2006; San Francisco, CA.

Krawinkler H. Editor, *Van Nuys Hotel Building Testbed Report: Exercising Seismic Performance Assessment*. Report No: PEER 2005/11, Pacific Earthquake Engineering Research Center, PEER, 2005.

Kunnath SK. *Application of the PEER PBEE Methodology to the I-880 Viaduct*. Report No: 2006/10, Pacific Earthquake Engineering Research Center, PEER, 2006.

Mitrani-Reiser J, Haselton CB, Goulet C, Porter KA, Beck J, Deierlein GG. Evaluation of the seismic performance of a code-conforming reinforced-concrete frame building - part II: loss estimation *8th National Conference on Earthquake Engineering* (100th Anniversary Earthquake Conference). 2006; San Francisco, CA.

Lee TH, Mosalam KM. Seismic demand sensitivity of reinforced concrete shear-wall building using FOSM method. *Earthquake Engineering and Structural Dynamics* 2005; 34(14): 1719-1736.

Lee TH, Mosalam KM. *Probabilistic Seismic Evaluation of Reinforced Concrete Structural Components and Systems*. Report No: PEER 2006/04, Pacific Earthquake Engineering Research Center, PEER, 2006.

Günay S, Mosalam KM. PEER performance-based earthquake engineering methodology, revisited. *Journal of Earthquake Engineering* 2013; 17(6): 829-858.

Mosalam KM, Moustafa MA, Günay S, Triki I, Takhirov S. *Seismic Performance of Substation Insulator Posts for Vertical-Break Disconnect Switches* California Energy Commission, Publication number: CEC-500-2012. , March 2012.

Der Kiureghian A. Non-ergodicity and PEER's framework formula. *Earthquake Engineering and Structural Dynamics* 2005; 34:1643-1652.

EERI, Loma Prieta Earthquake Reconnaissance Report, Chapter 8: Lifelines. *Earthquake Spectra* 1990; supplement to 6: 239-338.

Field EH, Jordan TH, and Cornell CA. OpenSHA: A Developing community-modeling environment for seismic hazard analysis *Seismological Research Letters* 2003; 74(4): 406-419.

Akkar S, Ozen O. Effects of peak ground velocity on deformation demands for SDOF systems. *Earthquake Engineering and Structural Dynamics* 2005; 34(13):1551-1571.

IEEE Standard 693-2005. *Recommended Practice for Seismic Design of Substations*. New York, NY, 2006.

Takhirov S, Fenves G, Fujisaki E, Clyde D. *Ground Motions for Earthquake Simulator Qualification of Electrical Substation Equipment*. Pacific Earthquake Engineering Research Center, Report 2004/07. PEER, University of California, Berkeley, 2005.

Beyer K, Bommer JJ. Selection and scaling of real accelerograms for bi-directional loading: A review of current practice and code provisions, *Journal of Structural Engineering* 2007; 11:13-45.

Shantz T. Selection and scaling of earthquake records for nonlinear dynamic analysis of first mode dominated bridge structures, *8th National Conference on Earthquake Engineering* (*100th Anniversary Earthquake Conference*). 2006; San Francisco, CA.

Watson-Lamprey J, Abrahamson NA. Selection of ground motion time series and limits on scaling, *Soil Dynamics and Earthquake Engineering* 2006; 26(5):477-482.

Chopra AK. *Dynamics of Structures: Theory and Applications to Earthquake Engineering*. 3rd edition Prentice Hall: Englewood Cliffs, New Jersey, 2007.

Dastous J-B. Guidelines for seismic design of flexible buswork between substation equipment. *Earthquake Engineering and Structural Dynamics* 2007; 36(2): 191-208.

Fujisaki E, Takhirov S, Xie Q, and Mosalam K. Seismic vulnerability of power supply: Lessons learned from recent earthquakes and future horizons of research, *9th International Conference on Structural Dynamics EURODYN*. 2014; Porto, Portugal.

FEMA. *HAZUS-MH technical manual* Federal Emergency Management Agency 2003; Washington, DC.

Western Area Power Administration (WAPA). *Facilities Study* Report No 2011-T6, 2011.

Barrett JS, Chisholm WA, Kuffel J, Ng BP, Sahazizian A-M, de Tourreil C. Testing and modeling hollow-core composite station post insulators under shortcircuit conditions. *Proceedings of IEEE Power Engineering Society General Meeting*, Toronto, Canada, 13-17 July, 2003.

Koch BC. Utilities and Lifelines - LADWP Power System (oral presentation), *Northridge 20 Symposium* 2014; Los Angeles, CA.

第六届论坛特邀报告论文作者简介

Sashi Kunnath，于 1989 年获纽约州立大学布法罗分校博士学位，现任加州大学戴维斯分校土木及环境工程系教授及系主任，湖南大学外专千人计划特聘专家。他的研究领域包括基于性能的抗震设计，结构的非线性模拟，以及极端荷载下结构响应的评估。Kunnath 教授现为美国土木工程师学会（ASCE）结构工程分会（Structural Engineering Institute）的执行委员会委员，以及多个 ACI（美国混凝土学会）和 ASCE 技术委员会委员。Kunnath 教授曾任 ASCE Journal of Structural Engineering 的主编，并曾担任 ASCE 动力效应技术管理委员会（Technical Administrative Committee on Dynamic Effects）主席以及地震效应技术委员会（Seismic Effects Technical Committee）主席。Kunnath 教授已发表 200 余篇期刊和会议论文，并曾多次在世界各地的重要学术会议/研讨会做学术报告，包括中国，意大利，日本，韩国，葡萄牙，斯洛文尼亚及泰国。

Kunnath 教授曾经多次获奖，包括 ASCE 的三个奖项：Norman 奖章（2012），Richard Torrens 奖（2009），Raymond Reese 研究奖（2008），以及得 ACI 结构研究奖（2001）. Kunnath 教授是 ACI（2007），ASCE（2010）和 ASCE 结构工程分会（2013）的资深会员。

傅学怡 全国工程勘察设计大师，悉地国际设计顾问有限公司总工程师，深圳大学土木工程学院研究员，博士生导师，深圳大学建筑设计研究院顾问总工程师。长期从事建筑结构设计及研究，主持完成上百项高层超高层大跨建筑工程结构设计，获中国土木工程詹天佑奖 6 项、全国优秀建筑结构设计一等奖 8 项，国家游泳中心获国家科学技术进步一等奖（第 1 人）、国际桥梁与结构工程协会杰出结构大奖，多哈塔获都市高层建筑委员会世界最优高层建筑大奖、国际结构混凝土协会杰出混凝土结构特别贡献奖，万科中心获美国建筑师学会建筑荣誉奖。提出创新建筑结构设计方法多项并应用于实际工程，获国家发明专利 6 项，被国家设计规范采纳 3 项。主持完成 4 项国家自然科学基金项目，发表 100 多篇学术论文，3 篇获 SCI、EI 收录杂志优秀论文奖，出版 3 本专著，1 本 5 次印

刷,获工程界广泛应用,1本获国家新闻出版总署"三个一百"原创图书出版工程奖;培养硕士 50 名、博士 8 名;国际学术会议演讲十余次,赴美、英、日等国专场学术报告多次;为我国建筑抗震设计规范、高层建筑结构技术规程主要编制人;获华夏建设科学技术一等奖 4 项。

郁银泉 中国建筑标准设计研究院副院长、总工程师,教授级高工、国家一级注册结构工程师,中国工程勘察设计大师,享受国务院政府特殊津贴专家。全国超限高层建筑工程抗震设防审查专家委员会委员、中国建筑学会资深会员、中国钢结构协会常务理事、中国建筑学会建筑结构分会理事、中国建筑学会抗震防灾分会高层建筑抗震专业委员会委员、中国钢结构协会稳定疲劳委员会委员。担任《钢结构》、《建筑结构》等学术期刊编委。长期从事标准规范编制和建筑结构工程设计工作。曾主编和参编了《高层民用建筑钢结构技术规程》、《门式刚架轻型房屋钢结构技术规范》、《建筑抗震设计规范》等 9 项标准规范。作为技术审定人,参与了十几项国家标准设计项目。主持和指导设计了多项有影响的工程项目,获国家优秀工程设计金奖 1 项、银奖 2 项、铜奖 2 项,全国工程勘察设计行业优秀工程勘察设计奖一等奖 1 项、二等奖 2 项、三等奖 3 项,全国优秀建筑结构设计奖二等奖 1 项,华夏科学技术进步一等奖 1 项、河北省科学技术进步一等奖 1 项,北京市科学技术进步二等奖 1 项,北京市优秀工程项目二等奖 2 项、三等奖 1 项等。作为课题负责人或主要参与人参与国家"十一五"支撑计划子课题等 10 项科研项目,发表学术论文 40 篇。代表作品:数字北京大厦、国家体育场(鸟巢)、中关村金融中心、北京东四 D1 区海洋石油办公楼、天津新华世纪金融中心、鞍山市体育中心、阿尔及利亚奥兰体育场、中国残疾人体育综合训练基地、济南泉城公园全民健身中心、石家庄火车站、中纪委办公楼、唐山新华文化广场等。

李国强 同济大学教授,国家土建结构预制装配化工程技术研究中心主任,教育部建筑钢结构工程研究中心主任。主要从事多高层钢结构与钢结构抗火设计研究。历年来主持完成国家自然科学基金、国家杰出青年科学基金、国家 863 计划、国家攀登计划、国家科技支撑计划、霍英东教育基金、中科院知识创新工程、教育部青年骨干教师重点跟踪培养计划、上海市优秀学科带头人计划等资助的重要科研项目 40 余项。在国内外学术刊物上发表学术论文 500 余篇(其中 SCI 收录论文 83 篇、EI 收录论文 207 篇),出版学术

著作 13 部，主编或参编 11 部工程建设国家标准或行业及地方标准。担 2 本国际学术刊物和 1 本国内学术刊物主编，以及其他 5 本国际学术刊物和 10 多本国内学术刊物编委；科研成果获国家科技进步二等奖 1 项、省部级科技进步一等奖 5 项、二等奖 6 项。

蔡克铨 台湾大学土木系教授，主要研究领域为钢结构、抗震设计、地震工程模拟实验与分析。曾任国家地震工程研究中心主任，国家灾害防救科技中心地震组召集人，台湾结构工程学会与地震工程学会理事长。发表学术论文约 300 篇，荣誉包括行政院杰出科学与技术荣誉奖、国科会杰出研究奖三次、经济部国家发明创作金牌奖、国科会杰出技术移转贡献奖二次、中技社科技奖、侯金堆杰出荣誉奖、国家实验研究院杰出科技贡献奖三次、台湾大学终身特聘教授、俄罗斯国际工程院通讯院士。获台湾结构工程学会、土木水利工程师学会、中国工程师学会三学会共 13 次论文奖。担任 Earthquake Engineering and Structural Dynamics，Journal of Earthquake Engineering 编辑委员

李爱群 东南大学特聘教授，国家级教学名师，国家杰出青年基金获得者，"新世纪百千万人才工程"国家级人选。东南大学土木工程学院教授、博士生导师。国务院学位委员会第五、六届学科评议专家组成员，全国土木工程学科专业指导委员会副主任委员，南京土木建筑学会理事长。主要从事土木工程结构抗震抗风与隔震减振、结构健康监测与安全评估等方向的研究。获国家科技进步二等奖 2 项，省部级科技进步一等奖 5 项、二等奖 3 项等科技奖项。主编国家行业标准《建筑消能阻尼器》，合作主编国家建筑设计标准图集《建筑结构消能减震设计》。获国家发明专利授权 30 余项；主编著作 6 本，发表 SCI、EI 收录论文 120 余篇。

任伟新 合肥工业大学土木与水利工程学院院长。1995 年破格晋升教授，1996 年批准为博士生导师。2006 年入选教育部"长江学者奖励计划"桥梁与隧道工程岗位特聘教授。2004 年入选人事部等七部委首批"新世纪百千万人才工程国家级人选"、教育部首批"新世纪优秀人才支持计划"和国务院享受特殊津贴专家。作为访问教授在日本、比利时、美国和澳大利亚工作 6 年之久。目前担任 4 个国际杂志（SCI）编委；中国振动工程学会理事，应邀为本专业 30 余种国际期刊的论文审稿人，30 余次应邀到 20 余个国

家的大学做学术报告和专题讲座。长期从事桥梁结构稳定和振动研究和教学，培养毕业博士后5人，博士18人，硕士69人。在桥梁/结构工作模态参数识别及其应用、结构损伤识别、小波有限元方法及模型修正、结构稳定极限承载力等方面，取得了创新性成果，为国际同行所承认。理论研究的同时，注重工程实际的结合，有关研究成果已纳入铁道部桥梁设计规范，并在一些重大桥梁工程中应用。已出版专著3部，主编国际学术会议论文集3部，发表各类学术论文280余篇，被SCI、EI、ISTP收录150余篇，其中国际一流期刊论文70余篇，本领域最著名的美国土木工程师学会（ASCE）会刊发表论文18篇。迄今SCI引用1000余次，个人SCI H-index＝18，Scopus数据库H-index＝23，Google Scholar引用3100余次，H-index＝28。已主持和参加完成科研项目50余项，2002年回国后主持国家自然科学基金7项。

肖 岩 湖南大学土木工程学院院长、"长江学者"特聘教授、中组部"千人计划"国家特聘专家；兼美国南加州大学教授。美国土木工程师学会ASCE《Journal of Structural Engineering》，《Journal of Bridge Engineering》副主编；Elsevier《Journal of Constructional Steel Research》编委；国际组合结构学会ASCCS理事；《自然灾害学报》副主任委员；《建筑结构学报》编委。美国混凝土学会会士ACI Fellow，美国土木工程师学会会士ASCE Fellow。发表学术著作300余篇、部，包括SCI检索论文60余篇，SCI论文他引2000多次（单篇最高引用率超过400），ISI H-index 18。拥有发明专利和实用新型专利十多项。负责设计建造的湖南耒阳导子乡行车现代竹桥被美国《Popular Science》杂志评为2008年最佳工程创新（The Best of What's New in 2008）。获2009年湖南省国际合作奖。

童根树 浙江大学教授、博士生导师。1983年浙江大学土木系毕业，1988年6月西安建筑科技大学研究生毕业获博士学位，留校任教，1995年12月起在浙江大学土木系任教。主要从事钢结构稳定和抗震方面的基础理论研究，发表论文300余篇，著有《钢结构的平面内稳定》、《钢结构的平面外稳定》（修订版）、《钢结构设计方法》、《薄壁曲梁的线性和非线性分析》等著作。参编《钢结构设计规范》、《冷弯薄壁型钢结构技术规范》、《高层建筑钢结构技术规程》、《门式刚架轻型房屋钢结构技术规范》、《农业温室钢结构技术规范》；是《香港及邻近地区钢结构设计规范》起草委员会顾

问成员。讲授《钢结构》《结构稳定理论基础》《薄壁结构稳定》《高等钢结构理论》等本科与研究生课程。主持设计 Dubai 跑马场停车场综合体、多个国内多层卖场综合体建筑，亲自设计昆明世纪广场、成都成达大厦等大量高层钢结构建筑；担任过西安迈科商业中心，昆明海航酒店，西子电梯试验塔等民用和工业高层建筑结构顾问。

陆新征 清华大学土木工程系教授，防灾减灾工程研究所所长，《工程力学》期刊副主编。主要从事数值计算和防灾减灾方面的研究工作。主持或参加国家、部委课题多项。发表论文百余篇，出版教材、专著等 8 部，论著累计被引用 5000 余次。曾获得国家自然科学二等奖，国家科技进步二等奖等奖励。研究成果被我国国家、行业标准及美国混凝土学会（ACI）指南采纳，并在多个重要工程中得到应用。

Khalid M. Mosalam，获开罗大学学士和硕士学位，1996 年获康奈尔大学结构工程专业博士学位。1997 年起执教于加州大学伯克利分校土木及环境工程系，现为该校的结构工程教授，同时也是结构工程、力学和材料部的负责人。Mosalam 教授的研究领域包括混凝土结构、砌体结构及木结构在极限荷载下的结构行为和健康评估。他还致力于重要基础设施（如桥梁、变电站等）的评估和加固研究。他的研究领域包括大型结构的计算分析（包括确定性的和随机性的分析）和试验研究（包括混合试验）。Mosalam 教授采用先进计算手段和大型试验相结合的方法，在解决结构实际工程问题方面作出了突出贡献，由此获得了 2006 年的美国土木工程师学会 Walter L. Huber 土木工程研究奖。为了表彰他在摩洛哥廉价房屋的地震安全性及相关法规制定的国际合作项目中的突出贡献，Mosalam 教授被授予 2013 年度的 UCB Chancellor Award for Public Services。Mosalam 教授还活跃在高能效建筑和可持续发展的土木工程研究领域。